Hard Rock Hydraulics

Hard Rock Hydraulics

An Introduction to Modeling

Fernando Olavo Franciss

CRC Press is an imprint of the
Taylor & Francis Group, an **informa** business

CRC Press/Balkema is an imprint of the Taylor & Francis Group, an informa business

Typeset by codeMantra

Library of Congress Cataloging-in-Publication Data
Names: Franciss, F. O., author.
Title: Hard rock hydraulics : an introduction to modeling / Fernando Olavo Franciss, Academia Nacional de Engenharia, Brazil.
Description: Boca Raton : CRC Press, [2021] | Includes bibliographical references and index. | Summary: "Hard rock hydraulics concerns arrangements of adjoining intact rock blocks, occurring down to a depth of hundreds of meters, where groundwater percolates within the gaps between these blocks. During the last decades, technical papers related to successful or failed attempts for mining groundwater from hard rocks, and achievements or failures of public or mining developments with respect to these rocks, increased the knowledge of their hydraulics. Examples of activities where the mechanical behavior of these rocks highly interacts with their hydraulics are projects under the sea or groundwater level, such as open pits or underground mines, galleries, tunnels, shafts, underground hydropower plants, oil and LPG storage caverns, and deep disposal of hazardous waste. This book dedicated to hard rock hydraulics assumes some prior knowledge of hydraulics, geology, hydrogeology, and soil and rock mechanics. Chapter I discusses the main issues of modeling; chapter II covers the fundamentals of hard rock hydraulics; chapter III presents concepts regarding approximate solutions; chapter IV discusses data analysis for groundwater modeling; chapter V focuses on finite differences and chapter VI provides examples of some particular unusual applications. This book will help civil and mining engineers and also geologists to solve their practical problems in hydrogeology and public or mining projects"— Provided by publisher.
Identifiers: LCCN 2020038811 (print) | LCCN 2020038812 (ebook) | ISBN 9780367376208 (hbk) | ISBN 9780429355325 (ebook)
Subjects: LCSH: Hydrogeological modeling. | Hard rock mines and mining.
Classification: LCC GB1001.72.H93 F73 2021 (print) | LCC GB1001.72.H93 (ebook) | DDC 551.4901/1—dc23
LC record available at https://lccn.loc.gov/2020038811
LC ebook record available at https://lccn.loc.gov/2020038812

Published by: CRC Press/Balkema
Schipholweg 107C, 2316 XC Leiden, The Netherlands
e-mail: Pub.NL@taylorandfrancis.com
www.routledge.com – www.taylorandfrancis.com

ISBN: 978-0-367-37620-8 (hbk)
ISBN: 978-0-367-69442-5 (pbk)
ISBN: 978-0-429-35532-5 (ebk)

DOI: 10.1201/9780429355325
https://doi.org/10.1201/9780429355325

Contents

Preface

My interest in hard rock hydraulics arose when, defied by practical problems, I developed in 1962 a variant of electric analogs drawn with specific inks on polyester drawing papers. Later, in the 1968–1970 period, during my thesis termination, I was captivated by second-order tensors' capabilities to describe the permeability of fracture systems. Since then, I have faced many technical challenges in my professional life involving the hydraulics of dam foundations, water reservoirs interaction, hydrothermal water resources, pits, underground mines, excavated caverns tunnels for water supply.

Thus, I decided to write my last book on hard rock hydraulic. It includes some unusual examples but concisely written with self-explanatory illustrations. Also, I inserted and rectified a few lacking information and lapses of my previous books.

I wish to express my gratitude to many students, professional colleagues, and clients from Sondotecnica and Progeo, who helped me with due criticism and sensible proposals. Finally, my thanks also go to Marjanne Bruin and Alistair Bright, from CRC Press/Balkema and Assunta Patrone from Codemantra, for their patient support and incentive during this book's gestation period.

Rio de Janeiro, December 2020

Chapter 1

About modeling

1.1 Introduction

Modeling is the art of translating relatively complex things into simpler ones. A useful model may be as simple as a modest and concise oral explanation, i.e., a recorded description (a verbal model), or obscure as a complex differential equation (a mathematical model). What is essential is that the chosen model fulfills its purpose. Thus, before building any model, it is crucial to know why a model is needed. For instance, a model may aid to understand the following:

- How may a particular grouping of pumping wells economically satisfy the water supply for irrigation?
- Why a particular layout is the most appropriate for an electric power distribution system?
- What is the predicted transient pore pressure at a concrete dam foundation after loading its water reservoir?

Another point to be clarified is how closely the model shall simulate the prototype performance to give an adequate answer to a query.

However, before developing some guidelines about models and modeling, it is convenient to restrict the present book's application domain, mostly devoted to hard rock models.

Hard rocks typically include all types of igneous rocks, like granites, basalts, and gabbros, as well as metamorphic rocks, like gneisses, schist, and slates, forming more than 95% of the Earth's crust. Acid igneous rocks include the continents' basements, attaining more than 70 km depth at the root of high mountain chains. Like petrified seafloor basalt extrusions, basic igneous rocks pile down to 7 km thickness under the oceans. However, despite its vast quantity, igneous and metamorphic rocks emerge as outcrops, only over 15% of the Earth's surface, frequently beneath layers of sedimentary rocks, like sandstones, siltstones, and limestone, up to hundreds of meters thick. However, all rock types are mostly covered by less resistant thinner veneers, such as weathered rocks and residual soils or unconsolidated sediments as gravels, sand, and clays.[1]

However, rock masses near the Earth surface, down to tens or hundreds of meters depth, are not massive. Distinct groups of almost planar discontinuities split all igneous and metamorphic rock masses into contiguous and practically impervious blocks of quite resistant intact rock. From a strictly mechanical point of view, even if genetically dissimilar, these rock masses and some types of quartzites, crystalline limestone, and dolomites can be collectively classified as *hard-rocks* or *fractured rocks*. The generic term fracture stands for all kinds of restricted partitions or discontinuities within rock masses.

Irrelevant and small amounts of groundwater, almost immobilized, accumulate inside the minute voids and micro-cracks of intact rock blocks and hardly drop under the action of gravity. That kind of water remains practically unavailable for exploitation. On the other hand, significant groundwater quantities may saturate open fractures and different sorts of discontinuities within rock masses. That kind of water, called *free water*, may drain and percolate under its self-weight. Only this type of "gravity-driven" groundwater is the focus of the present book.

Groundwater storage volume is proportional to the rock reservoir's porosity, including interconnected interstitial voids, fractures, and dissolution discontinuities. As porosities of hard-rocks seldom attain 2% of their total volume, they are not considered suitable aquifers. On the other hand, consolidated rocks composed of small fragments of other rocks have porosities ranging from 5 to 15% and up to 25% for unconsolidated sands and gravels and are considered praised aquifers.

However, during the last decades, the growing need for groundwater supply in hard-rock land, mainly in emerging countries, coupled with a better assessment of the seasonal groundwater recharge and corresponding temporary storage at the top of deep weathered hard rock in wet inter-tropical climate, had altered that point of view. Moreover, motivated by attempts to enhance the structural and environmental safety factors of more and more large underground mines and civil public works in hard rocks land, new investigation methods and new technologies based on more sound theoretical concepts have improved hard-rock models.

Public or mine facilities developed under the sea or the groundwater table are significant works where the excavated rock's mechanical behavior interacts with their hydraulics. Examples are access drifts, stopes, panels, galleries, tunnels, shafts, underground hydropower plants, oil and gas storage caverns, and nuclear waste deep disposal.

Natural rock fractures occur from exceedingly small to an enormous scale. A specific petrograph microscope allows the observation of very tiny fractures affecting the rock-forming minerals. In contrast, satellite images may show immense fractures, eventually separating huge continental pieces.

A water drop height summed to its inner pressure defines its potential hydraulic head, provided they have compatible energy units, usually in m. Then, interconnected systems of fractures within hard rocks may convey significant amounts of groundwater – *occasionally mixed with hydrocarbons* – from one place to another, but only if there are potential driving forces between origin and arrival points. Thus, correct appraisal of water transport and pressure transfer within various kinds of hard rocks, under different conditions, is vital to public and mining projects. Only proper geological and geotechnical investigation of the hydraulic properties of hard rocks, jointly with the monitoring of the groundwater hydraulic head's spatial and temporal variation, allows the prediction and control of the hydromechanical behavior of hard rocks. In short, to answer hydraulic problems, the answers to two interdependent questions are always needed:

1. How can a suitable fractured hard rock model be made?
2. How hydraulic charges and induced groundwater flow, for different conditions, can be adequately modeled based on this hard rock suitable model?

There are no simple answers to these questions because they are codependent. After all, the premises of the second answer logically follows the conclusions of the former. Moreover, the modeler must agree with the model user, how correct, and eventually, how accurate the model results must be.

This book, dedicated to hard rock hydraulics, aims to help civil engineers and mining and geologists solve their practical problems in hydrogeology, public works, and mining engineering.

1.1.1 Scope and level of this book

This book is proposed for readers having some knowledge of hydraulics, geology, hydrogeology, and soil and rock mechanics and is structured as follows:

Chapter 1 – About modeling: that introduction discusses some questions concerning modeling.
Chapter 2 – Fundamentals: this chapter covers the fundamentals of hard rock hydraulics under a tensor approach.
Chapter 3 – Approximate solutions: this concise chapter presents concepts regarding approximate solutions for practical purposes.
Chapter 4 – Data analysis: this chapter presents a few data analysis techniques for groundwater simulations.
Chapter 5 – Finite differences: this chapter reviews specific finite-difference algorithms for fractured media.
Chapter 6 – Applications: this final chapter exemplifies some particular unusual applications.

1.1.2 Modeling uncertainty

Before considering several aspects of modeling, it is advisable to review some concepts about modeling uncertainty and how simple statistical methods can help construct preliminary models. However, mathematical models, highly praised by young geologists and soil and rock engineers, are considered in other sections of this book.

Risks may result from sequential and contingent incidents – *i.e., contingent in the sense that they* ***may*** *or* ***may not*** *occur* – concurrently changing in time and space. Consequently, geologists, civil and mining engineers, when working at significant public works or mine complexes, do not ignore that they always are compelled to make decisions under the context of many uncertainties regarding the mechanical and hydraulic behavior of fractured rocks. These *points of no return* may arise at any moment of the development of these projects. For example, they may occur in the feasibility, pre-design, design, construction, and maintenance phase for a hydroelectric power facility. However, for a mining complex, *progressive exploitation* and *long-term safe closure* stages replace *construction* and *maintenance* phases. In both cases, modeling of fractured rock masses may be mandatory but with different approaches.

For a hydroelectric power facility, modeling must guide the design of an enduring safe structure since the early phases of the project. However, models for a mining complex must be routinely revised as geological scenarios change with the exploitation progress. Permanent public works require more sophisticated and more consistent models than those continuously adjusted during mine developments, except if significant mining risks are suspected.

1.1.3 Reducing uncertainty

The first step to reducing uncertainties is establishing an appropriate model for hard rocks based on field mapping, laboratory, and site investigations. Nevertheless, as any model yields no more than a simplified picture of "what is," it must be sufficiently defined to comply with the construction methods, safety requirements, and functional performance. These models combine many communication categories: written, graphical, numerical, but they must be able to translate – *keeping a minimal internal coherence* – all the geological and geotechnical information wanted for the project in hand.

Concerning the project area, these models must allow the following:

- To appraise the site risks for making design decisions from the beginning to the project's closure.
- To anticipate the hydromechanical constraints for construction and operational scenarios.
- To devise the monitoring plans for tracking future project performance.

However, as there are many kinds of models, it is recommended, before choosing the most convenient type, to answer the following:

- What kind of features must the proposed model abstract from the real world?
- Based on the previous answer, what type of model may satisfy the application's requirements in view?

Figure 1.1 displays how modeling may evolve during the growth of a project. To keep simplicity, one may distinguish only two model variants: *factual* and *predictive*. Objective models split into *attribute-based* or *rock-type-based* types. On the other hand, predictive models can be *deterministic* or *probabilistic*. Although a descriptive model is rarely topo-referred (or geo-referred), a predictive model must always be.

The simplest factual attribute-based model, but rarely topographic referred, must always be bounded in space and time. It describes the probable characteristics or features that can be seen or suspected in quasi-homogeneous rock subdomains. When constructing this type of model, the first thing to be done is to delineate the boundaries – *suitable or even ill-defined* – of these subdomains. Their attribute-based features may be the *standard geologic and structural features. T*hey may *be* specified by appropriate adjectives concerning their structural modes, weathering grades, apparent failure or quasi-failure styles, and morphology.

These subdomains must be correctly associated with their adjoining or detached subdomains and with the attributes relevant for the analysis desired. For example, one may want to analyze an aquifer groundwater's behavior during local dewatering, the stability of a natural slope after heavy rains, or the settlement prospect during the drought season. Whenever possible, quantify attributes by appropriate grades referred to their mean and variance and cross-correlated with other features. If needed, modes of occurrence and associated probabilities, even as first approximations, may be given.

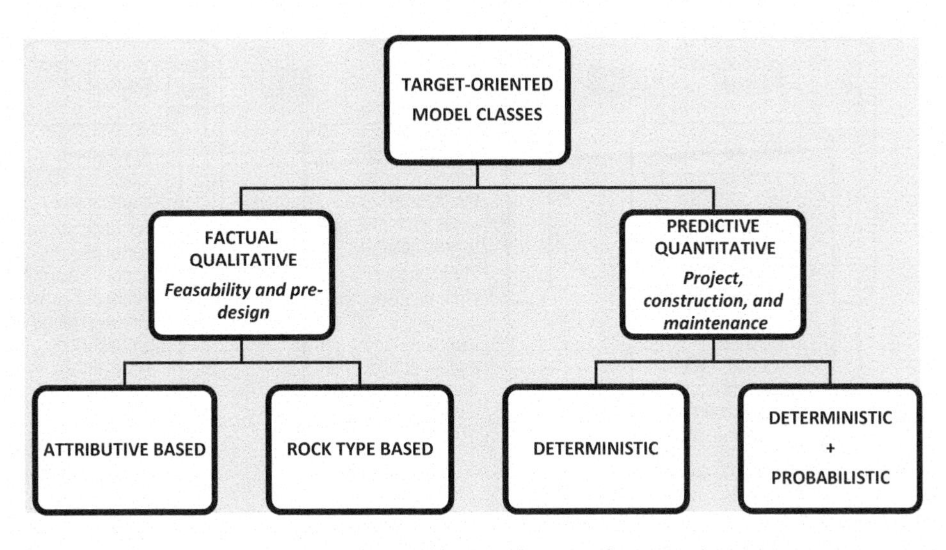

Figure 1.1 Simple target-oriented model classification for a civil engineering or mining project. For a mining complex, *exploitation* and *final safe closure* stages replace *project, construction,* and *maintenance* phases.

Rock-type-based models are condensed variants of attribute-based models to be only used in mature contexts. In this case, quasi-homogeneous subdomains are related to specific regional rock types having their attributes characteristics already statistically established in the past, as for rock classes applied to well-known geologic formations, as the Entrada Sandstones in the Arches National Park, Utah.

During the initial studies for feasibility reports for a hydroelectric facility or a mine complex, its quasi-homogeneous subdomains' boundaries can initially be vaguely located. As investigations proceed, these boundaries must be correctly referred by aerial-photo interpretation and detailed mapping, usually improved by standard topographic procedures, including drone-based photogrammetry. Its main attributes' central value and variances may also be identified with simple or more refined tests.

At the end of the pre-design phase, early 2D factual models are replaced by more exact ones, preferably in 3D. Select appropriate additional investigation techniques from the previous factual models' conclusions to improve the search for critical factual data. These critical occurrences may be previously unsuspected, as this customarily happens. In fact, according to Waltham,[2] *unforeseen ground conditions are, in most cases, only unforeseen because nobody had looked for them.*

An excellent factual model can only result from many subsequent interactions among field and desk works, as described in Figure 1.2.

Prediction models include many mathematical models based on previous factual models' results, now topographically referred. Simple mathematical models are essential to endorse the pre-design phase of big projects but are replaced by sophisticated mathematical models in the design phase. These models quantify the fractured rock masses' hydromechanical behavior under the extreme conditions devised for the project. They can have a traditional deterministic approach but supplemented from a probabilistic perspective.

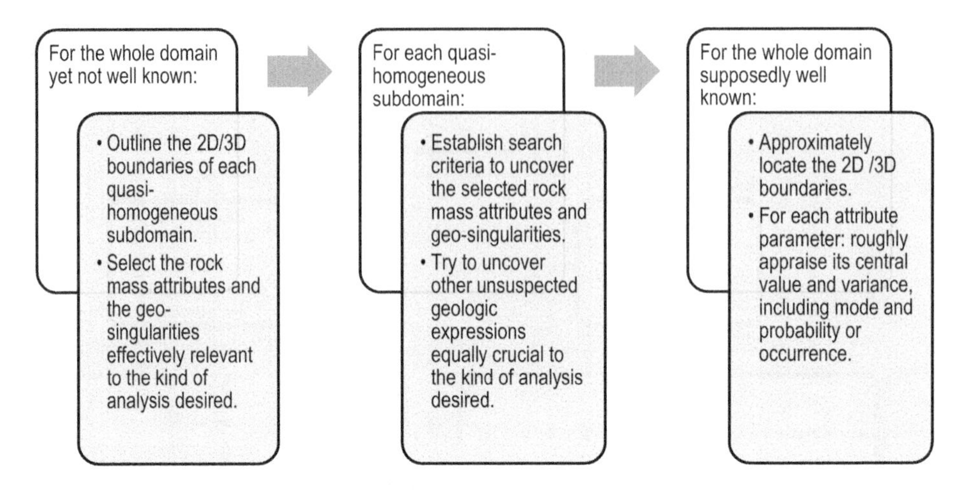

Figure 1.2 An excellent factual attribute-based model usually results from successive interactions between field and desk.

If limited to a deterministic approach, the mathematical model gives a unique answer to each relevant question: a simple **yes** or **no**. In this case, all uncertainties remain uncovered, hidden within the safety factor SF without any quantifier of its confidence. However, supplemented by a probabilistic approach, each possible SF value corresponds to a failure probability P_f, a reliable measure of its uncertainty.

All outputs of predictive mathematical models, either deterministic or probabilistic, have an imposing academic appeal! Nevertheless, without exception, their results rely much more on the quality of its previous factual models than on their algorithm cleverness. These constraints are usually ignored, as mathematical models' outputs often have a good-looking appearance, despite any occasional incoherence of their geological and geotechnical inputs.

As the quality of a factual model largely determines the predictive model's quality, the project's technical staff must conjointly analyze and judge all geological and geotechnical assumptions before passing to the predictive phase. The participants of this technical meeting must consider the following:

- To help spoken interaction during technical discussions, all information about relevant geological features and their significance to the project, including photos and mappings, can be condensed in a PowerPoint presentation as a baseline to organize the critical subjects, including their *pro* and *con* arguments. This presentation is essential to separate natural geomaterials and systematic fracture systems from specific geosingularities that can convey considerable amounts of groundwater and quickly transfer pore pressures over great distances. All geological and geotechnical features must accord to classification schemes proposed by credited professional associations, occasionally adapted to the project's natural environment. During these meetings, some doubts may naturally arise that can be verbally solved. However, additional explanations must be summarized and added to the original presentation for future meetings and pertinent doubts.

Finally, each geomaterial and geosingularity may bear a local generic name, as a tag, still genetically derived but easily memorized. A simple geological description must follow it but only applied to the concerned site.

- All geological structures are described at various observation scales but pointing out existing cross-correlations, for example, between small sigmoid wrinkles in a rock hand-sample and the dip and direction of sequential and similar prominent folds. It is also essential to describe the exterior appearance and each remarkable geological feature's internal structure, including its geometrical dimensions. Then, clarify how these features can be directly noticed with naked eyes and, if applied, in other observation scales: satellite image, air photo, outcrop, borehole sample, hand sample. Equally important is to explain to the non-specialists of the project staff how these features were inferred in depth by trained geologists or by an experienced consultant.
- Preliminary zoning of a fractured hard rock may result from many differentiation criteria. These criteria may be lithological variations, weathering grades, fracture density, granulometry. They can also be combining to define more distinct zones in the rock mass. For each defined zone, its geometrical parameters and other useful quantifiers, even if roughly estimated, must be communicated, preferably by their typical, significant, and expected values. As, during the development of a project, new information comes from the field, the original zoning, previously subdivided into quasi-homogeneous subzones, is reconsidered.
- Well-defined zones alternating, almost rhythmic, as for shales and marls, but having different physical properties, are duly signaled. However, when these characteristics gradually change in space, as weathering grades in an intertropical climate, without well-defined contacts, subzones limits are suitably chosen to allow their judgment given the project's technical requirements. In these meetings, alerts are given to suspicious mineral associations in certain zones, pointing to future problems.
- Exceptional and enormous singularities, such as some megafractures, dikes, sills, dissolution caverns, are described at different observation scales, for they may pass easily unnoticed in a field walk.
- Each group of a fracture system having peculiar properties must be individually considered, including the description of its mode of occurrence and its classical parameters: persistence, modal spacing, dip, dip-direction. Comments about their natural condition: sealed or open, type of fillings, moist or dry, weathering grade, some conjecture about its hydromechanical properties are also wanted.

To conclude, one must keep in mind that the quality of the predictive models' answers relies obviously on its inputs' quality. Also, it is possible to measure the uncertainty **i** of these answers by the Information Theory. That theory, not covered in this book, states what its metrics **i** communicates:

- **i** = 0: zero input uncertainty, then an exact predictive model may give a 100% dependable answer;
- 0 < **i** < 1: some input uncertainty, then an exact predictive model cannot give an absolute dependable answer; all project managers must always ask for inputs based on more reliable factual models to minimize the input uncertainty and get more positive predictive model results, notably when fatalities may occur;
- **i** = 1: total uncertainty, unable to answer **yes** or **no** to any question.

The probability of a specific feature occurrence or event can take any value between 0% and 100%. However, when answering **i** values, it is essential to remember that, according to information theory, quantifying a **yes** with X% confidence is the same as quantifying a **no** with (1 – X)% confidence because both answers are mutually exclusive. It follows that 0% or 50% probabilities imply *absolute belief* or *absolute doubt*. That remark is valid either for a **yes** or a **no**. That premise must always be in mind before starting any technical dialog involving geologists and civil or mining engineers. It is desirable to qualify an answer by its probability of occurrence instead of a plain **yes** or **no**.

Figure 1.3 shows an example where a straight answer at the start of a project was impossible. In this case, the project manager asked the civil engineer if two crossing subvertical diabase dikes near the shaft could be considered hydraulic barriers? The civil engineer was prone to decide based on quick answers **yes** or **no**, but the geologist doubted, and the question remains unanswered, and in a future decision round should involve risk concepts. Paying attention to Wahlstrom[3] guidance: *It is a mistake for a geologist associated with engineering planning and design to extend guesswork beyond reasonable limits but is equally a mistake for the engineer to ask the geologist to provide interpretations of bedrock based on anything less than a very careful geological investigation.*

However, how to acquire a minimum but almost sure knowledge in advance before doing complementary field investigations that take time and are usually costly?

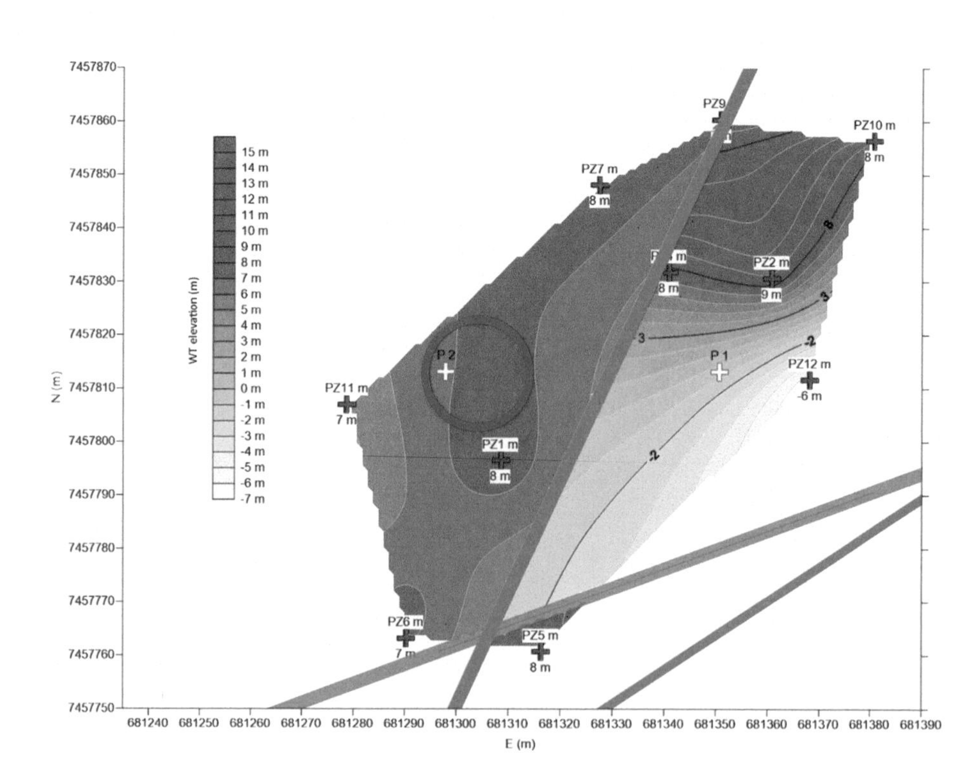

Figure 1.3 Probable configuration of the hydraulic heads (November 25, 2014) around a vertical shaft for an underground metro station but assuming the two subvertical dikes as hydraulic barriers for the groundwater.

To dramatically and rapidly reduce the uncertainty, make successive reviews of the project site's geological and geotechnical characteristics based on topo-referred factual models, as suggested in Figure 1.4,[4] but at different observation scales and specialized professionals, such as field geologists, geophysical experts, petrographers, and laboratory analysts.

Many geological features that may jeopardize the future functionality or even the structural stability of a planned project can only be detected if simultaneously analyzed and cross-correlated with other observations at appropriated scales.

The gain resulting from making geological investigations at many observation scales exceed the simple detection of a relevant feature but initially unsuspected at a particular scale. There always exists the possibility of predicting the occurrence of other types of significant features if directly or indirectly noticed in other observation scales, bigger or lower than the project scale itself.

Observations and cross-correlation at various observation scales produce a low preliminary uncertainty **i** and avoid costly and slow-paced field borings. Additionally, for a specific spatial domain, the complementary investigations must not rely on the results of a unique and exhaustive kind of prospection. Prescribe complementary investigations considering all previous experience and all cross-correlated information at other observation scales and cheap prospecting types.

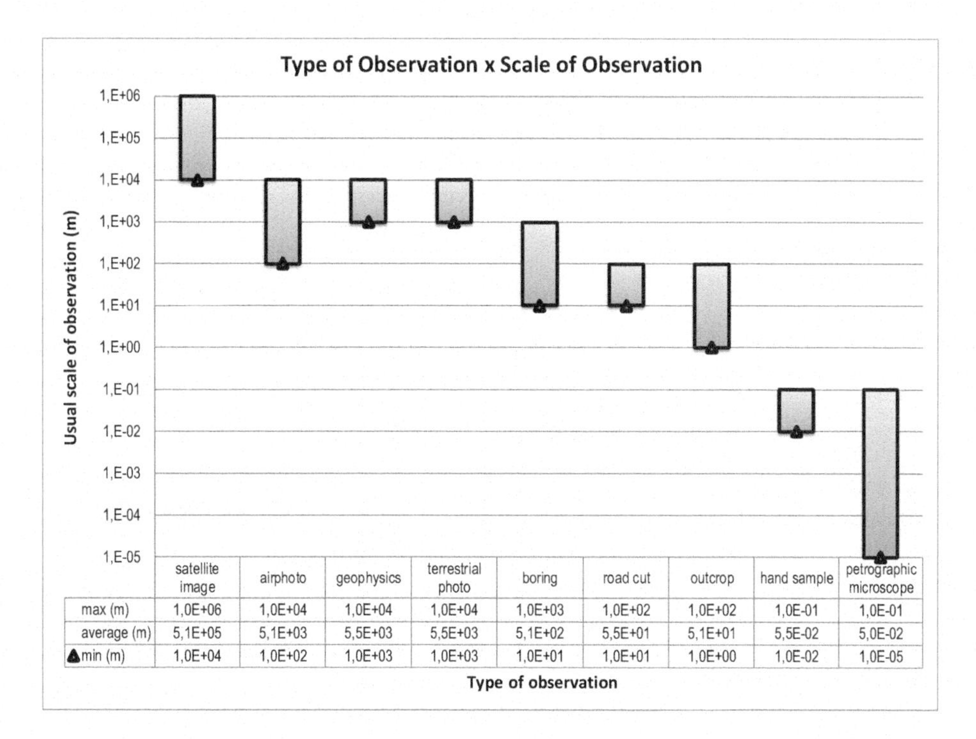

	satellite image	airphoto	geophysics	terrestrial photo	boring	road cut	outcrop	hand sample	petrographic microscope
max (m)	1,0E+06	1,0E+04	1,0E+04	1,0E+04	1,0E+03	1,0E+02	1,0E+02	1,0E-01	1,0E-01
average (m)	5,1E+05	5,1E+03	5,5E+03	5,5E+03	5,1E+02	5,5E+01	5,1E+01	5,5E-02	5,0E-02
▲min (m)	1,0E+04	1,0E+02	1,0E+03	1,0E+03	1,0E+01	1,0E+01	1,0E+00	1,0E-02	1,0E-05

Figure 1.4 Different observation scales to build topo-referred and factual-data models during preliminary geological investigations at relatively low costs. They include tele-detection, aerial-photographs, field maps, boring samples appropriately located, cheap field or laboratory tests, petrographic analysis.

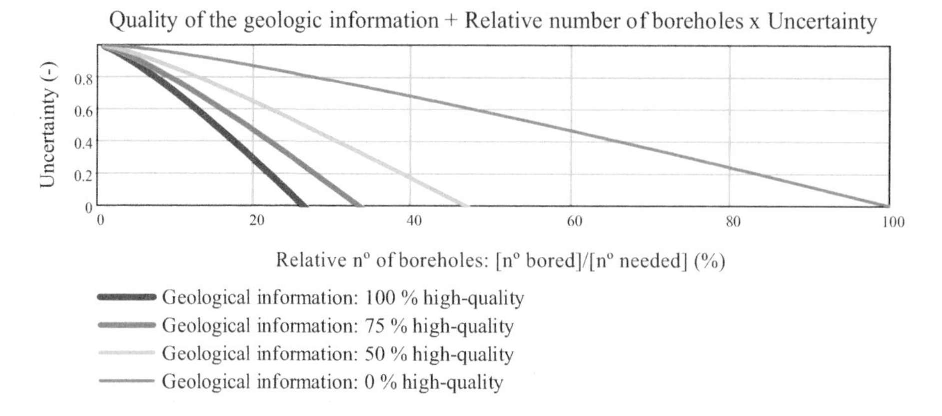

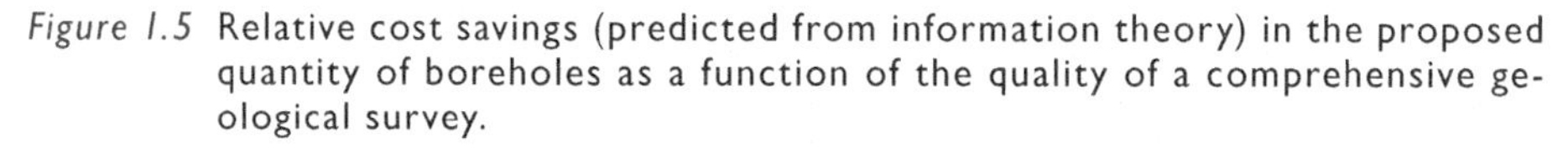

Figure 1.5 Relative cost savings (predicted from information theory) in the proposed quantity of boreholes as a function of the quality of a comprehensive geological survey.

For instance, appraise rock-mass characteristics along a tunnel by geophysical methods, but investigate the local bedrock's soundness in a tunnel portal by rock borings, never the contrary.

Finally, costs and time savings in rock borings for any civil or mining work rely on the quality of the previous comprehensive geological survey, as suggested in Figure 1.5.

The following item gives a few practical examples to show how preliminary factual models may be constructed, even from inadequate information. It also exemplifies how traditional but straightforward statistical methods can help answer questions about geological-geotechnical issues, usually in probabilistic terms, to analyze possible risks.

1.1.4 Practical examples

1.1.4.1 Example of the difficulty to detect karstic features

Figure 1.6 shows the estimated probability of spotting unsuspected highly permeable karstic dolomitic rocks within the gneissic foundation of a small-height but 5 km long earth dam.

In this case, as Figure 1.6 shows, to spot these very permeable dolomitic rocks with 90% chance would require 179 rock borings regularly spaced, a figure much less than the 28 executed. However, a previous field-walk around the dam site area before starting the borings would reveal some sparse outcrops of weathered pinnacles of karstic dolomite near the planned dam suggesting the possibility of the occurrence of similar rocks under the dam axis. That possibility was later confirmed by an electrical-resistivity geophysical survey along the dam axis.

1.1.4.2 Example of the value of published information[5]

Figure 1.7 shows presumed shear and extension faults in an extended geomorphic zone to be crossed by several tunnels running SE-SEE. Figure 1.8 indicates the probabilities

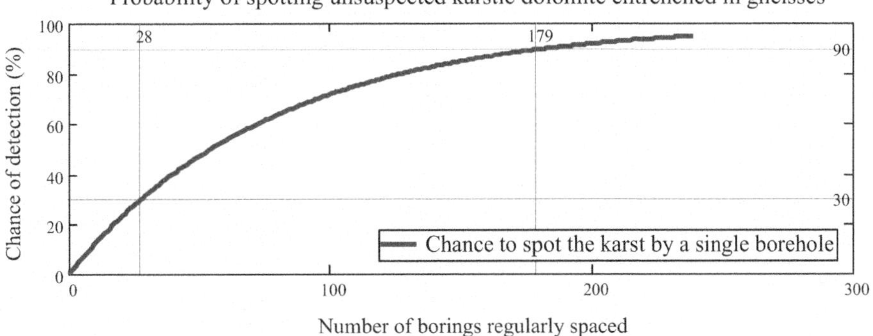

Figure 1.6 Chance of spotting an unsuspected 20 m wide karstic and very permeable dolomitic rocks within a sizeable gneissic basement at the foundation of a small-height earth dam about 5 km long. Estimated probability considered regularly spaced borings.

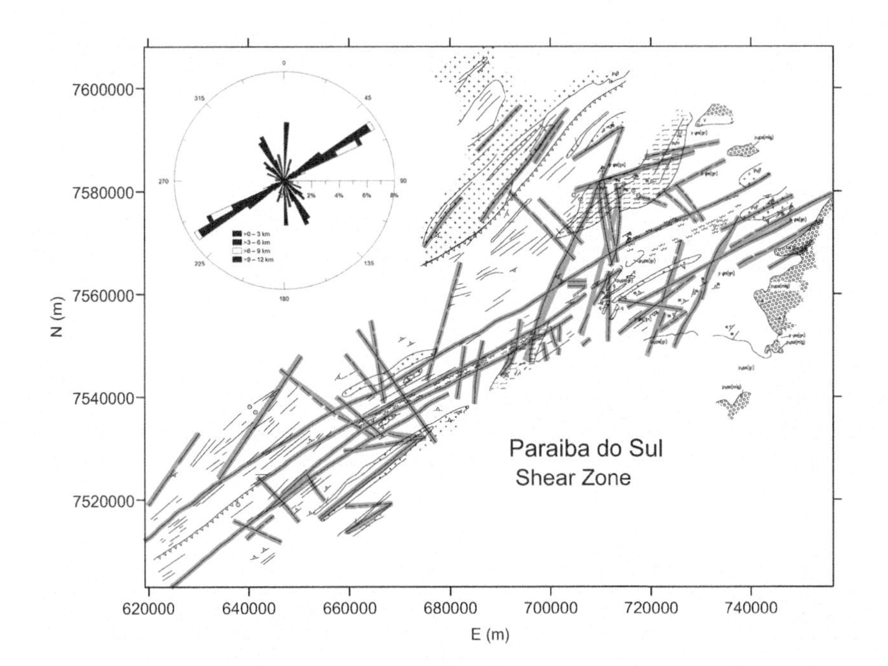

Figure 1.7 Portion of a 3 km wide zone of the Serra Geral gneissic basement in the Brazilian Plateau's eastern portion. It shows long tectonic lineaments recently reactivated during the Cenozoic-Mesozoic transcurrent tectonics, running NE-SW to NNE-SSW, WNW-ESE to E-W, N-S, and NW-SE. Regional seismicity points to mostly NW-SW compressive stresses with local deviations to E-W and NS. A few tunnels, 3–10 km long, were planned to cross that belt along NE-SW. These tunnel crossings could be occasionally tricky due to adverse faulted rocks properties and the natural state of stresses.

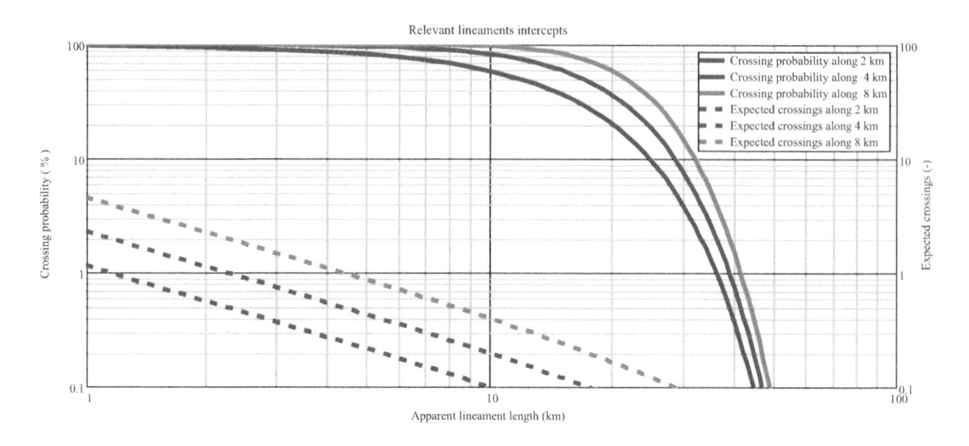

Figure 1.8 Probability of interception (curved lines) and the number of interceptions (straight dotted lines) for tunnel crossings of relevant lineaments. Read the apparent length (km) on the horizontal axis. The probability of interception (%) and the number of interceptions (–) are, respectively, indicated in the right and left vertical axis.

of crossing these faults during the future excavation of 2, 4, and 8 km long tunnels. Although all these presumed faults could not necessarily be considered adverse in a preliminary conjecture (but as someone proved to be during tunnel excavations), they deserved attention.

1.1.4.3 Example of the consequences of not detecting minor geologic features

To avoid common mistakes, always consider the implications of critical minor details detected in rock core samples. In this example, pieces of evidence of an unexpected risk for the stability of a roller-compacted concrete RCC dam only become unquestionable just before finishing its bedrock drainage galleries. Then, duly reappraisal of all stored old core samples by geologists yields the corrected bedrock model. Figure 1.9 shows diagonally crossed zones indicating subparallel planes of weak shear strength at elevations where the rock core samples revealed many interspersed graphite coatings: potential weak shear strength zones.

1.1.4.4 Example of anticipating future WT levels

For project planning, the project manager needed to presume the WT level at a monitored piezometer, but after the dry season. The only solution was to guess the WT level decay after ± 100 days of observations, but with only consecutive 26 observations (see Figure 1.10). As expected, this prediction's standard deviation was relatively high, suggesting a final WT level between 9.6 and 17.2 m at the end of these 100 days.

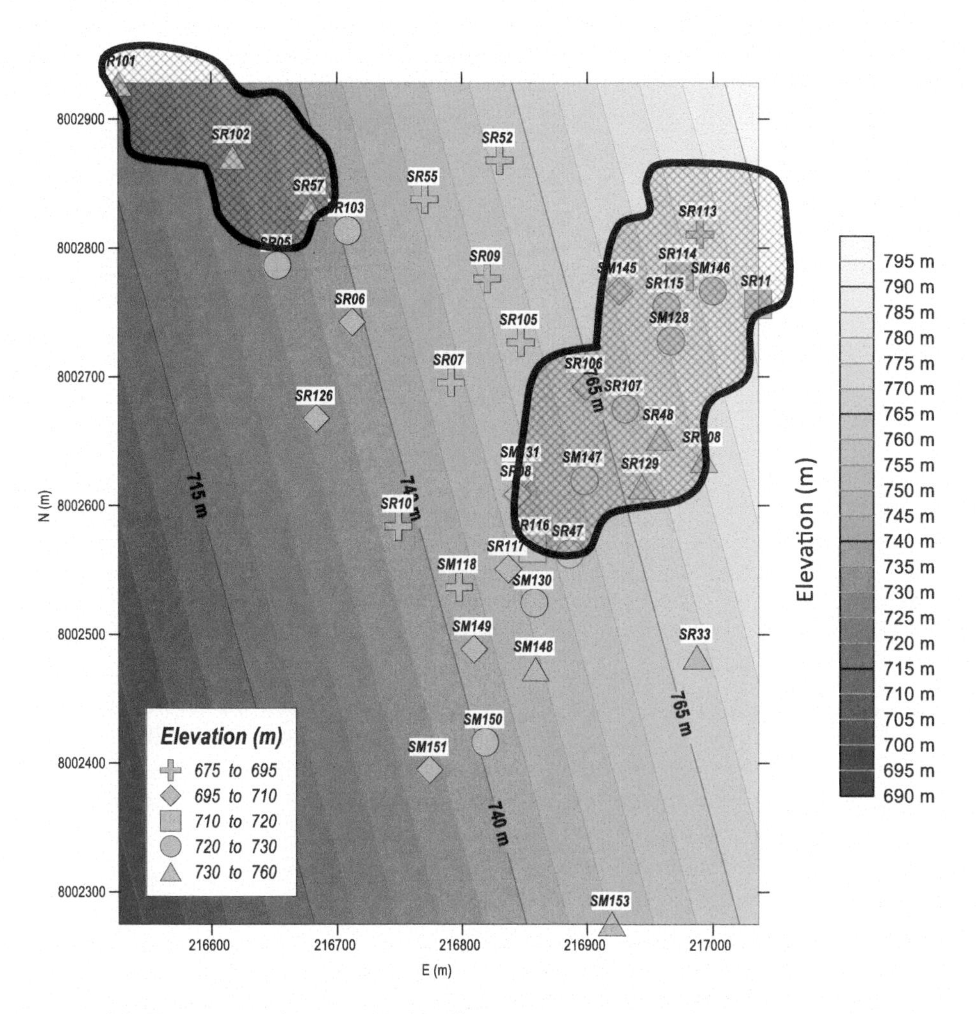

Figure 1.9 The cloud of points (E, N, Z) within the diagonally crossed areas correspond to a dense suite of weak shear strength planes, but only perceived after the foundation excavation. However, the reappraisal of old rock core samples disclosed many interspersed graphite coatings corresponding to these weak planes, defined by the figure's average parallel lines.

1.1.4.5 Example of pumping-rates predictions

This example concerns predictions, made according to Burg's method,[6] of daily and monthly pumping-rates associated with the gradual lowering of the water table above an underground mine, subject to seasonal fluctuations. The predicted yearly pumping-rate time-series, including trend evaluations, suggested a moderate growth trend related to mine's slow deepening. Given its importance to the excavations'

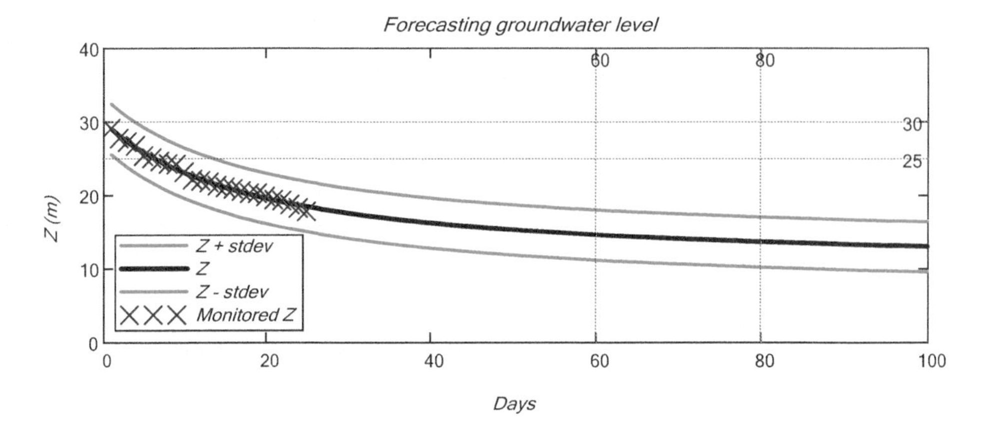

Figure 1.10 Prediction of the groundwater level at the end 100 days of drought (shorter historical period: 86 days) based on only 26 daily observations counted from the presumed beginning of the dry period. The expected values resulted from the adjustment of a decay function of type A·exp[1/(B + Ct)], where the independent variable t symbolizes days and A, B, and C the adjusted parameters by the least-squares method.

physical and financial constraints, these predictions, based on several past values, were made annually for every five years ahead, despite its low reliability. The deepening of the groundwater level, probably due to the climatic change and the rock permeability reduction at low levels, slightly modified the predicted seasonal cycles. Figures 1.11 and 1.12 exemplify the typical results of the predictions. Errors of five prediction years varied between 15% and 20% of the correct values.

1.1.5 Concluding remarks

Specialized geologists and soil and rock engineers, duly integrated, coordinated, and dominate appropriate expertise to cooperate with project managers to solve practical technical questions associated with public works and mining.[7]

At first, the geologist, backed by what he can see and uncover, is an essential professional to help do risk analysis because he can *qualitatively* model probable interactions between the future project and the soils and rocks at the project site. Later, the soil and rock engineer can *quantitatively* model these interactions but obeying all technical constraints for the construction and operational phases. Presently, both models allow reliable risk assessments of the integrity and functionality of a project under many anticipated situations.

However, if geological and geotechnical investigations follow simple passe-partout standards, they do not give consistent results or avoid future losses. It is the project that must obey the site constraints and not the contrary.

Sometimes, after deciding to go ahead, but only based on financial criteria, a project may be conceived in a hurry, almost irreversibly, without duly considering its site environment's main restrictions.

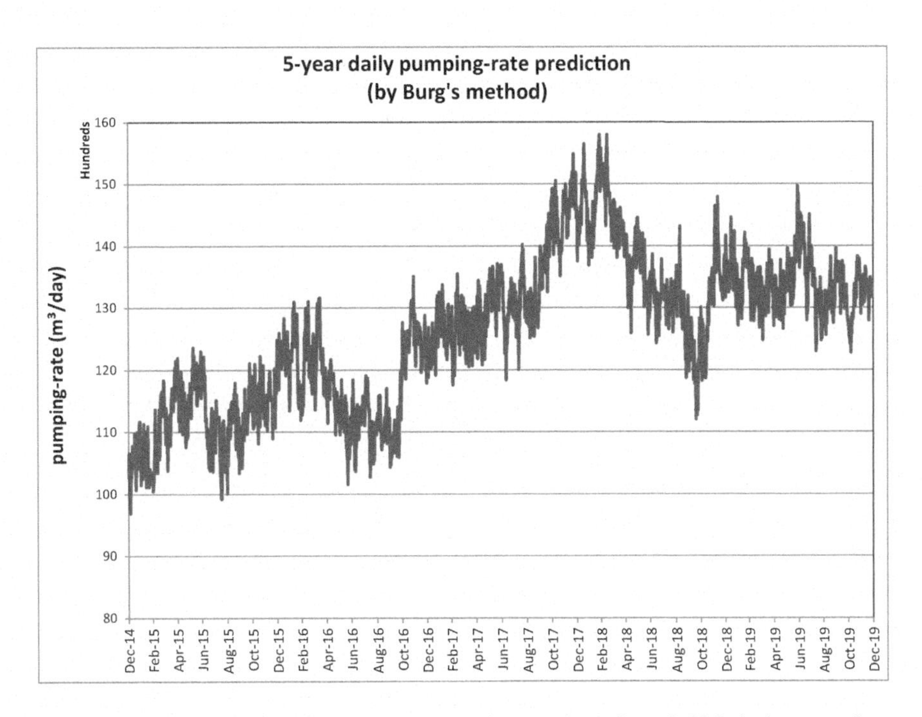

Figure 1.11 Five-year daily pumping-rate prediction based on 1,828 daily records.

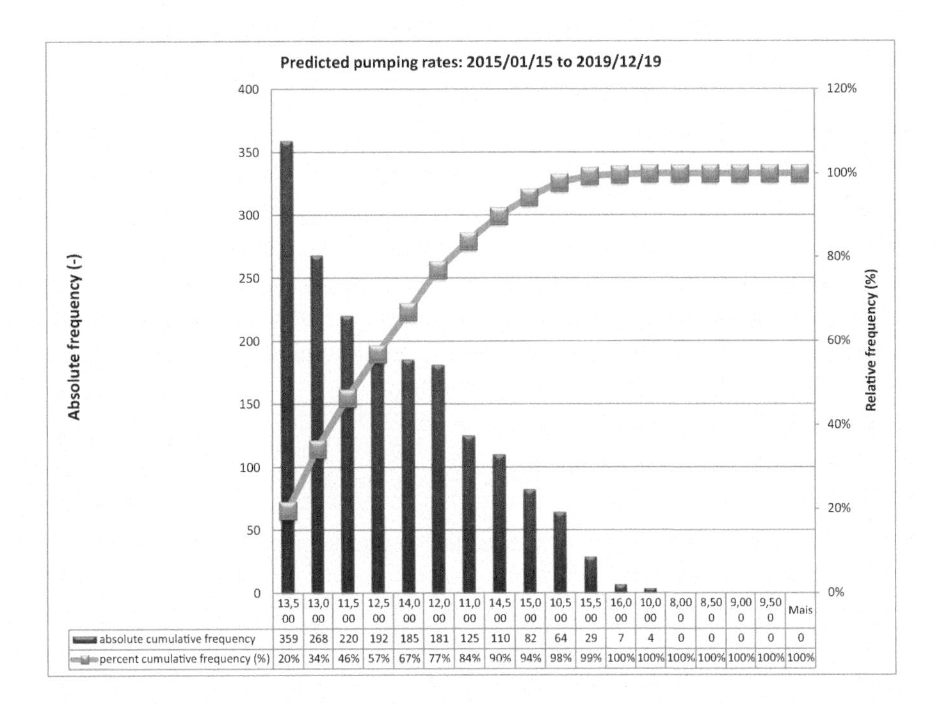

	13,500	13,000	11,500	12,500	14,000	12,000	11,000	14,500	15,000	10,500	15,500	16,000	10,000	8,000	8,500	9,000	9,500	Mais
absolute cumulative frequency	359	268	220	192	185	181	125	110	82	64	29	7	4	0	0	0	0	0
percent cumulative frequency (%)	20%	34%	46%	57%	67%	77%	84%	90%	94%	98%	99%	100%	100%	100%	100%	100%	100%	100%

Figure 1.12 Five-year monthly pumping-rate prediction based on 59 monthly records.

However, natural conditions are the only variables that Mother Nature imposes on the builder and from which he cannot escape. When unforeseen difficulties arise, and costs increase, and deadlines failed, it is almost impossible to change the project conception. It is only possible to accept the escalating costs and the urgent need to make unpredicted and undesirable adaptations. These unexpected extra costs are usually attributed to unforeseen "geological conditions" by the stockholders and the financial staff, unfamiliar with the geological inference's possibilities. However, the author maintains that most unexpected extra costs are foreseeable if the project risks are adequately analyzed, naturally with some error margin, by a competent technical staff that meticulously scrutinizes their potential adverse constraints. That activity is the core of all risk analyses leading to a safe and economical project.

New projects are built faster and frequently involve less favorable sites. Consequently, their risks grow at the same pace! Facing more and more aggressive financial markets but trying to ensure competitive profits, new civil works or mining extractions became increasingly industrialized. As a result, job planners approve equipment, materials, processes increasingly efficient and performant, presumably more suited to their new jobs' natural environments. Nevertheless, when unforeseen disasters happen, it is not easy to devise new technical solutions or to change costly equipment already allocated to the site. The extra costs of these adaptations may appreciably exceed the budgeted amount, considered initially profitable.

However, a poorly planned project may still go ahead, albeit at a slower pace. But at what price? In public works, invoking ambiguous contractual clauses may aid recovering part of the costs increase, usually imputed to ill-defined geologic risks in the construction contract. Nevertheless, these annoyances may result from a forced work schedule and insufficient investigation resources to face the natural environment's complexity at the chosen project site.

Undoubtedly, experienced technical staff can plan cost-effective investigations for large projects and forecast, with tolerable error margins, the chief difficulties to be overcome. Nevertheless, already known geological-geotechnical imperatives may be ignored – *if not neglected* – by the job organizers, giving rise to embarrassments when schedule delays and cost increases occur. In these moments, consultants, geologists, and soil and rock engineers are hurriedly assembled – *assuming the role of firefighters facing unexpected significant fire events* – to propose remedial solutions and avoid the worst. In a little time, generally insufficient, they should appraise the situation, evaluate the terrain, program, and perform a minimum of complementary studies, if possible, and solve the difficulties in question.

Then, to avoid, at least partially, future financial problems, owners and "stakeholders" must take into due consideration the advice of knowledgeable and experienced professionals during the planning, schedule, and budget preparation of any significant undertaking.

To conclude this item, it is worth considering that despite all the care observed during the acquisition, interpretation, interpolation, and extrapolation of all laboratory and field data, the resulting geological-geotechnical model must not conflict with the practical experience obtained with previous projects in the same area. If there are divergences, the final predictive model must be revised.

Thus, one of the essential activities to be conducted before proposing a factual model is to make a selective survey of the past geological-geotechnical issues that have occurred in the project region. That survey must include natural events, like earthquakes, floods, earthquakes, and those induced by the construction or operations of roads, canals, dams, underground works, open-pit, or underground mining. Check all sources of useful information: old newspapers, event registers, risk registers, as-built of old civil works and mine extraction, personal or group testimonials.

Notes

1 An excellent review on this subject can be found in "The new global lithological map database GLIM: A representation of rock properties at the Earth surface Jens Hartmann and Nils Moosdorf Institute for Biogeochemistry and Marine Chemistry", KlimaCampus, Universität Hamburg, Bundesstrasse 55, DE-20146 Hamburg, Germany (geo@hattes.de).
2 Waltham, T., *Foundations of Engineering Geology*. Spon Press, London, 2002, 2nd ed., chap. 37.
3 Wahlstrom, E. E., "Tunneling in Rock", *Developments in Geotechnical Engineering*, E. E. Wahlstrom, ed. Elsevier, Amsterdam, 1973.
4 Franciss, F. O., "Curriculum Planning Project for Fundamental Courses on Geotechnology", II International Congress of Engineering Geology, São Paulo, Brazil, 1974.
5 Suguio, K., "Geologia do Quaternário e Mudanças Ambientais"; fig. 127; Biblioteca Nacional, Câmara Brasileira do Livro, S.P., Brazil, 1999.
6 Burg, J. P., *A New Analysis Technique for Time Series Data*. NATO Advanced Study Institute on Signal Processing with Emphasis on Underwater Acoustics, Enscheda, The Netherlands, August 1968, reprinted in *Modern Spectrum Analysis*, D. G. Childers, ed. IEEE Press, New York, 1978.
7 Detailed duly considerations about that subject can be appreciated in Pierre Martin, *Géotechnique appliquée au BTP*. Editions Eyrolles, 61, bld. Saint-Germain, 75240, Paris, Cedex 05, 1997.

Chapter 2

Fundamentals

2.1 Introduction

Soils and rock hydraulics concern the fundamental differential relations for modeling groundwater movement. However, different models describe this movement with its advantages and disadvantages. Then, what sorts of models and description techniques this book present? For what kinds of practical applications this book considers?

Concerning the first question, this book mainly deals with subjects limited to the author's professional experience, mostly dealing with hard rocks. That rock type is usually inhomogeneous and anisotropic. Geological contacts, faults, fractures, and other discontinuities command their hydraulic behavior.

Concerning the second question, this book favors two approaches: an integrated and discretized analysis. The integrated analysis only requires the knowledge of the hard-rock domain's average properties to give satisfactory solutions, but the discretized analysis requires its partition into small subdomains. Moreover, each small domain needs information about the architecture of its discontinuities and related hydraulic properties.

This book presents specific 3D finite difference algorithms to deal with complicated problems concerning hard-rock hydraulics regarding discretized models, without restrictions about anisotropy, inhomogeneity, and shape of discontinuities.

2.2 Basic concepts

2.2.1 Pseudo-continuity

The governing equations of groundwater hydraulics adhere to the fundamental assumption of Continuum Mechanics: *the continuum concept of matter*. According to this concept, the groundwater movement description through pervious soils and rocks may be described from a "macroscopic" point of view, but taking implicitly into account all interactions, heterogeneities, and anisotropies, and discontinuities influencing the groundwater movement from a "microscopic" point of view. However, limits separating "macro-scale" from "micro-scale" depend on the analysis's purposes and the pervious media characteristics.

The *continuum* concept applied to a soil or rock domain assumes continuity of matter in space and time. Consequently, to model a groundwater system according to this concept, two critical conditions must be fulfilled:

- First, the domain subsystems' size is set above an almost practical limit, separating the macro from the micro observation scale.
- Second, *a pseudo-continuum* fictitious system replaces the real discontinuous stuff but having an almost similar hydraulic behavior.

If these two conditions are satisfied, it is possible to construct an acceptable numerical model by observing the following preliminary steps:

- *First*: select the observation scale for the pseudo-continuous system to assure compatibility between the model size and the desired details.
- *Second*: according to the most convenient observation scale, split the pseudo-continuous system into several smaller pseudo-continuous subsystems.
- *Third*: describe the hydraulic properties of these subsystems and formulate their corresponding governing flow equations to nearly replicate the system's actual hydraulic behavior at the elected observation scale.

For usual numerical models, the modal (*most observed or most frequent class of values*) linear dimension of the pseudo-continuous subsystems generally does not exceed 1/30 to 1/300 of the model's most significant linear dimension.

2.2.2 Porosity and permeability

From a physical point of view, earth and rock masses are three-phase mixtures, or assemblages, of mineral particles, water, and air combined with a small proportion of other gases. Figure 2.1 shows two samples of pervious soils and Figure 2.2 shows two samples of fractured hard rocks.

Figure 2.1 The above image shows two kinds of granular soil images: cobbles and gravel. Both types, when immersed below the water table, allow the movement of groundwater within their voids.

Figure 2.2 Two examples of hard rock images. At left, an almost horizontal exposure showing shear and extension fractures. At right, a vertical cut showing an almost rectangular fracture system. Both rock masses allow rainwater to infiltrate at all rock exposures and spread across within their fractures during rainy seasons.

Space and time variations of the state of stress acting on soil and rock masses may modify their solid phase structure and their liquid and gaseous phases' flow pattern. In fully saturated soils or rocks, the gaseous phase is almost inexistent, but tiny proportions of gases may be present, dissolved in the aqueous phase. These three-phase systems are considered pervious when groundwater freely percolates within their voids under the gravity action. This percolating water behaves like an almost Newtonian liquid of extremely low dynamic viscosity (1.310 cP at 10°C).[1] A small fraction of this water phase, called pellicle or adherent water, stays bonded to the solid phase due to physiochemical and capillary forces. The smaller the voids or thickness of the fractures, the more significant the adherent-water relative amount. This amount attains almost 100% for void size or fracture opening smaller than 10^{-2}mm. Between 10^{-4} and 10^{-2}mm, nearly all adherent water is due to capillarity, and below 10^{-4}mm, this water behaves like a non-Newtonian fluid, a viscoplastic liquid.

From the hydrogeological point of view, a simple classification of soils and rock is:

- Impervious but porous with non-connected voids: e.g., unfractured and non-weathered vesicular basalt.
- Poorly pervious, but with minimal interconnected voids and a significant amount of adherent water: e.g., coastal clays.
- Pervious, but with interconnected voids and a small amount of adherent water: e.g., aeolian silts, with capillary voids size between 10^{-4} and 10^{-2}mm, and fluvial gravels and cobbles, with non-capillary voids size between 10^{-2} and 10^{2}mm.

2.2.3 *Volumetric and gravimetric relations*

Practical mass–volume relations describe the relative amounts of the *solid*, *liquid*, and *gaseous* phases within three-phase mixtures. However, while a homogenous and isotropic medium (in a statistical sense) admits an average description, a heterogeneous medium may require a space-dependent description because its heterogeneity and anisotropy vary with direction, consequently, variations of its physical properties. Also, while simple scalar numbers can describe homogeneous and isotropic systems' properties, this is not the case for anisotropic ones.

2.2.3.1 Homogeneous and isotropic systems

Figure 2.3 shows the gravimetric and volumetric phases for a three-phase permeable system.

The ratio V_W/V_T symbolized by n denotes the total porosity; note that $0<n<1$. The effective porosity represented by n_e corresponds to the ratio $(V_W - V_{aW})/V_T$, where V_{aW} is the adherent-water volume. Then, $n_e<n$.

The ratio V_W/V_V, denoted by S, measures the liquid phase's degree of saturation, generally in percentage. Then, $0\% \leq S \leq 100\%$.

There exist standardized laboratory procedures to measure all these volumetric and gravimetric relations for soils and rocks. The evaluation of the porosity of cobbles, gravels, or other types of coarse granular rocks may be found by the division of the

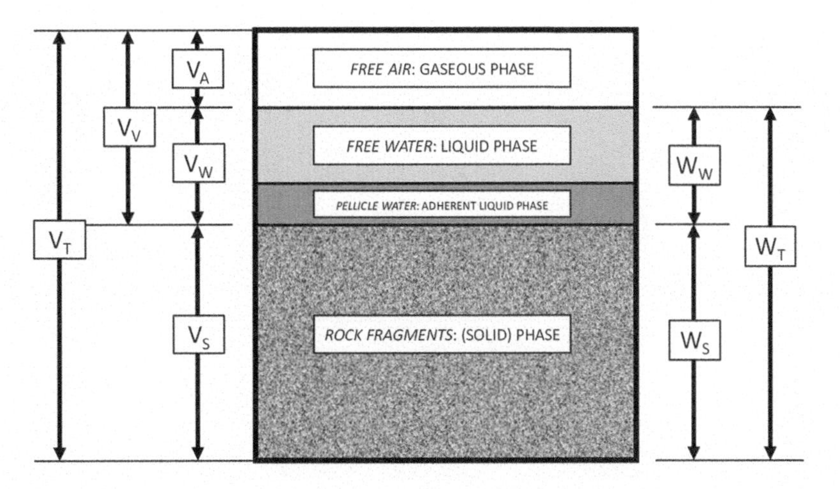

Figure 2.3 Relative proportions of air, water, and solid phases in soils and rocks; the symbols V and W stand for volume and weight; the subscripts A, W, and S stand for air, water, and solid phases, whereas T stands for mixtures of all phases. The general volumetric and gravimetric relations are $r_i = V_i/V_T$ and $R_i = W_i/W_T = \rho_i V_i/\rho_T V_T$, where ρ_i designate the mass-density of the i-phase.

sum of their void linear (or areal) dimensions measured on a scanning line (or planar surface) by the scan line total length (or the total area).

2.2.3.2 Homogeneous and anisotropic systems

Second-order tensors describe anisotropic properties, formally portrayed by nine numbers but reducible to three mutually orthogonal axes of an ellipsoid, as the well-known stress tensor.

Soils characterized by platy structures or rocks demarcated by groups of almost parallel fractures but of different thicknesses and spacing have anisotropic physical properties. Figure 2.4 exemplifies a typical anisotropic rock mass outcrop.

While scalar numbers can describe the porosity of homogeneous and isotropic soils and rocks, anisotropic systems require its description by a second-order tensor, as exemplified in Figure 2.5. This figure shows a cubic rock mass divided by three almost mutually orthogonal groups of joints, each group i defined by an average spacing d_i and an average opening e_i at each axis direction i. Then, for each cubic face of total area A_i, and total voids area a_i, its porosity n_i viewed along each axis direction i can be roughly estimated by:

$$n_i = \lim_{A_i \to 0} \frac{a_i}{A_i} = \frac{d_j}{e_j} + \frac{d_k}{e_k} \quad \text{for } (i \neq j \neq k)$$

Figure 2.4 Image of a succession of rocky slabs inclined at an angle of almost 45° (see the white dip symbol) where significant discontinuities are parallel to the exposed face while two other sets cut the plunging slabs (see their orientations depicted by the two arrowed white lines). In reality, these apparent massive slabs result from a dense package of minor slabs (see inside the white circle).

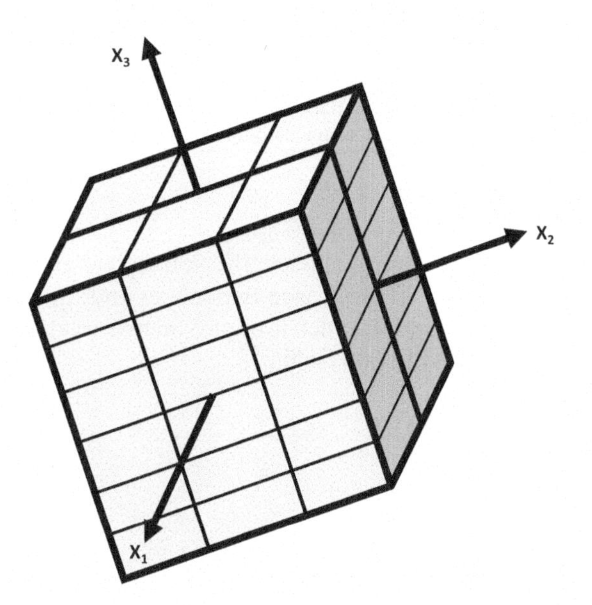

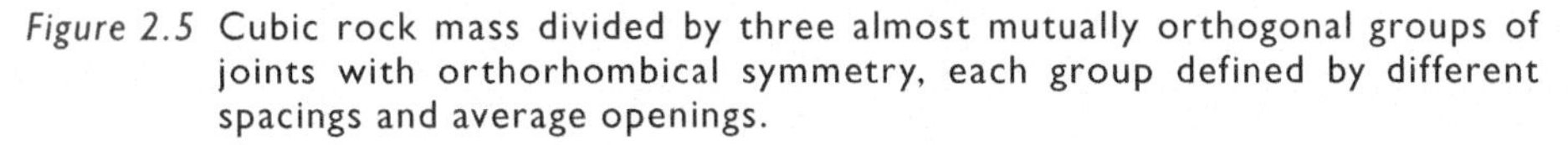

Figure 2.5 Cubic rock mass divided by three almost mutually orthogonal groups of joints with orthorhombical symmetry, each group defined by different spacings and average openings.

The three scalars n_i (n_1, n_2, and n_3) are the principal porosities of the second-order tensor [n] written as:

$$|n| = \begin{pmatrix} n_1 & 0 & 0 \\ 0 & n_2 & 0 \\ 0 & 0 & n_3 \end{pmatrix} = \begin{pmatrix} \frac{d_2}{e_2} + \frac{d_3}{e_3} & 0 & 0 \\ 0 & \frac{d_1}{e_1} + \frac{d_3}{e_3} & 0 \\ 0 & 0 & \frac{d_1}{e_1} + \frac{d_2}{e_2} \end{pmatrix}$$

However, these unique scalars n_i ($i = 1$, 2 and 3) are associated with a particular set of axes x_i, called its principal axes. Concerning another set of axes, the same tensor has nine symmetrical components, that is, $n_{ij} = n_{ji}$:

$$|n| = \begin{pmatrix} n_{11} & n_{12} & n_{13} \\ n_{21} & n_{22} & n_{23} \\ n_{31} & n_{32} & n_{33} \end{pmatrix}$$

With matrix notation, it is easy to describe the same second-order tensor [n] in any other system of coordinates, preserving its character and allowing calculations in any reference system. As the hydraulic conductivity for fractured hard rocks can only be described by second-order tensors, as clarified ahead, it follows that models for hard rocks can only give good results if worked accordingly.

2.2.2.3 Heterogeneous systems

For inhomogeneous soil and rock masses, the volumetric and gravimetric relationships presented in Section 2.2.3 are space-dependent and, in some cases, also time-dependent. In this case, general volumetric or gravimetric relations r_i or R_i can be only locally defined by the following expressions:

$$r_i = \lim_{\delta V \to 0} \frac{\delta V_i}{\delta V_t}$$

$$R_i = \lim_{\delta V \to 0} \frac{\delta W_i}{\delta W_t}$$

Relatively simple heterogeneous systems may allow functional approximations or permit the entire system's resolution into almost quasi-homogeneous subsystems.

2.2.4 Observation scales

Natural hard rock discontinuities delineate interconnected and hierarchical fracture nets. Based on their relevance, it is possible to distinguish some classes discontinuities families, from major lineaments, as extension fractures and shear zones, to minor joint systems. The significant discontinuities define contiguous hard rock mosaics, from high order size, between 10 and 100 km, to low order size, between 100 and 1,000 m. Several minor discontinuities, classified by its frequency and magnitude, still crack these giant mosaics into smaller and less relevant units, imparting structural complexity to hard rock domains (see Figure 2.6). As a result, it is difficult to construct adequate pseudo-continuous models for these intermingled lithologic and structural features. However, there are rough guidelines to set the minimum size for their pseudo-continuous subsystems. This minimum is related to the size of the model, the length and frequency of the discontinuities, and the desired model details; that is, it depends on the desired *observation scale*.

Observation scales for fractured rocks may be classified as follows:

Class I – "plugging scale": Consider, for example, a mega-fracture wide opening full of dense-packed pervious granular soil (see Figure 2.7). To simulate that confined layer's hydraulic behavior, the size of their pseudo-continuous subsystems must be greater than 10 to 30 times the soil modal pore diameter. Their hydraulic properties must reflect the combined effect of all individual grains and pores they contain. Civil engineers or hydrogeologists hardly ever perform this type of analysis. However, that very small-scale approach may help evaluate and predict the effects of burial pressure changes on significant oil discontinuities reservoirs' permeability and porosity.

Class II – "intact rock *scale"*: Consider, for example, the tension gashes associated with stylolites in carbonate rock blocks. These small structural features play an

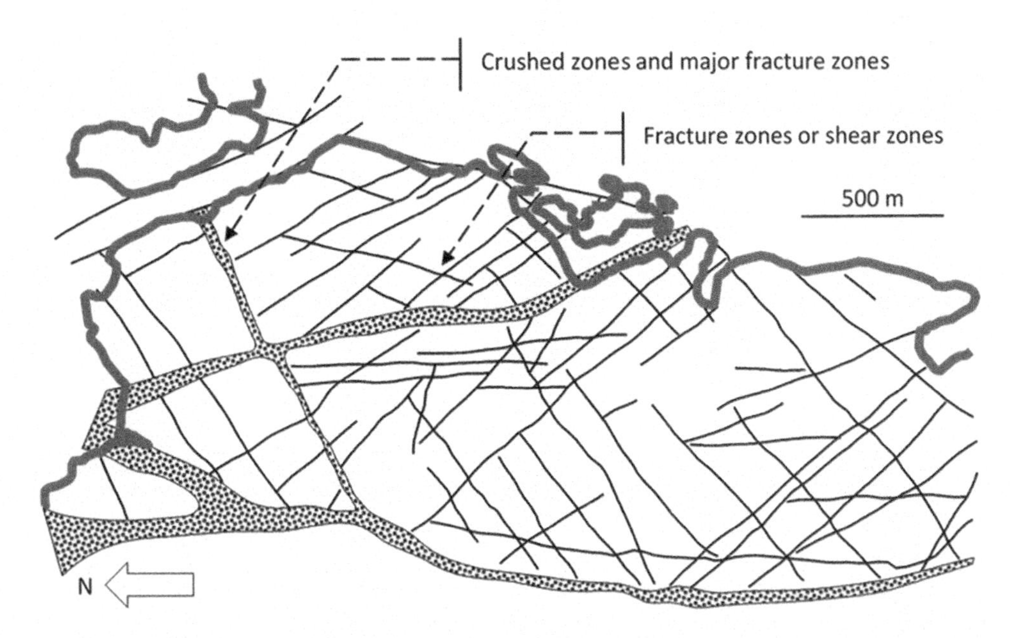

Figure 2.6 Typical mosaic pattern, delineated by shear zones and fractures, affect the underlying hard-rock basement of Norway, Sweden, Finland, and Northern Denmark, over 3.1 billion years old, dubbed Fennoscandia (For details, see Johansson, S., Large Rock Caverns at Neste Oy's Porvoo Works, Large Rock Caverns Proceedings of the International Symposium, Helsinki, 1986, structural features adapted from Fig. 2 of Volume 1.).

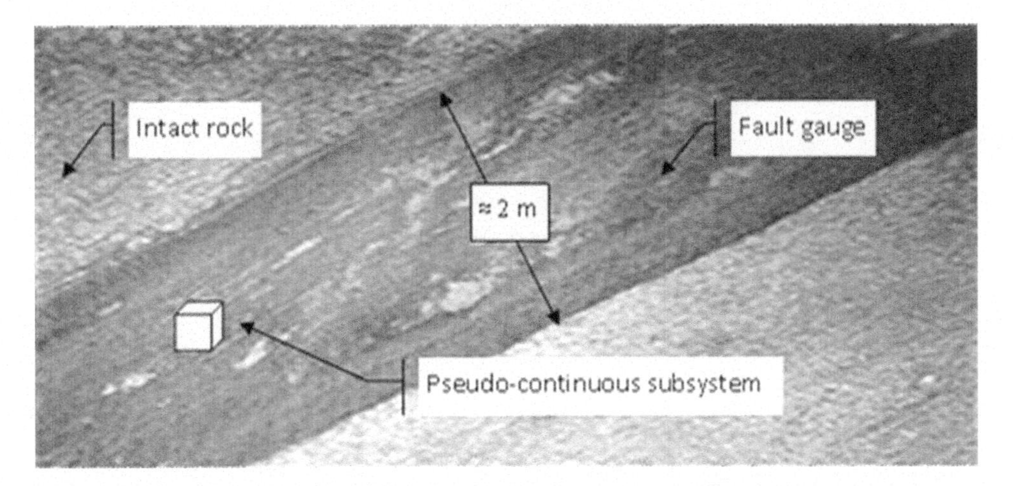

Figure 2.7 Some simulations require a detailed appraisal and hydraulic modeling of the hydraulic influence of the porous stuff jam-packed in their openings at the intersections of their major discontinuities, under high differential pressures. (Adapted illustration from NTNU-Project Report ID-98.)

essential role in some oil reservoirs (see Figure 2.8). Their hydraulic properties are sensibly anisotropic. The smallest size of their pseudo-continuous subsystems must be about 10 to 30 times the modal discontinuity spacing.

Class III – "discontinuity scale": Consider, for example, the observation of the hydraulic behavior of partially filled and partially sealed discontinuities taken one by one. In this case, the higher dimension of rectangular pseudo-continuous subsystems must be greater than 10 to 30 times the modal discontinuity thickness (see Figure 2.9). Their hydraulic properties reflect the combined effect of their incomplete plugging, poor sealing, and erratic voids. This type of analysis may be employed to predict discontinuities' hydraulic behavior near underground civil engineering works or groups of wells.

Class IV – "rock *mass scale*": Consider, for example, a pseudo-continuous cuboid subsystem enclosing a volume of rock mass subdivided by several discontinuities at random, separating intact rock blocks (see Figure 2.10). For one or more than one erratic discontinuities, the equivalent hydraulic property of that pseudo-continuous subsystem results from specific construction rules depending on every discontinuity's hydraulic properties, as discussed later. In this case, the smaller subsystem size must be greater than 3 to 10 times the most relevant discontinuity thickness. The same rule applies to quasi-systematic fracture sets with their hydraulic properties statistically inferred. In this case, their subsystems must be greater than 10 to 30 times the modal fracture spacing.

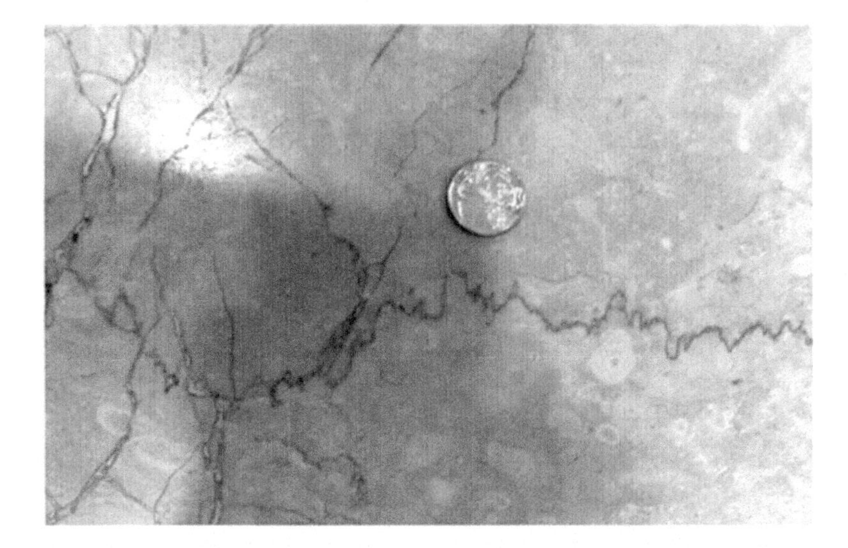

Figure 2.8 That image shows of stylolites associated with secondary fissures. These features are jagged discontinuities but are not structural fractures. They occur in limestone and other sedimentary rocks and typically delineate irregular and interlocked small columns, pits, and teeth-like projections, probably formed diagenetically by differential movement under pressure, accompanied by the partial dissolution of carbonates. They may enlarge by subsequent groundwater flow. Nelson (2001) presents a valuable discussion about the 3D effect of intersecting stylolites on the permeability of whole-rock cores.

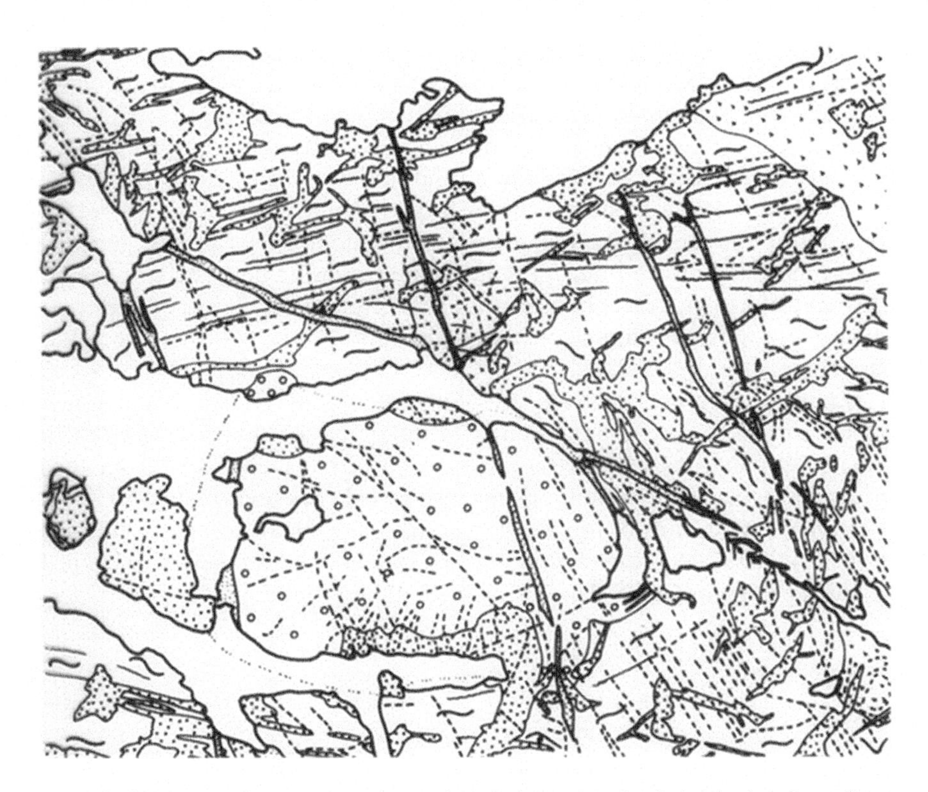

Figure 2.9 Some simulations require a detailed appraisal of the hydraulic influence of major discontinuities that crosses the flow domain. In these particular cases, an assemblage of pseudo-continuous subsystems with variable hydraulic properties may model the pervious discontinuity adequately. (Canadian Shield, figure adapted from Les Eaux Souterraines des Roches Dures du Socle, UNESCO, 1987, approximate scale 1:8000.)

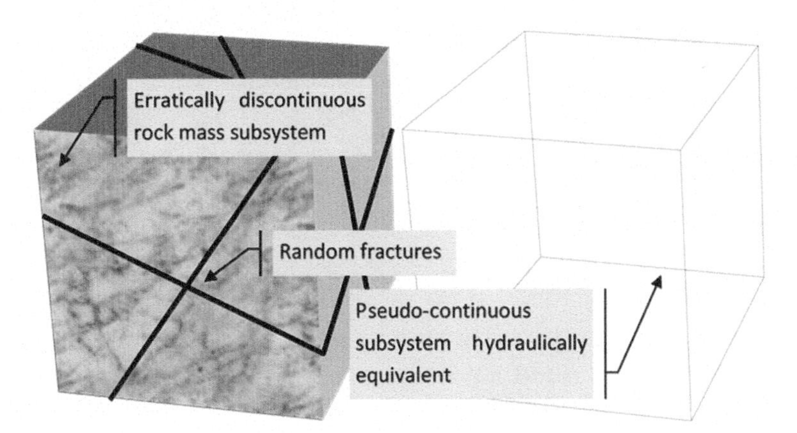

Figure 2.10 An erratically discontinuous rock mass subsystem may be substituted by a pseudo-continuous subsystem, having an almost similar hydraulic behavior.

Class V – "hard-rock basement scale": Consider, for example, a hard-rock basement showing mosaics separated by several lineaments. In this case, combine more than one observation scale into a single model: a smaller one, restricted to a group of lineaments, and a greater and integrated one for all mosaics (see Figure 2.11).

In practice, the minimum size of the pseudo-continuous subsystems for numerical simulations usually exceeds the minimum imposed by the above rules. There are no simple criteria to define the best one. Usually, select an acceptable limit after judging the most convenient size for the model, the costs of getting the geological data, the desired output details, the trade-off between truncation errors and round-of-errors, the computer time, and storage capacity. To reduce costs, it is always possible to refine a previous simulation only within some subdomains where this is essential.

It is important to note that the size of a pseudo-continuous subsystem may range from few centimeters (e.g., for a granular seam) to tens or hundreds of meters (e.g., for a vast aquifer). In transient processes, the system variables change from an unbalanced condition to a final state of equilibrium. In this case, the observation time interval may vary from a few seconds or minutes (e.g., during the equalization of neutral pressure transients in saturated cores tested in a triaxial chamber) to several days or months (e.g., during the progressive dewatering of an underground mine).

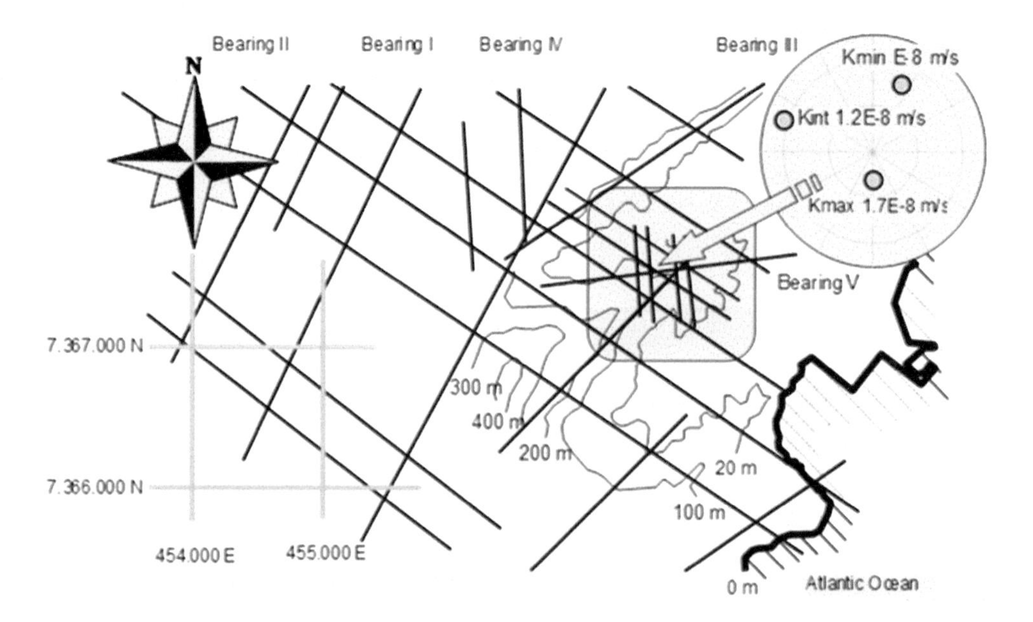

Figure 2.11 A mixed model may adequately simulate the hydraulic behavior of the small rectangular area enclosing well-delimited discontinuities but correctly coupled to the hard-rock mosaics (South Atlantic Brazilian coast).

2.2.5 Description at different scales

All variables describing the hydraulic properties and the physical state of a pseudo-continuous system, such as porosities, pressure heads, flow-rates, and hydraulic conductivities, not only must be referred to its center of gravity and to the mid-point of the observation time-interval but also must be, in some way, averaged adequately in space and time.

For quasi-homogeneous rock masses, the elegant concept of *representative elementary volume*, proposed by Bear,[2] abbreviated as REV, has a physical or a statistical sense but always associated with an elected class of lithologic or structural features. In a physical sense, REV corresponds to the minimum rock mass *test volume* that gives significant average large-scale test results. In a statistical sense, REV corresponds to the smallest *sampling volume* allowing a confident statistical description of their physical features.

Therefore, unless monitoring devices yield effective REV's average parameters, their outputs cannot be considered adequate space-dependent rock mass properties. Then, the REV's sizes affected by monitoring programs must be coherent to allow a consistent analysis of their results. These remarks must be kept in mind when calibrating numerical models against few field measurements or test results located in the flow domain's strategic points.

Depending on the chosen observation scale, the REV size and the pseudo-continuous subsystem size may be of the same order. However, for an erratically inhomogeneous fractured rock mass, the REV concept can hardly be applied.

2.3 Hydraulic variables

2.3.1 Scalars, vectors, and second-order tensors

At this point in this book, it is convenient to remember some basic concepts relating to vectors, tensors, matrix rotations, eigenvalues, eigenvectors, invariants of vectors, and symmetrical second-order tensors.

At any point defined by 3D-t Cartesian coordinates (x_1, x_2, x_3, t)[3] of a material continuum, one may associate a *scalar*, a *vector*, or a *second-order tensor*. A *scalar* is a single number N, generally associated with a specific physical unit that measures a variable's magnitude but unrelated to any coordinates system. Examples are the hydrostatic pressure or the temperature of a fluid at rest. A scalar is also named a zero-order tensor.

A *vector* **v** denotes a rectilinear "shift" of length $|v|$ in a 3D space, necessarily related to a right-handed Cartesian coordinate system that measures the magnitude and the orientation of a variable, usually associated with a physical unit. A vector has three Cartesian components (v_1, v_2, v_3). Examples are a force or a velocity at every point (r_1, r_2, r_3, t) of a projectile trajectory. For concision, one writes the group (r_1, r_2, r_3, t) as $(\mathbf{r}, t)$, where **r** is called the position vector. A vector is also called a first-order tensor.

A *second-order tensor* [T] associated with a point $(\mathbf{r}, t)$ describes the effect of three inclined vectors acting over three mutually perpendicular planes crossing at that point.

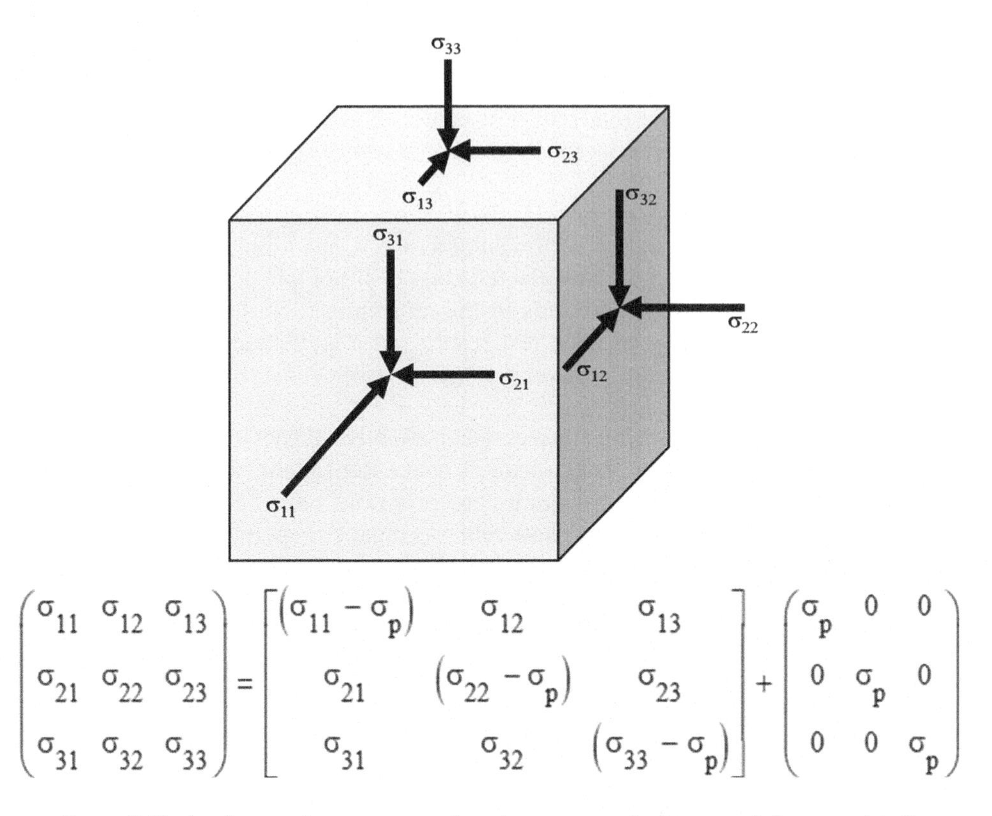

Figure 2.12 As shown above, a second-order symmetrical tensor [σ], as used in linear elasticity, has nine components that result, as shown in this figure, from the sum of a deviatoric tensor plus a spherical or hydrostatic tensor.

As vectors, second-order tensors are also usually related to right-handed Cartesian coordinate systems. Consequently, nine numbers (three cartesian projections related to each one of these planes) commonly portray second-order tensors. However, there exists a particular group of mutually orthogonal planes, called principal planes, where these vectors are at right angles to them. An example is the well-known stress tensor shown in Figure 2.12.

The vectors and tensors' components can be transformed by a right-handed rotation matrix [R] while the Zenithal coordinate system (E, N, Z) remains fixed. The general 3D rotation matrix [R] is defined by:

$$R = \begin{pmatrix} \cos(x,E) & \cos(y,E) & \cos(z,E) \\ \cos(x,N) & \cos(y,N) & \cos(z,N) \\ \cos(x,Z) & \cos(y,Z) & \cos(z,Z) \end{pmatrix} \tag{2.1}$$

In Equation 2.1, the three angular rotations, symbolized by (x, E), (x, N) … (z, Z), comply with the trigonometric rule, as shown in Figure 2.13.

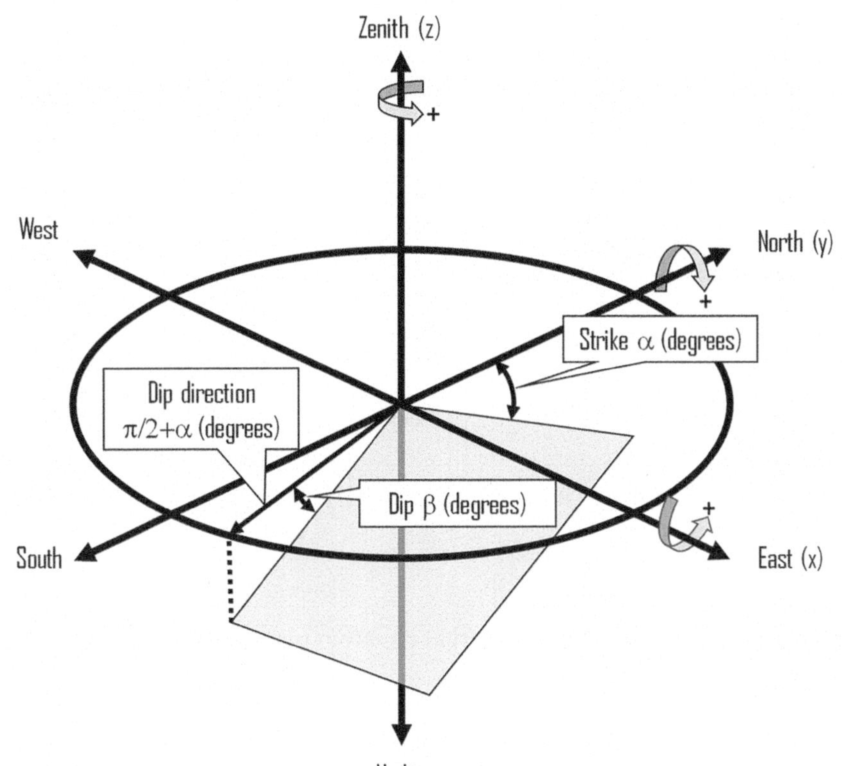

Figure 2.13 The right-handed Zenithal reference frame is an adequate coordinate system for groundwater modeling. For a positive axis pointing toward the eye of an observer, the curved arrows indicate the positive counter-clockwise rotations.

A rotation matrix [R] applied for a 3D vector **v** is:

$$\begin{pmatrix} v_{rot_1} \\ v_{rot_2} \\ v_{rot_3} \end{pmatrix} = \begin{pmatrix} r_{11} & r_{12} & r_{13} \\ r_{21} & r_{22} & r_{23} \\ r_{31} & r_{32} & r_{33} \end{pmatrix} \cdot \begin{pmatrix} v_1 \\ v_2 \\ v_3 \end{pmatrix}$$

A rotation omatrix [R] applied for 3D second-order tensors [k] is:

$$\begin{pmatrix} k_{rot_{11}} & k_{rot_{12}} & k_{rot_{13}} \\ k_{rot_{21}} & k_{rot_{22}} & k_{rot_{23}} \\ k_{rot_{31}} & k_{rot_{32}} & k_{rot_{33}} \end{pmatrix} = \begin{pmatrix} r_{11} & r_{12} & r_{13} \\ r_{21} & r_{22} & r_{23} \\ r_{31} & r_{32} & r_{33} \end{pmatrix} \cdot \begin{pmatrix} k_{11} & k_{12} & k_{13} \\ k_{21} & k_{22} & k_{23} \\ k_{31} & k_{32} & k_{33} \end{pmatrix} \cdot \begin{pmatrix} r_{11} & r_{12} & r_{13} \\ r_{21} & r_{22} & r_{23} \\ r_{31} & r_{32} & r_{33} \end{pmatrix}^T$$

In the above equations, the subscript rot concerns the transformed "scalars" components of the original matrix [k] (or vector **v**).

Referred to the Zenithal coordinate system (E, N, Z) – *commonly used for groundwater flow modeling* – a complete 3D right-handed rotation matrix [R] may result from the product of three successive independent rotations:

- First, apply to the original matrix the first rotation through an angle ζ about the coordinate Z-axis.
- Second, apply to the previously rotated matrix the second rotation through an angle ψ about the coordinate N-axis.
- Finally, apply to the previously rotated matrix the last rotation through an angle ξ about the coordinate E-axis.

The resulting rotation matrix [R], written as a function of these three angles, is:

$$R(\zeta,\psi,\xi)=\begin{pmatrix}\cos(\zeta) & -\sin(\zeta) & 0\\ \sin(\zeta) & \cos(\zeta) & 0\\ 0 & 0 & 1\end{pmatrix}\cdot\begin{pmatrix}\cos(\psi) & 0 & \sin(\psi)\\ 0 & 1 & 0\\ -\sin(\psi) & 0 & \cos(\psi)\end{pmatrix}\cdot\begin{pmatrix}1 & 0 & 0\\ 0 & \cos(\xi) & -\sin(\xi)\\ 0 & \sin(\xi) & \cos(\xi)\end{pmatrix} \tag{2.2}$$

Performing matrices multiplication, that equation turns to:

$$R(\zeta,\psi,\xi)=\begin{pmatrix}\cos(\psi)\cdot\cos(\zeta) & \cos(\zeta)\cdot\sin(\xi)\cdot\sin(\psi)-\cos(\xi)\cdot\sin(\zeta) & \sin(\xi)\cdot\sin(\zeta)+\cos(\xi)\cdot\cos(\zeta)\cdot\sin(\psi)\\ \cos(\psi)\cdot\sin(\zeta) & \cos(\xi)\cdot\cos(\zeta)+\sin(\xi)\cdot\sin(\psi)\cdot\sin(\zeta) & \cos(\xi)\cdot\sin(\psi)\cdot\sin(\zeta)-\cos(\zeta)\cdot\sin(\xi)\\ -\sin(\psi) & \cos(\psi)\cdot\sin(\xi) & \cos(\xi)\cdot\cos(\psi)\end{pmatrix}$$

It is important to remember that vectors and matrices transformed by the above rotation formula preserve the same L^2 norm.[4] This property allows for second-order tensors to obtain its eigenvalues (principal values) and associated eigenvectors (normalized principal direction cosines). Moreover, three important tensor invariants, I_1, I_2, and I_3, do not change for all possible second-order tensors transformations:

$$I_1 = k_{11} + k_{22} + k_{33}$$

$$I_2 = \begin{pmatrix}k_{22} & k_{23}\\ k_{32} & k_{33}\end{pmatrix} + \begin{pmatrix}k_{11} & k_{13}\\ k_{31} & k_{33}\end{pmatrix} + \begin{pmatrix}k_{11} & k_{12}\\ k_{21} & k_{22}\end{pmatrix}$$

$$I_3 = \begin{pmatrix}k_{11} & k_{12} & k_{13}\\ k_{21} & k_{22} & k_{23}\\ k_{31} & k_{32} & k_{33}\end{pmatrix}$$

2.3.2 *Groundwater flow description*

Pseudo-continuous groundwater models simulate the space-time variation of the *hydraulic gradient*, a metric of the water particles' space-time energy expenditure, and the *specific discharge*, which is a metric for the flow rate. The variation of these two parameters may be simulated for *steady* or *unsteady* conditions, and the *hydraulic conductivity* is the physical property connecting both variables. Together, the space-time variation of these three interrelated variables adequately describes the groundwater system's hydraulics but constrained by its external and internal boundary conditions. These three variables must always be referred at the center of each pseudo-continuous subsystem (see Figure 2.14).

Concerning groundwater flow, *scalar* quantities H describe the hydraulic heads at points (**r**, t) within the pervious soil or rock mass, but at the same points, *vector* quantities describe the hydraulic gradients **J** and associated specific discharges **q**.

A scalar quantity measures the hydraulic conductivity k of an isotropic pervious media. However, only a *second-order tensor* [k] correctly describes it for an anisotropic media.

2.3.3 *The specific discharge*

As for Soil Mechanics, the *void volume* V_v contained in *volume* V of a pseudo-continuous subsystem corresponds to the total volume occupied by the non-solid phase, that is, gas and water. Conduits, fractures, or other types of discontinuities, including minute openings, structure porous media voids. The "effective" *void volume* V_e excludes from the total void volume V_v not only all hydraulically isolated pores or

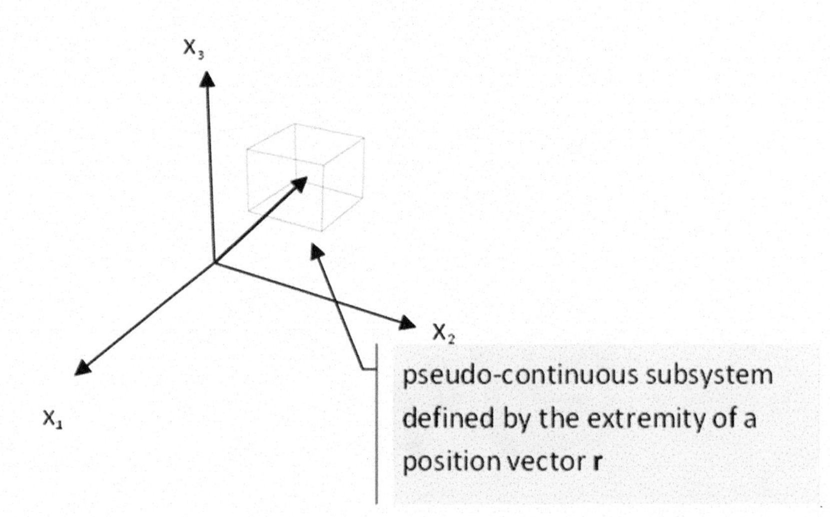

Figure 2.14 Coordinates (x_1, x_2, x_3) of the extremity of a position vector r defines the geometric center of a pseudo-continuous subsystem.

closed discontinuities but also all immobilized and thin water layers bonded to the solid phase surface. Consequently, V_e is somewhat smaller than V_V.

The *effective porosity* n_e equals the ratio between the "effective" volume V_e and the total volume V, that is, $n_e = V_e/V$. It has no units, and for hard-rocks, effective porosities seldom attain 1 to 2% of their total volume while may range from 5% to 15% for consolidated clastic rocks and up to 25% for unconsolidated sands and gravels (see Figure 2.15).

As formulated by Bear, the groundwater average "effective" velocity v_e ascribed to the center of the pseudo-continuous subsystem corresponds to the vector-average of the velocities of all water particles $\mathbf{v}_P$ percolating throughout its "effective" voids during an observation time interval $\boldsymbol{\delta}\mathbf{t}$ (Bear et alia, 1968). The specific discharge **q**, also called Darcy's velocity (Darcy, 1856), may be determined by the product of the "effective" velocity $\mathbf{v}_e$ by the effective porosity n_e, that is $\mathbf{q} = n_e \cdot \mathbf{v}_e$ (see the end of this chapter).

It is vital to retain that the specific discharge **q** is always smaller than the "effective" velocity $\mathbf{v}_e$. For isotropic media, the "effective" velocity $\mathbf{v}_e$ and the specific discharge **q** are collinear vector quantities, carrying the units of a velocity L/T. Both are tangent vectors to the pseudo-streamlines. They describe that vector fields **q**(**r**, t) and $\mathbf{v}_e$(**r**, t) are related to an arbitrary Cartesian frame, functionally linked to the space-time coordinates (x_1, x_2, x_3, t) also written as (**r**, t). However, as discussed ahead, the reader must always remember that v_e and q are non-collinear vector quantities for anisotropic media.

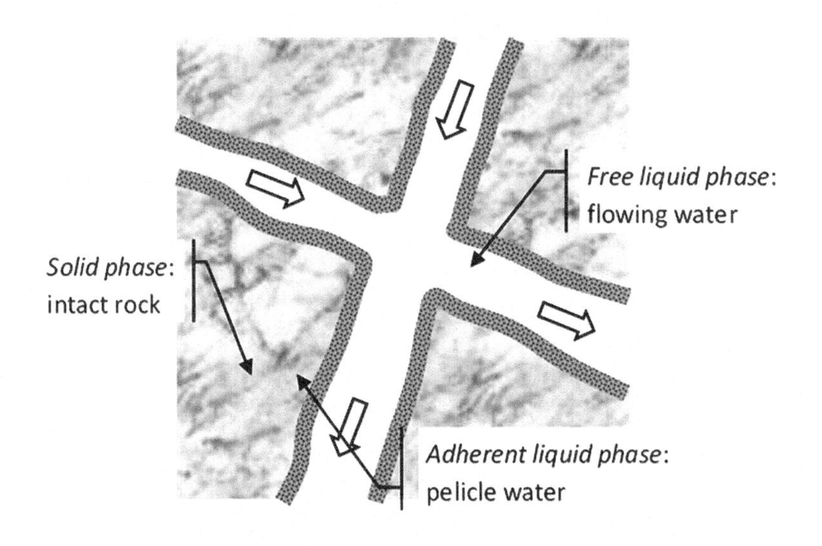

Figure 2.15 This figure schematizes the link between the intact rock, the percolating water, and the immobilized pellicle water. Free water particle seeps with varying speeds. The average "effective" velocity v_e of the moving water ascribed to the center of the pseudo-continuous subsystem corresponds to the vector-average of the velocities of all water particles v_P percolating throughout its "effective" voids during an observation time interval δt.

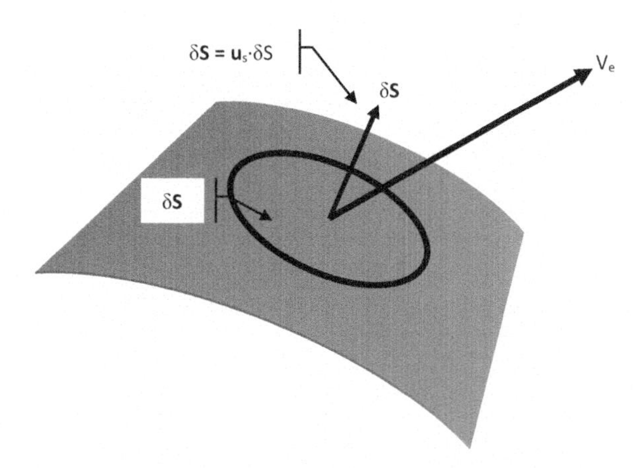

Figure 2.16 Any area element δS of a pseudo-continuous subsystem boundary S is a vector quantity defined by the product of its area δS and its "exterior normal" u_S.

Percolating water, entering or leaving the boundaries of a pseudo-continuous system, seeps across the voids of the boundary elements $\delta\mathbf{S}$ (see Figure 2.16). Then, the elementary discharge δQ traversing an area element $\delta\mathbf{S}$ results from the scalar product of its "effective" voids area $n_e\delta S$ by the "effective" velocity $\mathbf{v}_e$:

$$\delta Q = \mathbf{v}_e n_e \delta \mathbf{S} = |\mathbf{q}|\,|\delta \mathbf{S}|\,\cos\left(\mathbf{v}_e, \mathbf{u}_S\right)$$

The expression $(\mathbf{v}_e, \mathbf{u}_S)$ denotes the angle between the vector $\mathbf{v}_e$ and the "exterior normal" $\mathbf{u}_S$. By the convention adopted in this book, δQ is positive for *inflows* and negative for *outflows*.

2.3.4 The hydraulic gradient

A percolating water particle consumes part of its energy as it moves and as time elapses. The total stored energy associated with a percolating water particle, called the *hydraulic head*, customarily symbolized by H, is a scalar quantity that results from the sum of three kinds of energy: *gravitational*, *piezometric*, and *kinetic*. The *gravitational head* corresponds to the potential energy acquired by a water particle if raised from an arbitrary reference level (e.g., the mean sea level) to an elevation Z. The *piezometric head* measures the elastic energy stored in a water particle if compressed to a confining pressure p above the atmospheric pressure p_0. The *kinetic head* measures the additional energy due to its motion. As typical "effective" velocities $\mathbf{v}_e$ are minimal, usually, from 10^{-10} to 0.1 m/s, kinetic energies carried by percolating water particles in soils and rocks are commonly neglected. In this case, the total head H only has a *potential* character. To be homogeneously written, as shown in Table 2.1, these energies must be normalized by the water *density*, or water *unit weight*, or *unit volume* of water.

A *transient* or *unsteady-state* head field H(**r**, t) characterizes a flow domain, where the potential head H varies both in space and time. For any point in this field, a functional relationship involving its space-coordinate **r** and time-coordinate t may describe its

Table 2.1 Common normalized expressions of the total hydraulic head H

Type of energy	*Usual hydraulic head H units*		
	Normalized to the density of water ρ (L^2/T^2)	*Normalized to the unit weight of water* ρg *(L)*	*Normalized to the unit volume of water* L^3 $(M/L^1/T^2)$
Gravitational	gZ	Z	ρgZ
Piezometric	p/ρ	$p/\rho g$	p
Kinetic	$v_e^2/2$	$v_e^2/2g$	$\rho v_e^2/2$

potential head H. A *permanent* or a *steady-state* head field $H(\mathbf{r})$ characterizes a flow domain, where H only varies in a space domain and is described by the space-coordinates $\mathbf{r}$. Points in both fields but storing the same potential head H configure *equipotential* surfaces: moving for transient flows or fixed for permanent flows.

In a *zenithal* right-handed Cartesian coordinate system (E, N, Z), the vertical Z-axis points to the zenith. When written in L units, the head H at a generic point (E, N, Z) results from the sum of the gravitational head Z plus the piezometric head $p/\rho g$:

$$H = Z + \frac{p}{\rho g}$$

In another reference frame x_i $(i = 1, 2, 3)$ with its origin at the level z_0, the head H at a generic point (x_1, x_2, x_3) results from:

$$H = z_0 + x_1 \cos(Z, x_1) + x_2 \cos(Z, x_2) + x_3 \cos(Z, x_3) + \frac{p}{\rho g} \tag{2.3}$$

In the above expression, z_0 denotes the elevation origin O of the new reference frame, and cos (Z, x_i) corresponds to direction cosines of the axes x_i related to the zenithal axis Z, measured by the conventional trigonometric usage (see Figure 2.17).

The hydraulic head's spatial gradient measures head decay per unit length of a moving water particle, its *spatial head decay-rate* (see Figure 2.18). Now, consider the straight-line AB within the head field $H(\mathbf{r}, t)$ defined by two position vectors $\mathbf{r}_A$ and $\mathbf{r}_B$ and two heads H_A and H_B. Then, the vector-distance $(\mathbf{r}_B - \mathbf{r}_A)$ measures the distance $\delta\mathbf{r}$ between A and B, and the scalar-difference $(H_B - H_A)$ measures their differential head δH. If the straight-line AB is vertical to the equipotential surface H_A at point A, the *scalar-to-vector* ratio $\delta H/\delta\mathbf{r}$ measures the average spatial head decay-rate between the points A and B. This ratio is a vector quantity and has the same direction as $\delta\mathbf{r}$ but pointing towards B if $H_B > H_A$ and vice-versa.

Note that although mathematically inappropriate to derive a scalar quantity H by a vector quantity $\mathbf{r}$, or, more generally, to derive a tensor $\mathbb{P}$ by another tensor $\mathbb{Q}$, it is convenient, to simplify notation, to define the symbolic partial derivative $\partial\mathbb{P}/\partial\mathbb{Q}$ as a tensor $\mathbb{R}$ whose rank equals the sum of the ranks of $\mathbb{P}$ and $\mathbb{Q}$ (Koerber, 1962). Then, as point B gradually recedes toward point A all along the perpendicular AB, this vector rate-of-change continuously tends to a limit, and when B finally converges to A it defines the local spatial head decay-rate.

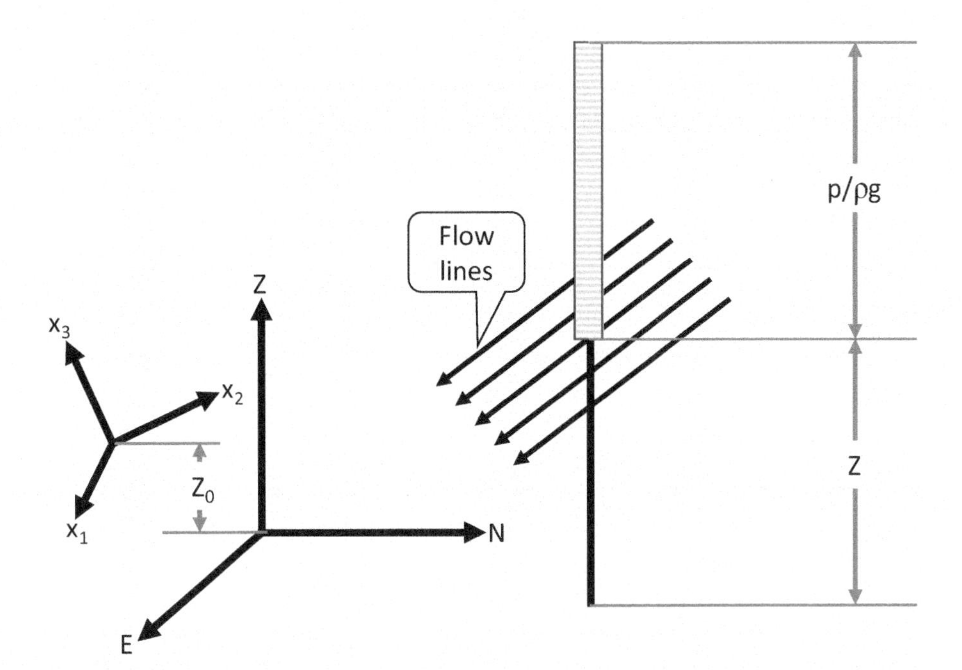

Figure 2.17 When written in L units, the head H at a generic point (E, N, Z) results from the sum of the gravitational head Z plus the piezometric head p/ρg.

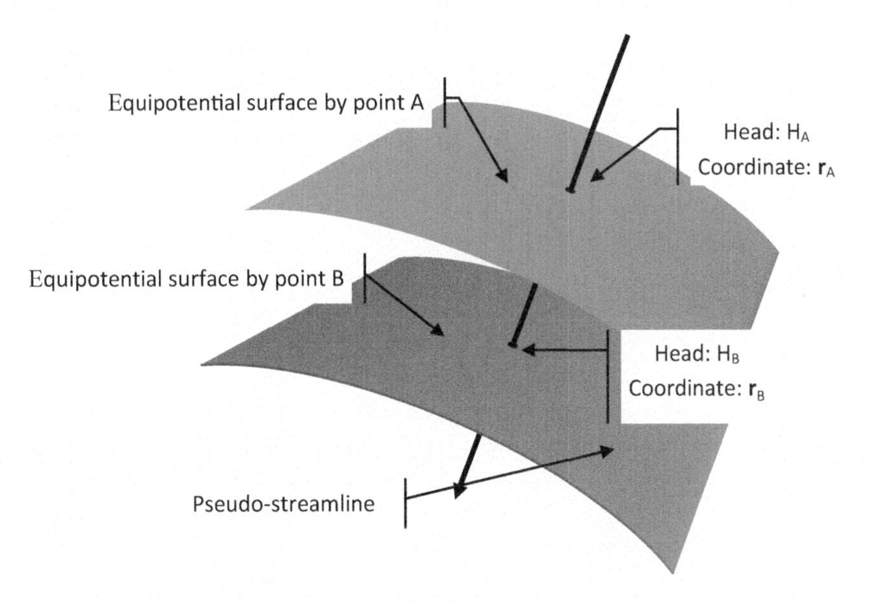

Figure 2.18 For homogeneous and isotropic pseudo-systems, all pseudo-streamlines traverse all equipotential surfaces at right angles. In this case, the limit of the symbolic ratio $-\delta H/\delta r$ when B recedes to A defines the hydraulic gradient $\mathbf{J}_A$ at A, tangent to the pseudo-streamline.

Practical groundwater formulas substitute $-\partial H/\partial \mathbf{r}$ (note the negative sign) for the *hydraulic gradient* **J** (see) to agree with the flowrate's orientation.

For any other point B′ of any other straight-line AB′ but not perpendicular to the equipotential surface at point A, it is also possible to apply similar reasoning and define another directional gradient $\mathbf{J}'_A$. However, the absolute value of $\mathbf{J}'_A$ is always smaller than the absolute value of $\mathbf{J}_A$. Indeed, $\mathbf{J}'_A = \mathbf{J}_{A1}\cos(\alpha)$ where α is the angle BAB′. Then, $\mathbf{J}_A > \mathbf{J}'_A$.

Being a vector, **J** results from the vector-sum $\left[(\partial H/\partial x_1)\mathbf{u}_1 + (\partial H/\partial x_2)\mathbf{u}_2 + (\partial H/\partial x_3)\mathbf{u}_3\right]$, where the base or unit vectors ($\mathbf{u}_1$, $\mathbf{u}_2$, $\mathbf{u}_3$) are associated to an arbitrary reference frame (x_1, x_2, x_3). Then:

$$\begin{pmatrix} J_1 \\ J_2 \\ J_3 \end{pmatrix} = \begin{pmatrix} -\dfrac{\partial H}{\partial x_1} \\ -\dfrac{\partial H}{\partial x_2} \\ -\dfrac{\partial H}{\partial x_3} \end{pmatrix} = \begin{pmatrix} -\left(\cos(x_1, Z) + \dfrac{1}{\rho g} \cdot \dfrac{\partial p}{\partial x_1}\right) \\ -\left(\cos(x_2, Z) + \dfrac{1}{\rho g} \cdot \dfrac{\partial p}{\partial x_2}\right) \\ -\left(\cos(x_3, Z) + \dfrac{1}{\rho g} \cdot \dfrac{\partial p}{\partial x_3}\right) \end{pmatrix} \tag{2.4}$$

In Equation 2.4, the components $-\cos(x_i, Z)$ and $-\left[1/\rho g(\partial p/x_i)\right]$, respectively, denote the *gravitational* and *piezometric* gradient. If the hydraulic gradient **J** is related to the zenithal reference frame ENZ, its components are:

$$\begin{pmatrix} J_E \\ J_N \\ J_Z \end{pmatrix} = \begin{pmatrix} -\dfrac{\partial H}{\partial E} \\ -\dfrac{\partial H}{\partial N} \\ -\dfrac{\partial H}{\partial Z} \end{pmatrix} = \begin{pmatrix} -\left(\cos(E, Z) + \dfrac{1}{\rho g} \cdot \dfrac{\partial p}{\partial E}\right) \\ -\left(\cos(N, Z) + \dfrac{1}{\rho g} \cdot \dfrac{\partial p}{\partial N}\right) \\ -\left(\cos(Z, Z) + \dfrac{1}{\rho g} \cdot \dfrac{\partial p}{\partial Z}\right) \end{pmatrix} = -\begin{pmatrix} \dfrac{1}{\rho g} \cdot \dfrac{\partial p}{\partial E} \\ \dfrac{1}{\rho g} \cdot \dfrac{\partial p}{\partial N} \\ \dfrac{1}{\rho g} \cdot \dfrac{\partial p}{\partial Z} + 1 \end{pmatrix} \tag{2.5}$$

As the axes x_1, x_2, and x_3 match that of E, N, and Z, cos (E, Z) and cos (N, z) equal zero and −cos (Z, Z) equals −1, denoting the Z-component of the gravitational pull. Note that for heads H in L units, the hydraulic gradient has no units.

The head decay per unit time of a moving water particle, its temporal head decay-rate, may be measured by the hydrodynamic gradient DH/dt resulting from the summation of the time-gradient $\partial H/\partial t$ and the product of spatial gradient $\partial H/\partial r$ by the specific discharge **q**, that is, $DH/dt = \partial H/\partial t + (\partial H/\partial r) \cdot \mathbf{q}$ (see Figure 2.19).

The scalar product $(\partial H/\partial r) \cdot \mathbf{q}$ measures the head spatial decay as the water particle moves during a unit time. Also, for $\mathbf{q} = 1 \cdot L/T$ the negative value of this product turns to be the hydraulic gradient **J**, a standard measure of the consumption of the potential

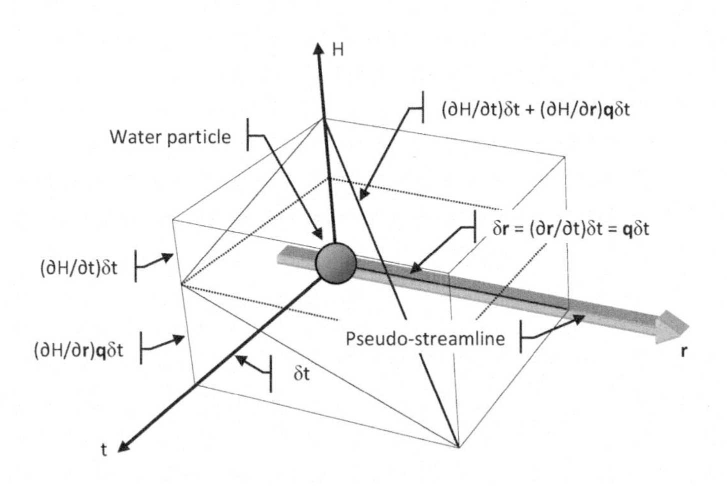

Figure 2.19 This schematic figure shows a transient 4-D head field H(r, t) and helps to understand the meaning of $DH/dt = \partial H/\partial t + (\partial H/\partial r)\cdot q$. Since head decays in both time and space, the observer must keep track of two kinds of motion: he must stay in a fixed point to watch the head time-decay, or he must go after the moving water particle to watch the head space-decay.

energy stored in a percolating water particle when it travels a unit length during a unit time with a unit apparent velocity (more details at the end of this chapter).

Numerical approximations of hydraulic gradients are relatively simple but require attention. For 3D cubic lattices referred to an arbitrary frame xyz (see Figure 2.20 left), devised to simulate the hydraulic behavior of pseudo-continuous pervious media,

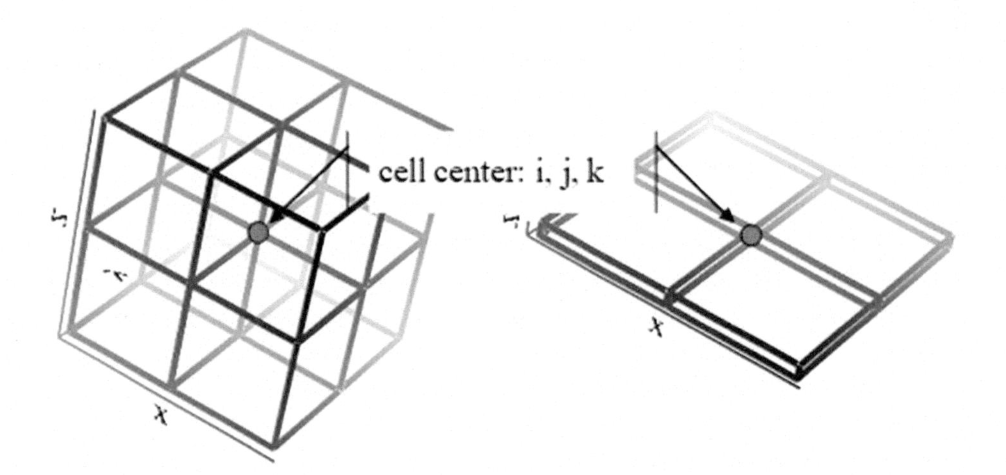

Figure 2.20 3D cubic cell (left) and 2D slab lattices (right) referred to an arbitrary coordinate system xyz. Hydraulic gradients **J** at the center of these cell types are approximated as linear functions of the total head values in its neighborhood.

the hydraulic gradient **J** components at the lattice center, specified by the subscripts (i, j, k), can be estimated by the head values H at its neighborhood:

$$(J_x)_{i,j,k} = -\frac{H_{i+1,j,k} - H_{i-1,j,k}}{x_{i+1,j,k} - x_{i-1,j,k}}$$

$$(J_y)_{i,j,k} = -\frac{H_{i,j+1,k} - H_{i,j-1,k}}{y_{i,j+1,k} - y_{i,j-1,k}}$$

$$(J_z)_{i,j,k} = -\frac{H_{i,j,k+1} - H_{i,j,k-1}}{z_{i,j,k+1} - z_{i,j,k-1}}$$

As the head H (written in m units) corresponds to the sum of the piezometric head p (written in m units, i.e. normalized by rg) and the gravitational head Z (elevation measured in the zenithal frame, but written in m units):

$$(J_x)_{i,j,k} = -\left(\frac{p_{i+1,j,k} - p_{i-1,j,k}}{x_{i+1,j,k} - x_{i-1,j,k}} + \frac{Z_{i+1,j,k} - Z_{i-1,j,k}}{x_{i+1,j,k} - x_{i-1,j,k}}\right)$$

$$(J_y)_{i,j,k} = -\left(\frac{p_{i,j+1,k} - p_{i,j-1,k}}{y_{i,j+1,k} - y_{i,j-1,k}} + \frac{Z_{i,j+1,k} - Z_{i,j-1,k}}{y_{i,j+1,k} - y_{i,j-1,k}}\right)$$

$$(J_z)_{i,j,k} = -\left(\frac{p_{i,j,k+1} - p_{i,j,k-1}}{z_{i,j,k+1} - z_{i,j,k-1}} + \frac{Z_{i,j,k+1} - Z_{i,j,k-1}}{z_{i,j,k+1} - z_{i,j,k-1}}\right)$$

Expressing, in compatible m units, the components of the gravitational gradient by the direction cosines of the axes x_i related to the zenithal axis Z:

$$(J_x)_{i,j,k} = -\left(\frac{p_{i+1,j,k} - p_{i-1,j,k}}{x_{i+1,j,k} - x_{i-1,j,k}} + \cos(x, Z)\right)$$

$$(J_y)_{i,j,k} = -\left(\frac{p_{i,j+1,k} - p_{i,j-1,k}}{y_{i,j+1,k} - y_{i,j-1,k}} + \cos(y, Z)\right)$$

$$(J_z)_{i,j,k} = -\left(\frac{p_{i,j,k+1} - p_{i,j,k-1}}{z_{i,j,k+1} - z_{i,j,k-1}} + \cos(z, Z)\right)$$

The components of the gravitational gradient, that is, cos (x, Z), cos (y, Z), and cos (z, Z), only depend on the direction of the z-axis concerning the Z-axis of the zenithal frame. If, instead of an arbitrary xyz coordinate system, the spatial lattice is consistent with the ENZ zenithal frame, these components, if expressed in compatible m units, are:

$$(J_E)_{i,j,k} = -\frac{p_{i+1,j,k} - p_{i-1,j,k}}{E_{i+1,j,k} - E_{i-1,j,k}}$$

$$(J_N)_{i,j,k} = -\frac{p_{i,j+1,k} - p_{i,j-1,k}}{N_{i,j+1,k} - N_{i,j-1,k}}$$

$$(J_Z)_{i,j,k} = -\left(\frac{p_{i,j,k+1} - p_{i,j,k-1}}{Z_{i,j,k+1} - Z_{i,j,k-1}} + 1\right)$$

For 2D slab-lattices referred to an arbitrary xyz frame, as thin flattened cells attached to the plane xy, as shown in Figure 2.20 right, it is assumed that the confined water pressure P in between the fracture walls are invariant in the z-direction. Indeed, the fracture aperture δz is relatively exceedingly small if compared to its x and y dimensions. In this case, the slab lattice simulates the hydraulic behavior of a planar fracture attached to the plane xy, the hydraulic gradient **J** components at the cell center (i, j, k), if expressed in m units, are:

$$(J_x)_{i,j,k} = -\left(\frac{p_{i+1,j,k} - p_{i-1,j,k}}{x_{i+1,j,k} - x_{i-1,j,k}} + \frac{Z_{i+1,j,k} - Z_{i-1,j,k}}{x_{i+1,j,k} - x_{i-1,j,k}} \right)$$

$$(J_y)_{i,j,k} = -\left(\frac{p_{i,j+1,k} - p_{i,j-1,k}}{y_{i,j+1,k} - y_{i,j-1,k}} + \frac{Z_{i,j+1,k} - Z_{i,j-1,k}}{y_{i,j+1,k} - y_{i,j-1,k}} \right)$$

$$(J_z)_{i,j,k} = -\left(\frac{p_{i,j,k+1} - p_{i,j,k-1}}{z_{i,j,k+1} - z_{i,j,k-1}} + \frac{\delta Z}{\delta z} \right)$$

Expressing in compatible m units all its terms, the components of the gravitational gradient again by the direction cosines of the axes x_i referred to the zenithal axis Z:

$$(J_y)_{i,j,k} = -\left(\frac{p_{i+1,j,k} - p_{i-1,j,k}}{x_{i+1,j,k} - x_{i-1,j,k}} + \cos(x, Z) \right)$$

$$(J_z)_{i,j,k} = -\left(\frac{p_{i,j+1,k} - p_{i,j-1,k}}{y_{i,j+1,k} - y_{i,j-1,k}} + \cos(y, Z) \right)$$

$$(J_z)_{i,j,k} = -\cos(z, Z)$$

For linear tubes conceived to simulate the hydraulic behavior of linear elements of solution conduits, shafts and galleries, and even rivers, adapt all previous procedures.

A hydraulic gradient **J** referred to an arbitrary frame xyz may be described in the zenithal frame ENZ by the following matrix operation:

$$\begin{pmatrix} J_E \\ J_N \\ J_Z \end{pmatrix} = R \begin{pmatrix} J_x \\ J_y \\ J_z \end{pmatrix} \tag{2.6}$$

The rotation matrix [R] is defined by:

$$R = \begin{pmatrix} \cos(x, E) & \cos(y, E) & \cos(z, E) \\ \cos(x, N) & \cos(y, N) & \cos(z, N) \\ \cos(x, Z) & \cos(y, Z) & \cos(z, Z) \end{pmatrix} \tag{2.7}$$

The nine angular measures (x, E), (x, N) … (z, Z) comply with the trigonometric rules. The magnitude and direction of **J** remain invariant under that rotation.

2.3.5 The hydraulic conductivity

2.3.5.1 Introduction

For any pervious pseudo-continuous subsystem, a physical property called its *hydraulic conductivity* measures its ability and efficiency to transfer groundwater from one point to another through its interconnected pores, fractures, conduits, and other open discontinuities. For laminar flow, typical of groundwater percolation, except nearby sinks and holes, the specific discharge **q**, the hydraulic gradient **J**, and the hydraulic conductivity [k] are linearly interrelated:

$$[\text{hydraulic conductivity}] \propto [\text{specific discharge}]/[\text{hydraulic gradient}]$$

That linear relationship was empirically discovered in 1856 by the French engineer Henry Philibert Gaspard Darcy and since then is known as Darcy's Law. It is a scalar quantity, a zero-order tensor, for homogeneous and isotropic pervious media. However, as the specific discharge **q** and the hydraulic gradient **J** are vectors, first-order tensors, their *vector-to-vector* ratio implies second-order tensors, generally written as a matrix [k] (see Figure 2.21).

If referred to an arbitrary orthogonal frame xyz, the relationship between the specific discharge **q**, the hydraulic gradient **J**, and the general hydraulic conductivity [k] is written, in full matrix lettering as:

$$\begin{pmatrix} q_x \\ q_y \\ q_z \end{pmatrix} = \begin{pmatrix} k_{xx} & k_{xy} & k_{xz} \\ k_{yx} & k_{yy} & k_{yz} \\ k_{zx} & k_{zy} & k_{zz} \end{pmatrix} \begin{pmatrix} J_x \\ J_y \\ J_z \end{pmatrix} \tag{2.8}$$

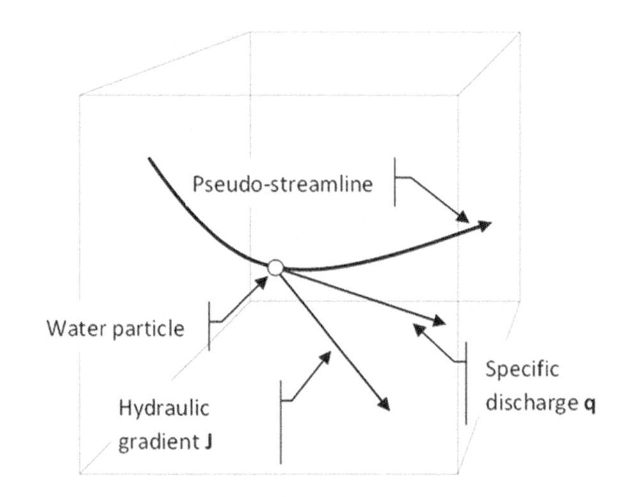

Figure 2.21 Specific discharge **q** and hydraulic gradient **J** at an arbitrary point of a pseudo-continuum pervious anisotropic media. Except for homogeneous and isotropic pervious soils and rock masses, both vectors q and J are non-collinear, except if the direction of **J** parallels one of the second-order tensor principal directions.

Expanding the right-hand product:

$$\begin{pmatrix} q_x \\ q_y \\ q_z \end{pmatrix} = \begin{pmatrix} k_{xx} J_x + k_{xy} J_y + k_{xz} J_z \\ k_{yx} J_x + k_{yy} J_y + k_{yz} J_z \\ k_{zx} J_x + k_{zy} J_y + k_{zz} J_z \end{pmatrix} \tag{2.9}$$

Note that each component q_i of the specific discharge **q** comes from the sum of three summands $q_{ij}=k_{ij}J_j$, each one parallel to an i-direction. However, each group of the three summands parallel to the same direction, driven by all components of the hydraulic gradient **J**:

$$\begin{pmatrix} q_x \\ q_y \\ q_z \end{pmatrix} = \begin{pmatrix} q_{xx} & q_{xy} & q_{xz} \\ q_{yx} & q_{yy} & q_{yz} \\ q_{zx} & q_{zy} & q_{zz} \end{pmatrix} \tag{2.10}$$

The matrix [k] is symmetric, $k_{ij}=k_{ji}$ (for i or j=x, y, z). The diagonal elements k_{ii} are always positive, but the off-diagonal k_{ij} elements can be positive, negative, or zero for reasons explained below. The *magnitude* of each component k_{ij} is proportional to the size of the interconnected discontinuities (pores and fractures and conduits) of the pervious soil or rock mass. The quantity, value, algebraic sign, and the location of each off-diagonal element k_{ij} depend on the geometric symmetry of the 3D pattern and the size of these discontinuities.

As the main diagonal components k_{ii} are always positive quantities, all diagonal ii-summands q_{ii} point to the same direction of J_i. However, to match the correct flow direction, all off-diagonal ij-summands q_{ij} make$\pm 90°$ to J_j depending on the algebraic signal of the off-diagonal components k_{ij} (+90° for positive k_{ij} and –90° for negative k_{ij}). A positive signal indicates that the fracture's strike and dip constrain the flowrate orientation to match the hydraulic gradient orientation. A negative signal indicates the opposite (as a sail-boat sailing against the prevailing winds).

In a frame defined by the *principal axes*, only the elements k_{ii} of the main diagonal are not equal to zero. These non-zero elements are called the *eigenvalues* of [k] (usually lettered as k_i, that is, as k_1, k_2, and k_3), and they measure the maximum, the intermediate, and the minimum hydraulic conductivities along the principal axes of [k]. A set of three direction cosines (a_{i1}, a_{i2}, a_{i3}), called *eigenvectors*, defines the spatial direction i of each eigenvalue k_i (see matrix *diagonalization* techniques for finding *eigenvalues* and corresponding *eigenvectors*). Using matrix notation:

$$\text{eigenvalues of } [k] = \begin{pmatrix} k_1 & 0 & 0 \\ 0 & k_2 & 0 \\ 0 & 0 & k_3 \end{pmatrix} \tag{2.11}$$

$$\text{eigenvectors of } k_i\,[i = 1,2,3] = \begin{pmatrix} a_{i1} \\ a_{i2} \\ a_{i3} \end{pmatrix} \tag{2.12}$$

2.3.5.2 *Fractures and conduits*

The *transmissivity* of a unit-wide strip of pervious sedimentary strata is $T=kH$, with dimension L^2/T. The same concept applies to an isotropic discontinuity. By simple premises, it is possible to presume that the magnitude of a fracture's transmissivity depends on its "equivalent" hydraulic aperture e, the interstitial permeability k of its filling, and the relative fracture extent κ over the fracture area. The parameter $\kappa=A_\kappa/A$ measures the ratio between the partially filled area A_κ and the total area A. As the partially filled area A_κ varies between 0 and A, the parameter κ varies between 0 and 1.

For wholly filled fractures, defined by $\kappa=1$, their average transmissivity T_1 for a filling permeability k may be estimated by:

$$T_1 = ke \tag{2.13}$$

For wholly clean fractures, that is for κ=zero, and their average transmissivity T_0 may be estimated by:

$$T_0 = Ce^n \tag{2.14}$$

In the last formula, C and n are empirical coefficients that must be dimensionally consistent with their units (see Louis, 1969; Quadros, 1982).

As frequently observed at leaking rock outcrops or underground excavations, groundwater usually percolates within a network of irregular and braided flow paths inside the fracture itself. Fracture thickness varies from place to place, from 0.01 to 1 mm for singular fractures and from 1 to 100 mm for narrow fracture groups, but occasionally higher. Consequently, the so-called "equivalent" hydraulic aperture e of a singular fracture or fracture set or a group of braided and crisscrossed fractures is an ideal concept that may yield approximate flowrates $\mathbf{q}$ by the expression $\mathbf{q}=[k_{equivalent}]\mathbf{J}$. The above formulas may be cautiously employed to estimate their average transmissivity T.

For partially filled fractures, that is, for $0<\kappa<1$, the Maxwell mixture rule may approximate their average transmissivity T as:

$$T_\kappa = (T_0)^{1-\kappa}(T_1)^\kappa \tag{2.15}$$

However, relatively reliable estimates of fracture equivalent transmissivities for laminar flows require ingenious field tests accurately performed, as discussed ahead.

The hydraulic regime of the groundwater flow throughout a pervious rough and geometrically perfect fracture depends on the magnitude of the hydraulic gradient $\mathbf{J}$, on the fracture average aperture e, and on its relative roughness $\delta e/e$, where δe is the modal height of the fracture asperities (see Figure 2.22). For clean fractured rocks, usually defined by "equivalent" hydraulic apertures $e<1$ mm and $\delta e/e$ around 1/3, the groundwater flowing far from wells or springs may generally be taken as laminar because the natural hydraulic gradients $\mathbf{J}$ seldom exceed 1 (Franciss, 1970; for a detailed account about these influences, see Louis, 1969).

The hydraulic transmissivity of a non-horizontal and non-isotropic fracture is a second-order tensor [T]. For example, the anisotropic hydraulic transmissivity [T] of

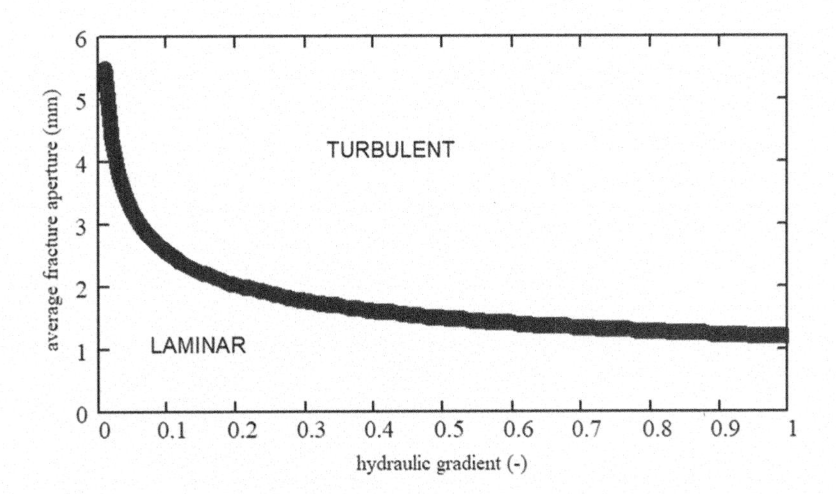

Figure 2.22 Approximate empirical boundary, as a function of the average fracture aperture e and the hydraulic gradients **J**, separating the laminar from the turbulent flow domain for relative roughness $\delta e/e$ around 1/3. (After numerical data from Louis, 1969.)

an almost planar fracture (related to an orthogonal coordinate system xyz, whose axis z is perpendicular to the fracture plane and the axes x and y coincide with the maximum and minimum transmissivities directions) is (see Figure 2.23):

$$|T| = \begin{pmatrix} T_x & 0 & 0 \\ 0 & T_y & 0 \\ 0 & 0 & 0 \end{pmatrix} \tag{2.16}$$

In this case, related to the fracture cartesian axes xyz, the relationship between the specific discharge **q**, the hydraulic gradient **J**, and the hydraulic transmissivity [T] is:

$$\begin{pmatrix} q_x \\ q_y \\ q_z \end{pmatrix} = \begin{pmatrix} T_x & 0 & 0 \\ 0 & T_y & 0 \\ 0 & 0 & 0 \end{pmatrix} \begin{pmatrix} J_x \\ J_y \\ J_z \end{pmatrix}$$

As intuitively expected, the z-component T_z of the tensor [T] – *but only when related to the fracture "proper" frame xyz* – must be zero. In this case, the z-component q_z of the discharge vector referred to this frame must also be zero, even though the z-component of the gravity driving pull J_z is not zero:

$$\begin{pmatrix} q_x \\ q_y \\ q_z \end{pmatrix} = \begin{pmatrix} J_x T_x \\ J_y T_y \\ 0 \end{pmatrix}$$

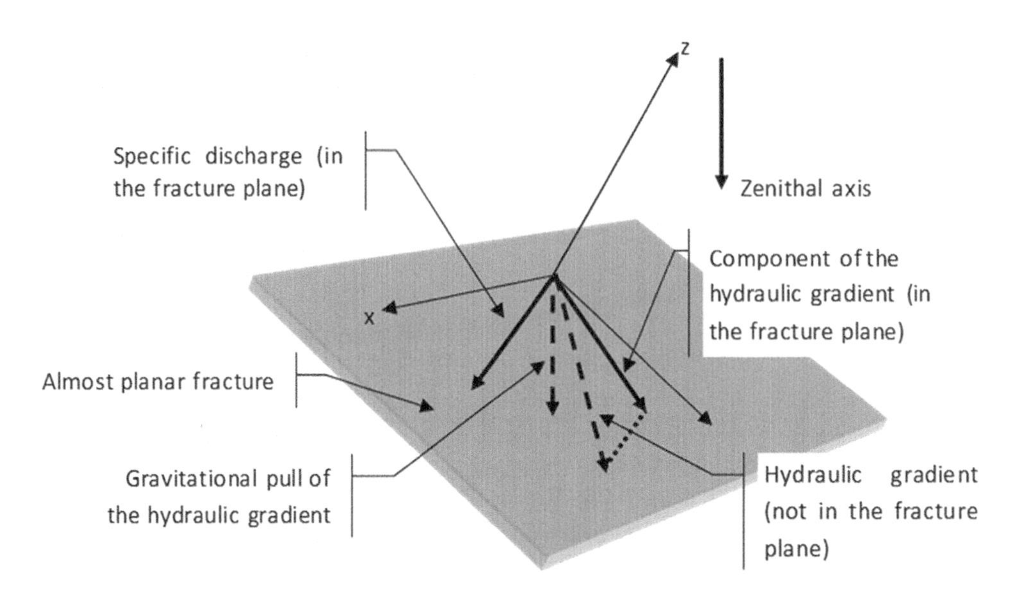

Figure 2.23 Inclined view of an almost planar and hydraulically anisotropic fracture. The fracture proper xyz orthogonal frame has its z-axis normal to the fracture plane. The axes x and y are the maximum and minimum transmissivities directions. The specific discharge **q**, only driven by the component of the hydraulic gradient **J** projected on the fracture plane, stays confined to the fracture plane. However, only for a hydraulically isotropic fracture, the specific discharge and the projected gradient are aligned. Note the out-of-plane gravitational components of the hydraulic gradient referred to the proper axes xyz and the zenithal axis Z.

2.3.5.3 Pseudo-continuity and pervious fractures

The components of a transmissivity tensor [T] referred to its proper axes may be defined for the zenithal frame ENZ from the following matrix rotation:

$$\begin{pmatrix} T_{EE} & T_{EN} & T_{EZ} \\ T_{NE} & T_{NN} & T_{NZ} \\ T_{ZE} & T_{ZN} & T_{ZZ} \end{pmatrix} = R \begin{pmatrix} T_x & 0 & 0 \\ 0 & T_y & 0 \\ 0 & 0 & 0 \end{pmatrix} R^T \tag{2.17}$$

The resulting matrix is symmetric, that is, $T_{EN} = T_{NE}$, $T_{EZ} = T_{ZE}$, and $T_{NZ} = T_{ZN}$.

As already known, the rotation matrix [R] is an orthogonal square matrix, and the elements of its three columns correspond, respectively, to the direction cosines of the unit vectors of the x-axis, y-axis, and z-axis to the zenithal reference frame ENZ. Then, its transpose equals its inverse, that is, $R^T = R^{-1}$. Consequently, the matrix transformation $R\{q\} = R[T]R^T R\{J\}$ holds. As a result, referred to the zenithal frame ENZ, the relationship between the specific discharge **q**, the hydraulic gradient **J,** and the hydraulic transmissivity [T] is (Rocha and Franciss, 1977):

$$\begin{pmatrix} q_E \\ q_N \\ q_Z \end{pmatrix} = \begin{pmatrix} T_{EE} & T_{EN} & T_{EZ} \\ T_{NE} & T_{NN} & T_{NZ} \\ T_{ZE} & T_{ZN} & T_{ZZ} \end{pmatrix} \begin{pmatrix} J_E \\ J_N \\ J_Z \end{pmatrix} \tag{2.18}$$

Modeling groundwater flow but considering the interplay of many arbitrary fractures applying for each one, a relationship $\mathbf{q} = [T] \cdot \mathbf{J}$ is somehow tricky. However, if less accuracy and less detail are accepted and adapting Darcy's law to discontinuous systems, it is relatively simple to construct a pseudo-continuous model for a rock mass implicitly incorporating several arbitrary discontinuities (Franciss, 1970). Then, the transmissivity [T] of an arbitrary discontinuity traversing a subsystem of volume $\delta V = \delta E \cdot \delta N \cdot \delta Z$ is by a hydraulically equivalent tensor [k] defined by:

$$|k| = \begin{pmatrix} k_{11} & k_{12} & k_{13} \\ k_{21} & k_{22} & k_{23} \\ k_{31} & k_{32} & k_{33} \end{pmatrix} = \begin{pmatrix} \frac{T_{EE}}{\sqrt{\delta N \, \delta Z}} & \frac{T_{EN}}{\sqrt{\delta Z \, \delta E}} & \frac{T_{EZ}}{\sqrt{\delta E \, \delta N}} \\ \frac{T_{NE}}{\sqrt{\delta E \, \delta Z}} & \frac{T_{NN}}{\sqrt{\delta Z \, \delta E}} & \frac{T_{NZ}}{\sqrt{\delta E \, \delta N}} \\ \frac{T_{ZE}}{\sqrt{\delta N \, \delta E}} & \frac{T_{ZN}}{\sqrt{\delta N \, \delta E}} & \frac{T_{ZZ}}{\sqrt{\delta E \, \delta N}} \end{pmatrix} \tag{2.19}$$

The next example helps to understand the above concepts.

Example 2.1

Consider the transmissivity T of a planar fracture referred to its "proper" 2D coordinate system xy and the directions of its principal transmissivities T_{max} and T_{min} agreeing to the x and y directions. Then, its transmissivity is described by a second-order tensor having T_{max} and T_{min} as its eigenvalues:

$$T = \begin{pmatrix} 5 \times 10^{-5} & 0 & 0 \\ 0 & 10^{-5} & 0 \\ 0 & 0 & 0 \end{pmatrix} m^2/s$$

It is then possible to sequentially rotate this matrix to conform its elements to the real fracture attitude in the field. Indeed, after two successive rotations: a $+30°$ rotation about Z and a $+30°$ rotation about X, the lower hemisphere stereographic projections of the fracture-pole and the T_{max} and T_{min} directions depicted in Figure 2.24 agree with the geological data.

The rotation operator R was:

$$R = \begin{pmatrix} 0.866 & -0.433 & 0.25 \\ 0.5 & 0.75 & -0.433 \\ 0 & 0.5 & 0.866 \end{pmatrix}$$

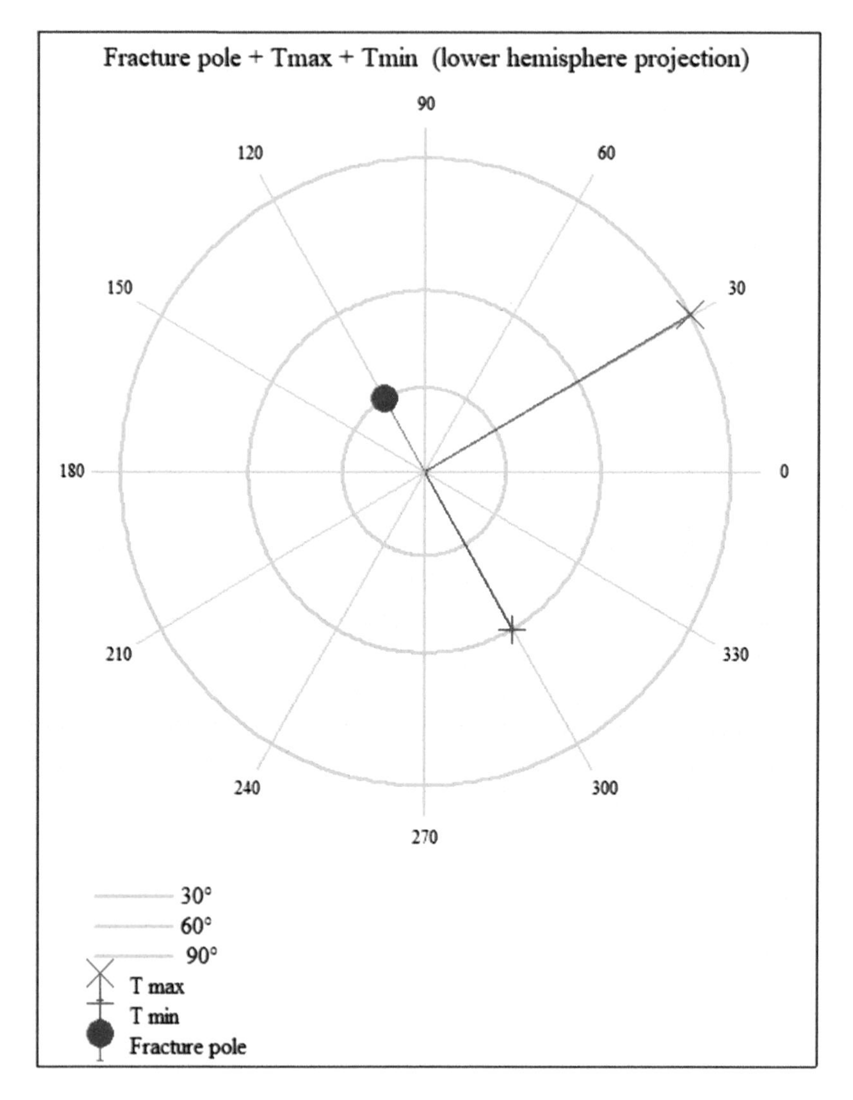

Figure 2.24 The above stereo-plot shows the result of two successive rotations of the tensor T that initially had its anisotropic transmissivities T_{max} and T_{min}, respectively, agreeing to x and y directions of the fracture proper 2D coordinate system.

The rotation operator $R[T]R^T$ gave the rotated hydraulic transmissivity T_{rot}:

$$T_{rot} = \begin{pmatrix} 3.938\times10^{-5} & 1.84\times10^{-5} & -2.165\times10^{-6} \\ 1.84\times10^{-5} & 1.812\times10^{-5} & 3.75\times10^{-6} \\ -2.165\times10^{-6} & 3.75\times10^{-6} & 2.5\times10^{-6} \end{pmatrix} m^2/s$$

The eigenvalues and eigenvectors of the rotated hydraulic transmissivity are preserved, as expected:

$$\text{eigenvalues}\left(T_{rot}\right) = \begin{pmatrix} 5\times10^{-5} \\ 1\times10^{-5} \\ 0 \end{pmatrix} m^2/s$$

$$\text{eigenvectors}\left(T_{rot}\right) = \begin{pmatrix} -0.866 & -0.433 & 0.25 \\ -0.5 & 0.75 & -0.433 \\ 0 & 0.5 & 0.866 \end{pmatrix}$$

Now the attitude of the rotated planar fracture conforms to the field observation, as sketched in Figure 2.25.

According to Equation 2.19 and considering a pseudo-continuous cuboid subsystem of volume $\delta V = \delta E \cdot \delta N \cdot \delta Z$, defined by edge lengths $\delta E = 2\,m$, $\delta N = 2\,m$, and $\delta Z = 2\,m$, the rotated hydraulic transmissivity tensor T can be substituted for an equivalent hydraulic conductivity [k]:

$$\text{equivalent\_k} = \begin{pmatrix} 1.969\times10^{-5} & 9.2\times10^{-6} & -1.083\times10^{-6} \\ 9.2\times10^{-6} & 9.06\times10^{-6} & 1.875\times10^{-6} \\ -1.083\times10^{-6} & 1.875\times10^{-6} & 1.25\times10^{-6} \end{pmatrix} m/s$$

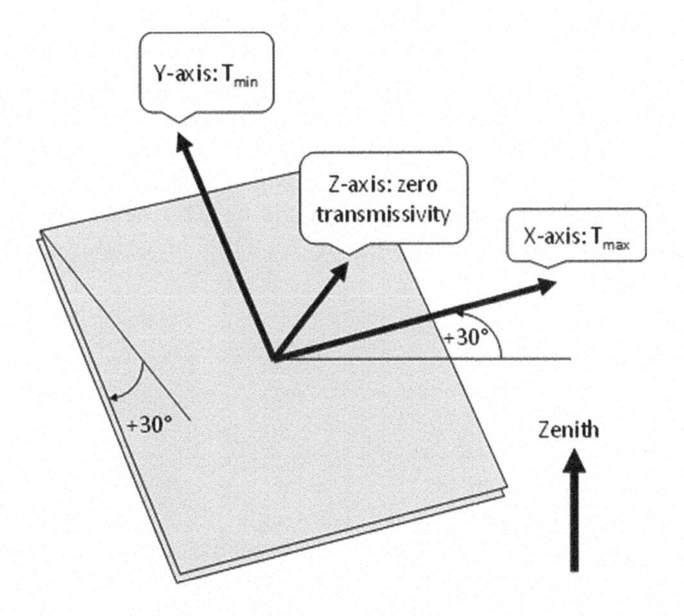

Figure 2.25 The attitude of the twice-rotated planar fracture conforms to its field observations. Its eigenvalues remain invariable.

The physical meaning of the positive or negative signs of the off-diagonal parameters of transmissivity tensor k are as follows:

- $k_{EN} = 9.2 \times 10^{-6}$: positive JN induces positive flow $q_{EN} = k_{EN} \cdot J_N$ along the E-axis (and vice-versa for negative J_N).
- $k_{EZ} = -1.083 \times 10^{-6}$: negative J_Z induces negative flow $q_{EZ} = k_{EZ} \cdot J_Z$ along the E-axis (and vice-versa for positive J_Z).
- $k_{NZ} = 1.875 \times 10^{-6}$: positive J_Z induces positive flow $q_{NZ} = k_{NZ} \cdot J_Z$ along the N-axis (and vice-versa negative J_Z).

However, it may happen that the original fracture principal transmissivity T_{max} was not horizontal. In this case, to replicate the observed fracture field characteristics, a previous plane rotation of 15° about the z-axis is needed. In this case, the initial planar hydraulic transmissivity matrix turned to be:

$$T = \begin{pmatrix} 4.732 \times 10^{-5} & -10 \times 10^{-6} & 0 \\ -10 \times 10^{-6} & 1.268 \times 10^{-5} & 0 \\ 0 & 0 & 0 \end{pmatrix} m^2/s$$

After the same successive rotations: +30° rotation about Z and a +30° rotation about X, the new fracture-pole and the T_{max} and T_{min} new directions conform to Figure 2.26.

The same arguments previously applied to planar fractures hold for planar fractures having open tubular conduits filled or partially filled with a pervious matrix. Also, assume that the groundwater transport capacity Q of this conduit is always laminar, linearly proportional to the hydraulic gradient **J**. Moreover, considering that contrasting to the conduit's transport capacity Q, its fracture flowrate is insignificant. In this case, the discharge capacity Q of the tubular conduit can be replaced by an equivalent planar T_x pseudo-transmissivity by simply dividing Q by the fracture width. Then, the fracture equivalent hydraulic transmissivity for a tubular conduit can be described by (see Figure 2.27):

$$|T| = \begin{pmatrix} T_{XX} & 0 & 0 \\ 0 & 0 & 0 \\ 0 & 0 & 0 \end{pmatrix}$$

These tubular tensors may also model rivers segments or inclined shafts and galleries but avoid nonlinearity using appropriate and specific coefficients, approximate solutions, and adequate inner boundary conditions, as exemplified in Chapter 4.

As shown below, the same theoretical principles applied to fractures apply to describe the transmissivity tensors of tubular conduits.

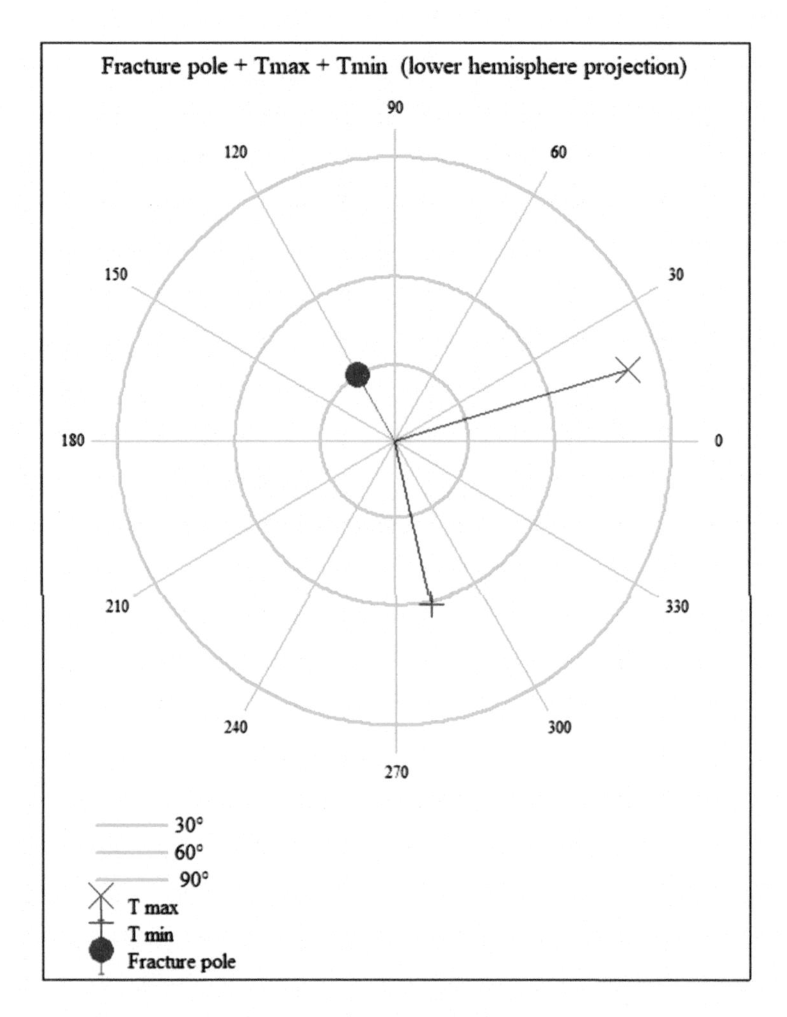

Figure 2.26 The above stereo-plot shows the result of two successive rotations of the tensor T having principal original anisotropic transmissivities T_{max} and T_{min}, respectively, parallel to the X and Y axes.

Example 2.2

For a fracture width of 2 mm, the equivalent transmissivity tensor [T] of a tubular conduit attached to the y-axis of the plane xy is described by (see Figure 2.28):

$$|T| = \begin{pmatrix} T_{XX} & 0 & 0 \\ 0 & 0 & 0 \\ 0 & 0 & 0 \end{pmatrix} = \begin{pmatrix} 10^{-2} & 0 & 0 \\ 0 & 0 & 0 \\ 0 & 0 & 0 \end{pmatrix} m^2/s$$

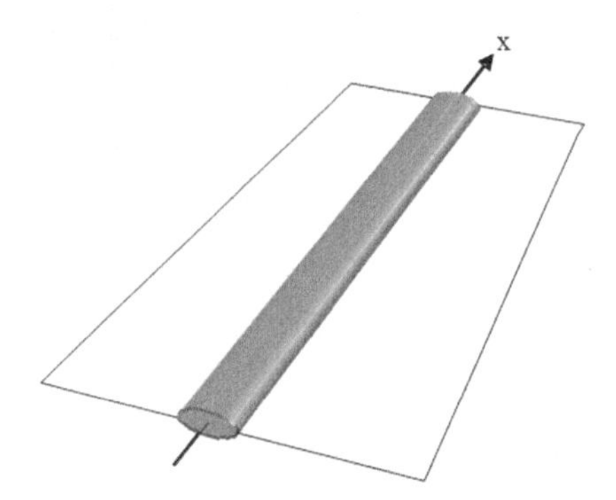

Figure 2.27 A tubular conduit along the x-axis and attached to a xy-planar pseudo-fracture.

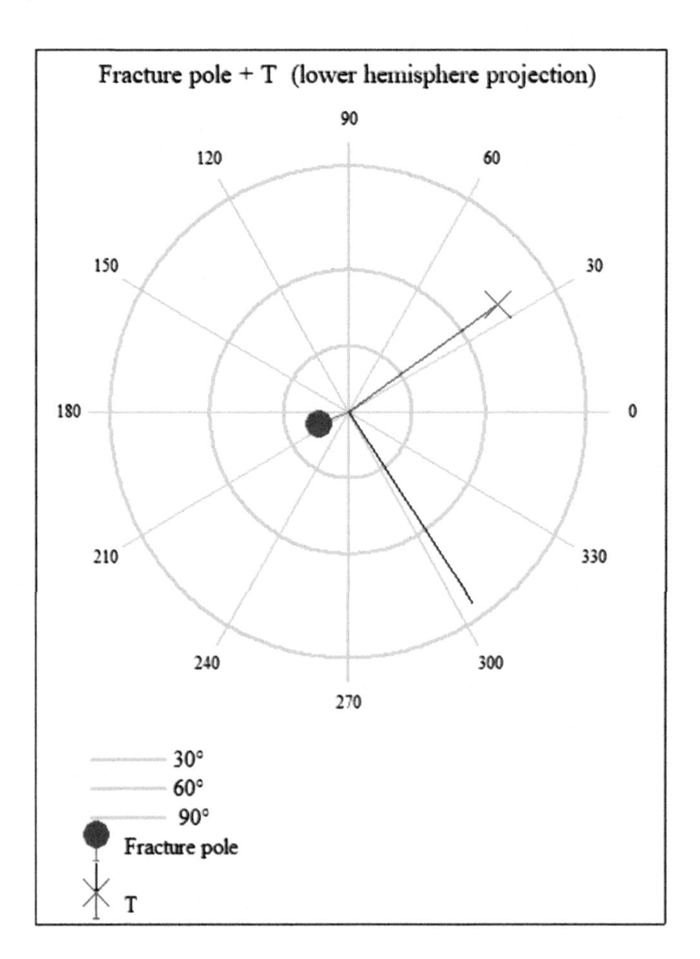

Figure 2.28 The above stereo-plots shows the result of two successive rotations of the pseudo-tensor T that simulate the hydraulic behavior of a single conduit having its original alignment along the X-axis.

Then, as in the previous example, it is possible to rotate this matrix to conform its elements to the real pseudo-fracture attitude in the field. Indeed, after two successive rotations: a + 35° rotation about Z and a + 15° rotation about Y, the lower hemisphere stereographic projections of the fracture-pole and the T_{max} and T_{min} directions depicted agreeing with the geological data.
The transmissivity tensor of the pseudo-fracture, now referred to the ENZ frame, is:

$$T_{rot} = \begin{pmatrix} 6.261\times10^{-3} & 4.384\times10^{-3} & -2.048\times10^{-3} \\ 4.384\times10^{-3} & 3.07\times10^{-3} & -1.434\times10^{-3} \\ -2.048\times10^{-3} & -1.434\times10^{-3} & 6.699\times10^{-4} \end{pmatrix} m^2/s$$

2.3.5.4 *Valid pseudo-continuity assumption*

For fractured rocks, pseudo-continuity is acceptable taken into account certain limits established with the help of simplified 2D electro-analogs models developed and employed by the author from 1.962 to 1.969.[5] In these models, discrete R-analogs sketched over polyester drawing paper (thickness < 0.1 mm) with an electro-resistive ink (linear specific resistance measured in ohm/m) simulate steady-state flows. However, the simulation of transient flows required discrete RC-analogs, having planar C elements made of congruent rectangles at both faces of the drawing paper, but painted with electro-conductive ink (specific planar capacitance measured in F/m^2; see Figure 2.29).

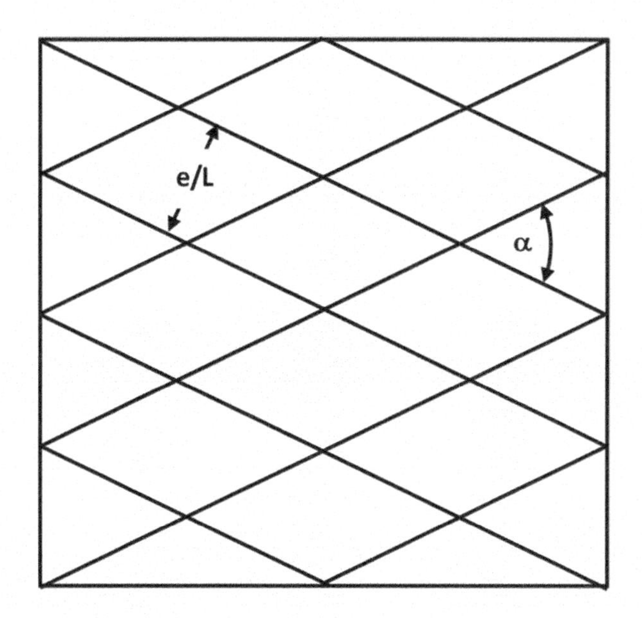

Figure 2.29 Simplified electro-analog to set an acceptable pseudo-continuity limit.

The test results indicated that acceptable pseudo-continuity might be settled for ratios $e/L \geq 0.1$ and $30° \leq \alpha \leq 150°$ (see Figure 2.30). Outside these bounds, single fractures must be individually modeled and correctly incorporated in the pseudo-continuous model constructed as explained in Section 2.3.5.5.

2.3.5.5 Erratic discontinuities

For erratic discontinuities labeled a, b, c ... defined by their equivalent hydraulic conductivities $\left[k^a\right],\left[k^b\right],\left[k^c\right]$... the resulting equivalent hydraulic conductivity for a pseudo-subsystem can be approached by a "parallel coupling" if the almost identical

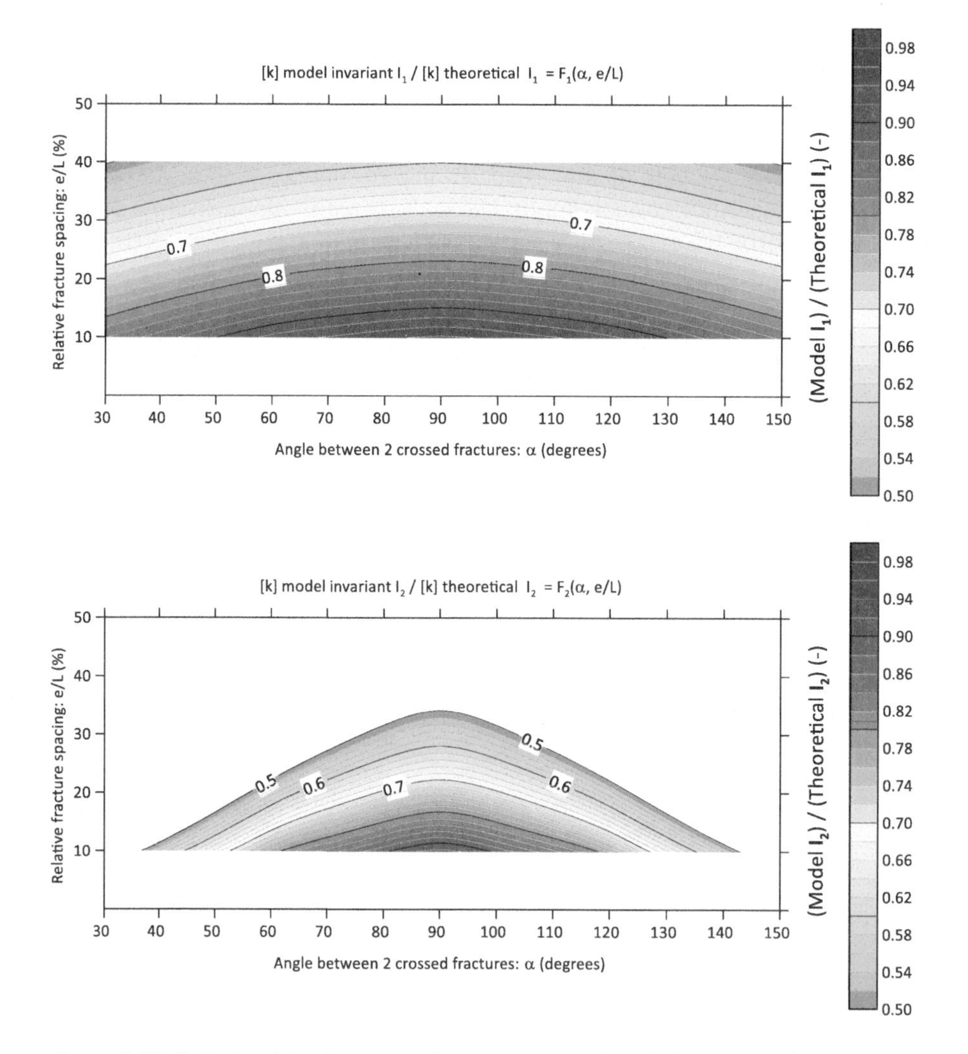

Figure 2.30 Relationships between the experimental and theoretical first I_1 and second I_2 invariants of the conductivity tensors [k]. Results from 34 sets of electro-analog models.

hydraulic gradients within this pseudo-subsystem may be substituted by an average value (see Section 3.3.8):

$$|k| = |k^a| + |k^b| + |k^c| + \ldots$$

Written as matrices:

$$|k| = \begin{bmatrix} \left[(k_{11})^a + (k_{11})^b + (k_{11})^c + \ldots\right] & \left[(k_{12})^a + \ldots\right] & \left[(k_{13})^a + \ldots\right] \\ \left[(k_{21})^a + (k_{21})^b + (k_{21})^c + \ldots\right] & \left[(k_{22})^a + \ldots\right] & \left[(k_{23})^a + \ldots\right] \\ \left[(k_{31})^a + (k_{31})^b + (k_{31})^c + \ldots\right] & \left[(k_{32})^a + \ldots\right] & \left[(k_{33})^a + \ldots\right] \end{bmatrix}$$

This summation rule only applies to uninterrupted discontinuities that cut the subsystem from side to side but neglects minor hydraulic head losses at discontinuities crossings (see Figure 2.31). It presumes that the "modeler" accept the simulation accuracy at the chosen observation scale. If the magnitude of hydraulic conductivity tensor $[k_r]$ of the intact rock exceeds 1/10 to 1/30 those for the fractures, they must be added to summation matrix.

Occasionally an almost impervious discontinuity, e.g., a thin diabase dike, a fracture completely sealed with calcite or clay fillings, can restrain the percolation. In this case, approximately construct the eigenvalues of the permeability tensor of the rock block as follows:

- *First hypothesis*: all pervious discontinuities that traverse the subsystem maintain their connectivity as they cross the impervious discontinuity. In this case, only

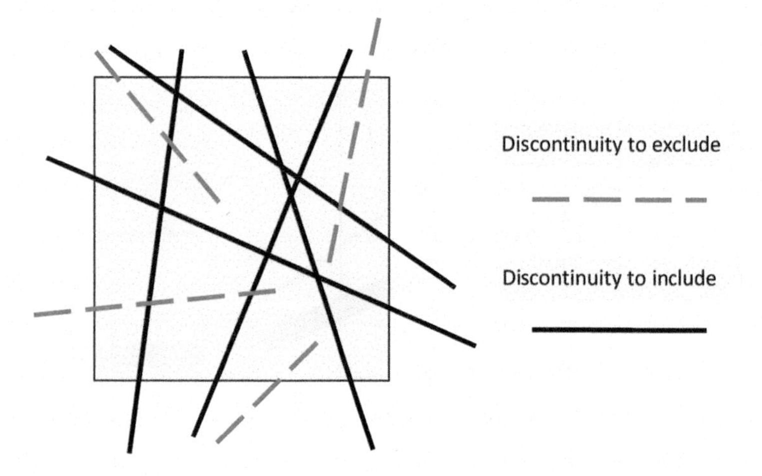

Figure 2.31 Exclude from the summation matrix all discontinuities that do not traverse the subsystem from side to side. If necessary, approximately incorporate its influence in the intact rock permeability.

reduce the eigenvalues of the permeability tensor $[k_r]$ of the intact rock matrix before applying the summation rule. Do this multiplying $[k_r]$ by $\left(k_f/k_r\right)^p/k_r$, where k_r is the average rock matrix original permeability, $k_f \ll k_r$ is the permeability of the quasi-impervious structure and $p \ll 1$ is its volumetric proportion (roughly estimated by its thickness divided by the subsystem side).

- *Second hypothesis*: some of the pervious discontinuities that traverse the subsystem do not retain their connectivity as they cross the impervious discontinuity. In this case, all blocked discontinuities are excluded from the summation matrix (see Figure 2.32). Then, reduce the rock matrix permeability tensor $[k_r]$, as explained above, but by only applying the summation rule to the remaining pervious discontinuities $\left[k^a\right], \left[k^b\right], \left[k^c\right]$.

Based on the geometry and hydraulic properties of the lithologic and structural features of a flow domain, it is possible to estimate the hydraulic conductivities [k] over that domain. However, by solving "inverse problems" (based on large scale field tests or time-series of piezometric and discharge data), the resulting parameters are much more reliable (Hsieh et al, 1985). That procedure, presented in Chapter 3, implicitly take all hydraulic boundary conditions, and automatically include all unseen discontinuities' hidden contributions.

2.3.6 Assessment of the hydraulic conductivities

2.3.6.1 Introduction

According to Zwikker[6] but here simply explained and applied to linear processes, the physical condition of a physical system in equilibrium with its environment – *for example, a rock mass* – may be stated by pairs of "intensive" **i** and "extensive" **e** parameters,

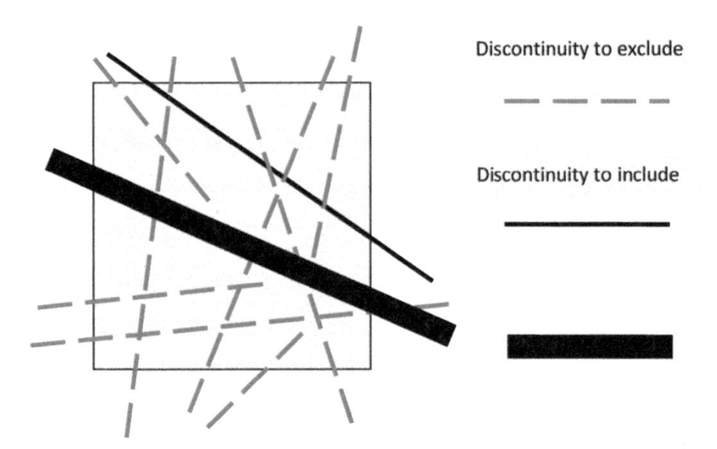

Figure 2.32 A thick impervious seam intercepts and restrains seepage of some pervious discontinuities. Some pervious discontinuities lose their connectivity as that discontinuity crosses them.

Table 2.2 Examples of pairs of intensive-extensive parameters

"Intensive" parameter **i** *(unaltered if the body mass is separated)*	*"Extensive" parameter* **e** *(altered if the body mass is separated)*
Stress	Strain
Temperature	Entropy
Electrical field	Displacement

as exemplified in Table 2.2, whose products have the dimensions of energy per unit volume. An incremental change $\partial \mathbf{i}$ in an "intensive" parameter **i** may induce a responsive change $\partial \mathbf{e}$ in one or more than one "extensive" parameter **e**. For example, a differential change $\partial \boldsymbol{\sigma}$ in the state of stress at a point of a rock mass induces a strain deformation $\partial \varepsilon$. A partial differential equation $[C] = \partial \varepsilon / \partial \sigma$ links the strain ε to the stress $\boldsymbol{\sigma}$ by elastic compliance [C] provided that other types of intensive parameters, like temperature **T**, are kept constant.

Applying a gradient to an *intensive* parameter causes a *steady-state* flow of the corresponding *extensive* parameter, whose product now has the dimension of an *energy dissipation density*. Then, *compliances*, *extensive* and *intensive parameters*, respectively, substitute for conductivities, gradients, and flow rates. In the case of groundwater flow through a rock mass, its hydraulic conductivity [k] is formally defined by the linear differential equation $[k] = \partial \mathbf{q} / \partial \mathbf{J}$, where [k] is a second-order tensor,[7] usually written as a matrix, reduced to a scalar for homogeneous and isotropic pervious media. The reciprocal of [k] are generically called *resistivity*.

The material composition and structural arrangement of the rock mass, including all their open fractures, determine the magnitude and the symmetry of the matrix [k]. The lower its structural complexity, the lower its anisotropy and simplicity as depicted by the matrix [k] (see Figure 2.33). Consequently, the magnitude and the anisotropy of the hydraulic conductivity [k] of a rock mass reveal the size and shape of the intact rock's interconnected porosity and the 3D geometry of its open fractures.

Also, the character of the hydraulic conductivity [k] of a complex rock mass follows Neumann's and Curie's principles. For structural arrangement specified by symmetry operations, Neumann's principle states that the tensor components of [k] remain invariant for these symmetry transforms. For complex structures constructed by grouping simpler ones, Curie's principle states that the resulted assemblage's symmetry cannot be lower than the higher symmetry of the collected tensors. Consequently, first inferences of the complex hydraulic conductivity [k] of a rock mass deduced from its litho-structural features, including weathering effects, constitute a valid theoretical attempt, to be further re-scaled based on the results or significant field tests. However, other factors, including changes of the intensity and the time-extent of environmental actions, may change the composition and structure of a rock mass, altering the character of [k].

Another difficulty concerns the appreciation of inhomogeneous and anisotropic hydraulic conductivity [k] of a rock masses marked by gradual or abrupt 3D variation of its litho-structural features. Its partition into statistically quasi-homogeneous zones and the identification of its singular features as dikes, faults, contacts may simplify its analysis. However, the distinction between heterogeneous and homogeneous masses always relies on the chosen observation scale.

Symmetry class	*Fracture orientation and spacings*	*Matrix [k]*
Triclinic a≠b≠c $\alpha \neq \gamma \neq \beta \neq 90°$	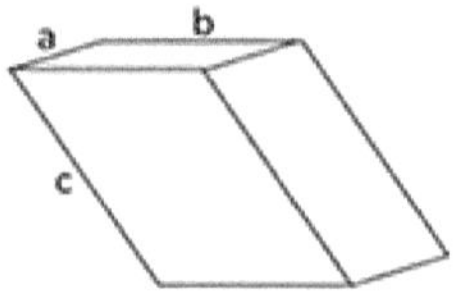 a, b, c	$\begin{pmatrix} k_{11} & k_{12} & k_{13} \\ k_{21} & k_{22} & k_{23} \\ k_{31} & k_{32} & k_{33} \end{pmatrix}$
Monoclinic a≠b≠c $\alpha = \gamma = 90° \neq \beta$	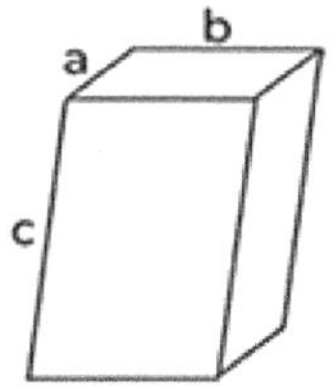 a, b, c	$\begin{pmatrix} k_{11} & 0 & k_{13} \\ 0 & k_{22} & 0 \\ k_{31} & 0 & k_{33} \end{pmatrix}$
Orthorhombic a=b=c $\alpha \neq \beta \neq \gamma \neq 90°$	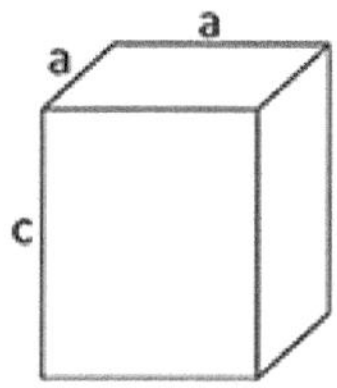 a, a, c	$\begin{pmatrix} k_{11} & 0 & 0 \\ 0 & k_{22} & 0 \\ 0 & 0 & k_{33} \end{pmatrix}$
Tetragonal a=b≠c $\alpha = \beta = \gamma = 90°$	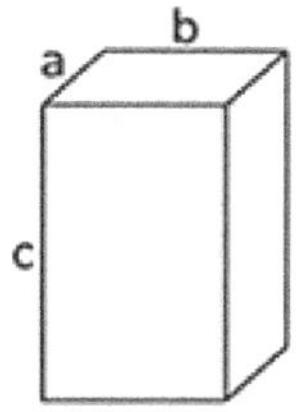 a, b, c	$\begin{pmatrix} k_{11} & 0 & 0 \\ 0 & k_{22} & 0 \\ 0 & 0 & k_{33} \end{pmatrix}$
Cubic a=b=c $\alpha = \beta = \gamma = 90°$	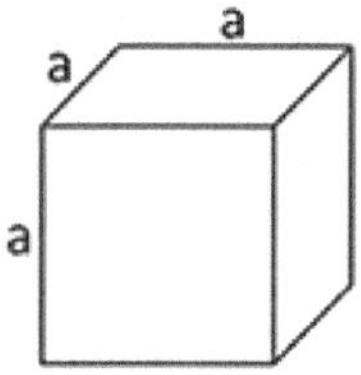 a, a, a	$\begin{pmatrix} k_{11} & 0 & 0 \\ 0 & k_{11} & 0 \\ 0 & 0 & k_{11} \end{pmatrix}$

Figure 2.33 Table of the maim symmetry classes × fracture spacing and orientation × matrix [k] character. For triclinic, monoclinic, and orthorhombic systems, not one plane accuses isotropic properties; for tetragonal system, all plans perpendicular to the symmetry axes accuse isotropic properties. For the cubic system, all planes accuse isotropic properties.

Inhomogeneous rock masses may be assimilated to quasi-homogeneous rock masses when they accuse irregular fluctuations of their K parameters at periods of variation below 1/10 to 1/30 of their average dimensions. Field tests may give reliable average [K] parameters for significant rock-mass volumes. However, if extensive, they may conceal the substantial effects of their singular litho-structural features. Hence, the need to inspect these rock masses at various observations scales, considering that quasi-homogeneous zones do not imply [k] invariance, only statistical homogeneity, marked by greater or lesser dispersion around [k] central values. A dependable knowledge of average values and standard deviations of the first I_1, second I_2, and third I_3 invariants of [k] may be valuable.

Finally, one must pay attention that estimates of [k] by several extensive field tests in fractured rock masses require a dependable solution for the differential equation $[k] = \partial \mathbf{q}/\partial \mathbf{J}$ but constrained by the appropriate recognition of the boundary conditions for each test, so often neglected. Bad choice of adequate tests and wrong boundary assumptions can lead to large estimation errors. To avoid complicated solutions (theoretical limitations) or to circumvent the difficulty of assessing the temporal and spatial evolution of the boundary conditions (practical limitations), the modeler may decide to adopt more or less simplifying hypotheses (homogeneity, isotropy) but compatible with the amount of time previously fixed to make the tests and the technical and economic importance of the project. As reliable estimates of [k] through appropriate several extensive field tests always involve significant time and energy consumption, the decision to do these tests depends on its cost-benefit for the whole project.

2.3.6.2 *Exploring at the "sample scale": back to 1930*

The extraction of the volume of oil trapped within fractured and porous sandstone or limestone reservoirs depends on the degree of interconnection of several small discontinuities' groups a few mm apart. Indeed, the commercial extraction of the hydrocarbons trapped in their interstitial porosity stops for almost clogged fissures by secondary mineralization. However, if these small fractures remain pervious, extraction may be very efficient and effective. Permeability pressure tests made on small samples, cored from large diameter samples of prospecting wells, gave an idea of how such fractures will behave during the exploitation. Moreover, the visual inspection of these small plugs, mainly, the nature and extent of their fillings and extents, gave an idea of its potential productivity (see Figure 2.34).

Laboratory tests measure the matrix permeability and the corresponding "effective" porosity of the small sample. If the rock matrix is anisotropic, the evaluation of its hydraulic conductivity requires measuring the permeabilities of many oriented "plugs" extracted from the large rock core. The oriented permeability $K_d(\mathbf{n})$ for a given direction, defined by its unit vector $\mathbf{n}$, is related to the tensor $|\mathbf{K}|$ by the relationship $k_d(\mathbf{n}) = 1/\left[\mathbf{n}^T \cdot |\mathbf{K}|^{-1} \cdot \mathbf{n}\right]$. Measuring more than six directional permeabilities in different directions allows the determination of the components of $|\mathbf{K}|$ by inverse methods (see references[8, 9] for details).

2.3.6.3 *Exploring old Lugeon tests: 1958 to 1968*

Lugeon tests were developed by the Swiss geologist Maurice Lugeon in 1933 for judging the water-tightness of bedrock foundation for concrete dams. They measure the

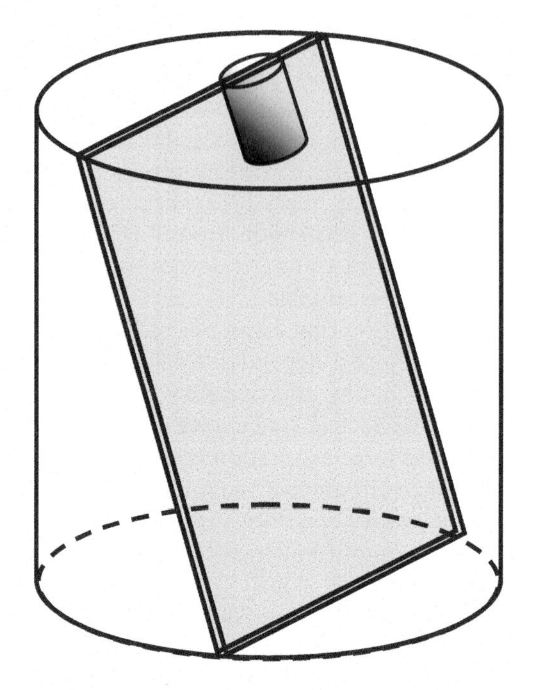

Figure 2.34 The image shows how to cut an oriented small sample from a large diameter core of a fractured rock. This small plug can be visually examined and subjected by permeability pressure-testing.

intake of pressurized water – *in L/min/m/atm* – into usually 1–3 m borehole segments under hydraulic heads that presumably would act at the dam foundation after finishing loading the reservoir.

From 1958 to 1968, the author made several Lugeon tests – *but at significantly lower test pressures* – at 19 river valleys in Brazil, basically including the following bed-rock types, but in an intertropical context: 53% in gneisses and granites; 16% in siltstones and slates; 11% in basalts; 10% in arkoses; 5% in conglomerates and siltstones; 5% in tillites. Average indicators and corresponding standard deviation at these dam valleys were: incoherent top strata depth: 7.1 ± 3.8 m; water table depth: 8.4 ± 8.9 m; "old rock-recovery percent" R for AX boreholes: $71.5\% \pm 23.0\%$; average valley slopes H/V: 0.20 ± 0.14 (see dam-site in areas Figure 2.35).

The "equivalent" isotropic hydraulic conductivity **k** of these bedrock foundations was inferred from the results of several Lugeon tests by the modified formula proposed by Babushkine and Guirinsky[10]:

$$k = \frac{Q}{2\pi HL} \cdot \ln\left(\frac{0.66L}{r}\right)$$

In the above formula, **k** (m/s) denotes the "equivalent" isotropic hydraulic conductivity, **Q** (m^3/s), the water intake during the test, H (m), the water pressure head, L (m), the length of the borehole segment tested, and r (m), the borehole radius.

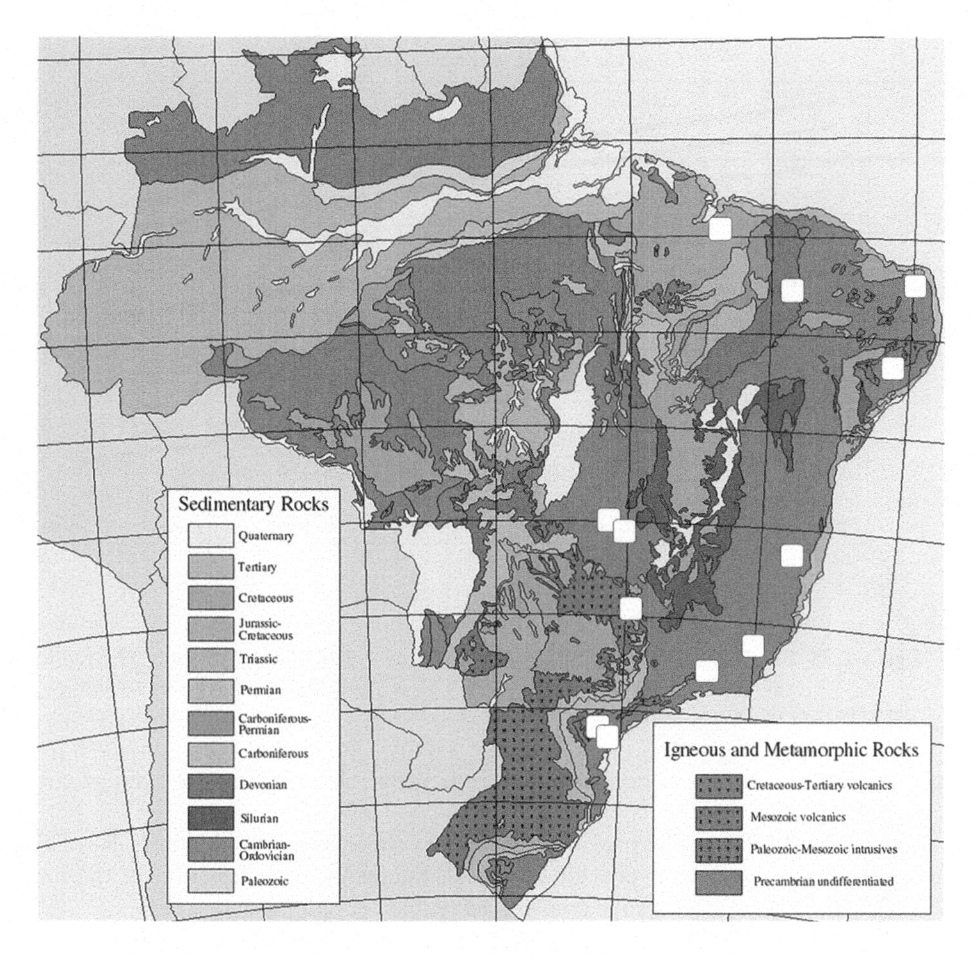

Figure 2.35 White squares indicate the location of one or more than one of the investigated dam sites.

For all the investigated sites, it was possible to estimate a polynomial and a logarithmic regression relating the equivalent hydraulic conductivity k, the depth P, and the old percent core-recovery R. The best global regression k = F(P, R), including all 1,625 Lugeon tests results[11], was logarithmic, as shown in Figure 2.36, defined by:

$$k(P, R) = e^{-\left(a + b \cdot R^3 + c \cdot P\right)}$$

For each dam site location, the equivalent hydraulic horizontal conductivity K_H, from the bedrock top up to 50 m depth, was estimated by the expression:

$$K_H = \frac{1}{P_f - P_i}\left(\int_{P_i}^{P_f} k \, dP\right)$$

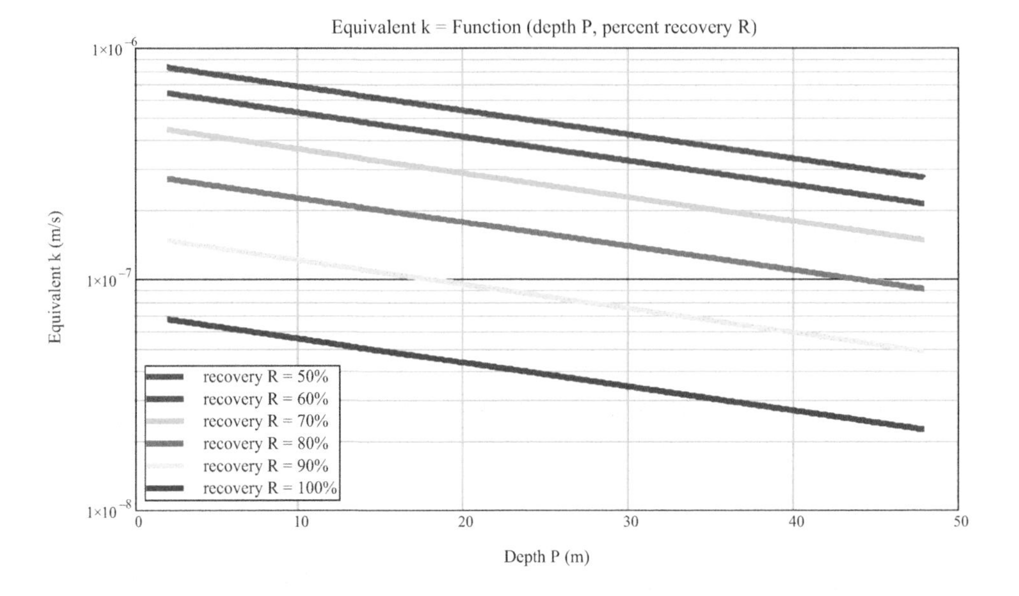

Figure 2.36 Best global regression $k = F(P, R)$ after 1,625 Lugeon tests at 19 Brazilian valleys (correlation coefficient $r = 0.57$ and log(standard dev.) $= 0.85$).

Figure 2.37 shows the graph expressing the average horizontal conductivity K_H at these 19 dam sites as a function of the average valley slope D and the average core-recovery R:

Figure 2.38, similar to the previous one, shows the average conductivity K_f of the fractured rocks, but only at pervious borehole segments, as a function of the dam valley's mean D slope the average core-recovery R.

Figures 2.39–2.41 show estimates, according to the methodology proposed by Snow,[12] of the average spacing s, average porosity n, and average equivalent fracture opening 2a of the bedrock foundation at the 19 investigated dam sites.

As collectively shown in Figure 2.42, as functions of the average valley slope **D** and the average core-recovery **R**, the above results show clear connections between the stage of geomorphological evolution of the investigated valleys and the bedrock fractures' character.

Relaxing stresses and growing deformation gradients follow these dam valleys' geomorphic evolution, roughly explained by gradual decreasing D values, increasing "equivalent" K_H values, and decreasing "equivalent" K_f values. These trends imply a continuous decreasing of fracture spacings s, decreasing open fracture apertures 2a, weathering contiguous fragmented rock masses, and increasing rock mass porosities n. These observations are evident if K_H and K_f are viewed as functions of the average valley slope D, an indirect rough aggregate measure of the weathering and denudation cycles (see Figure 2.42).

As the relief degradation pace diminishes in an intertropical environment, rock mass disintegration and weathering increase over time, followed by progressive groundwater saturation, giving rise to weaker rocks and soils accompanied by stress

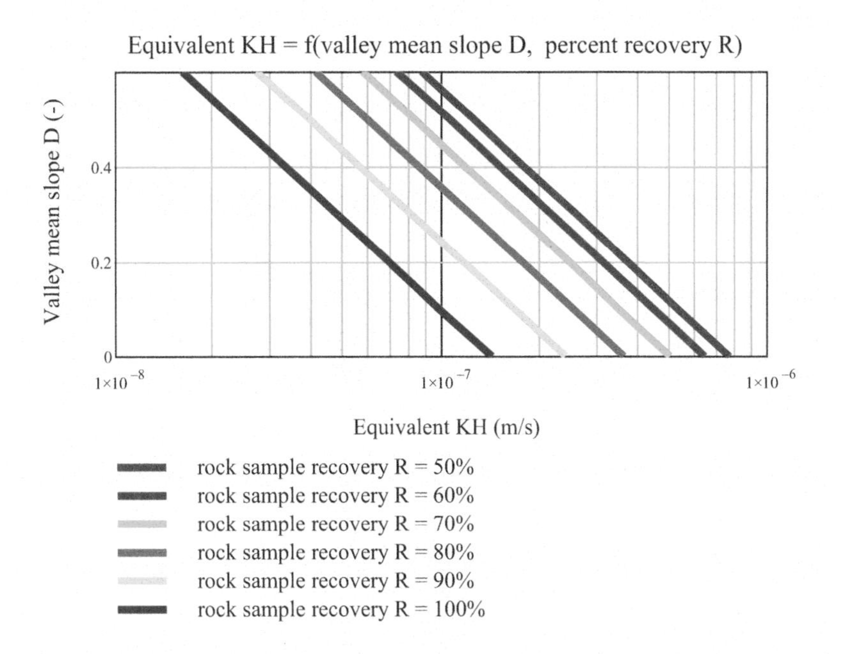

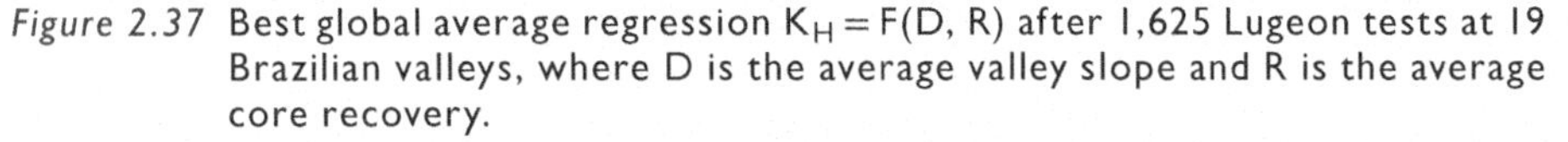

Figure 2.37 Best global average regression $K_H = F(D, R)$ after 1,625 Lugeon tests at 19 Brazilian valleys, where D is the average valley slope and R is the average core recovery.

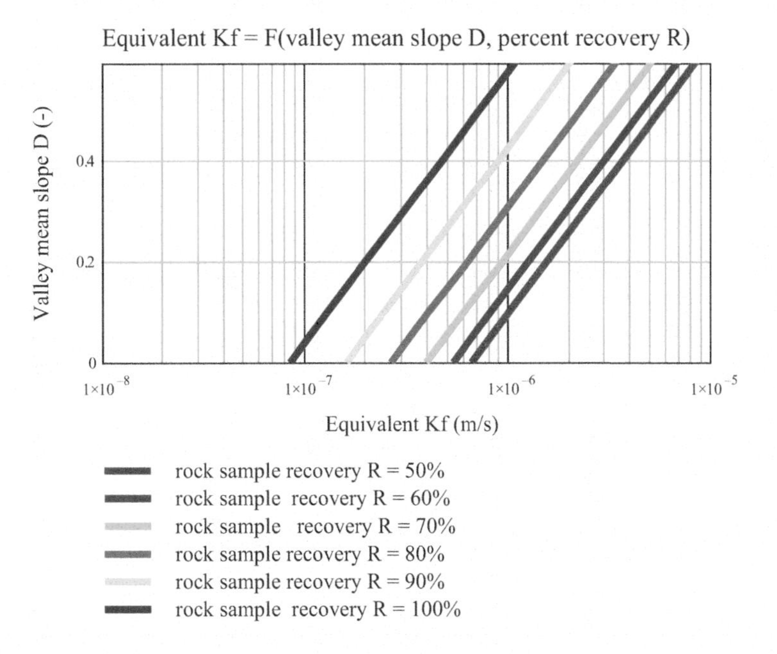

Figure 2.38 Best global average regression $K_f = F(D, R)$ after 1,625 Lugeon tests at 19 Brazilian valleys where, D is the average valley slope and R is the average core recovery.

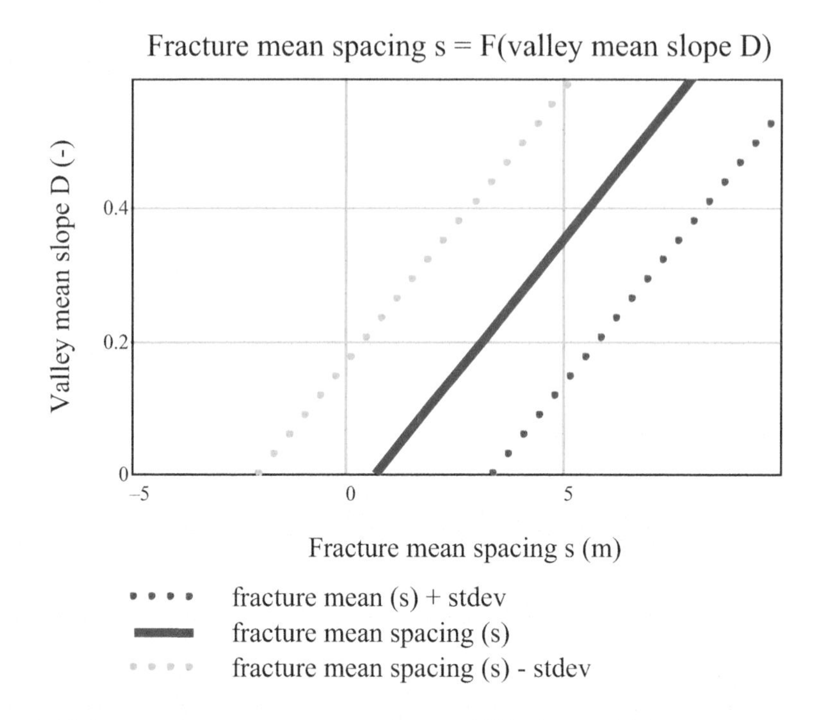

Figure 2.39 Best global average regression s = F(D) after 1,625 Lugeon tests at 19 Brazilian valleys, where D is the average valley slope.

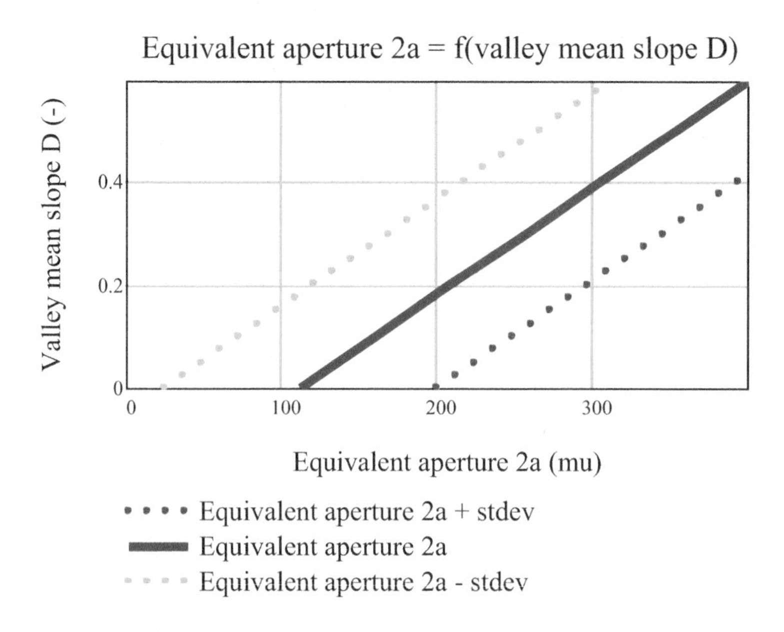

Figure 2.40 Best global average regression 2a = F(D) after 1,625 Lugeon tests at 19 Brazilian valleys, where D is the average valley slope.

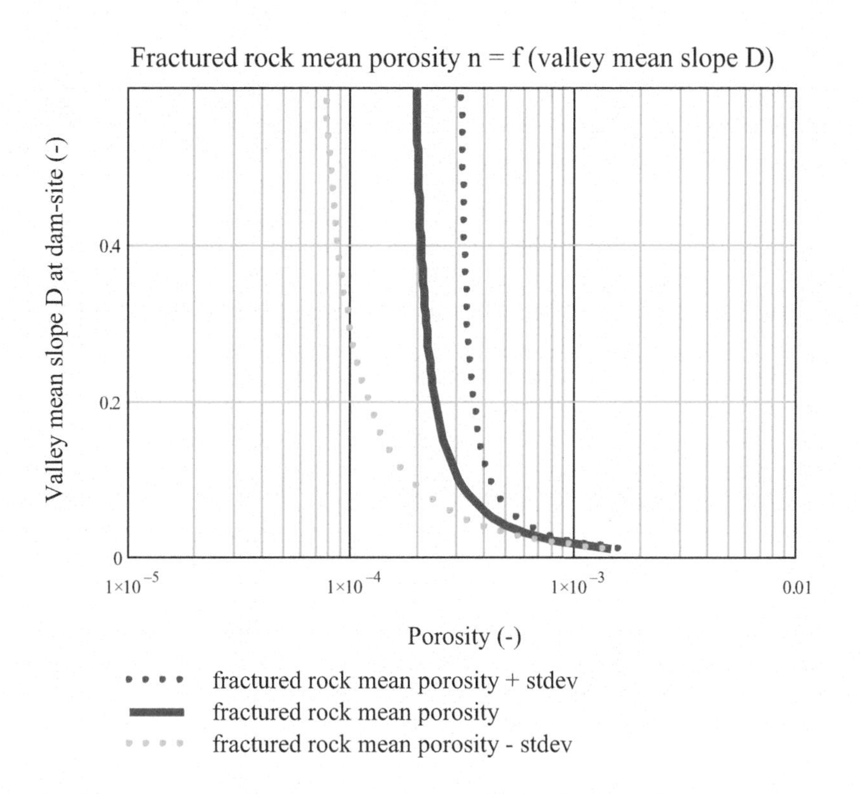

Figure 2.41 Best global average regression n = F(D) after 1,625 Lugeon tests at 19 Brazilian valleys, where D is the average valley slope.

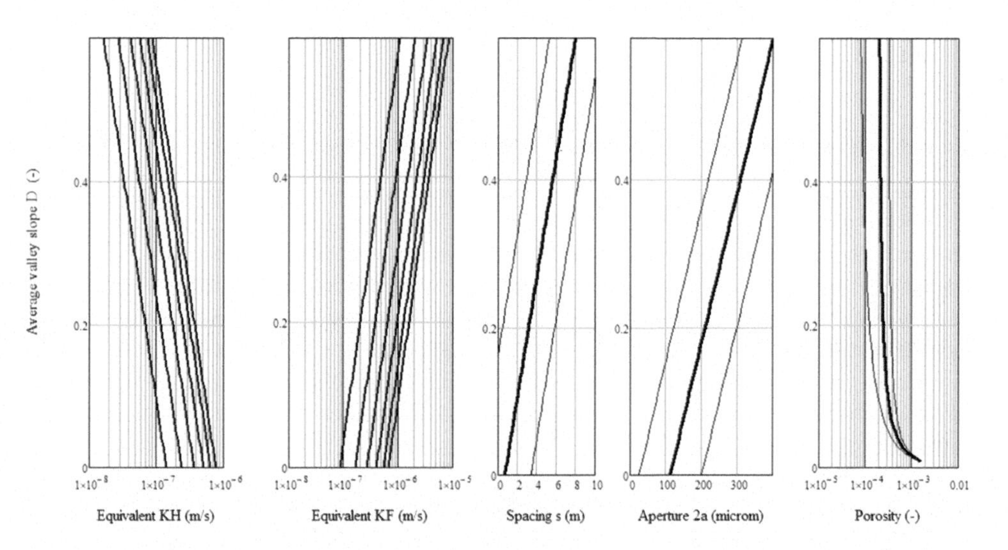

Figure 2.42 Putting together all five preceding graphs reveals common trends after 1,625 Lugeon tests at Brazilian valleys, where D is the average valley slope.

relaxation near the fractures and attaining full bedrock weathering down to several 10–100 m depths.

The time-lag between physical and chemical weathering effects brings in an inevitable delay to the fractures' gradual clogging by weathering products, almost impervious clay minerals, followed by a gradual reduction of the rock mass permeability (which explains the decrease of K_f with D). Consequently, the hydraulic impact of the original open fractures slowly disappears as the relief degrades. On the other hand, the rock mass strength diminishes and its deformability increases, both markedly. An identical effect appears for fractures of clearly tectonic origin (generally older than the fractures and joints due to the stress relief of the stored potential elastic energy within intact rock pieces), which usually have a pronounced weathering stage. However, if the permeability of the original clean fractures decreases over time, the frequency of the more recent but small fractures increases sharply due to stress relaxation, occasionally up to the point of increasing the permeability to the whole rock mass (increase in K_H with a decrease in D).

Emery's viscoelastic model simulates the intact rock's gradual fragmentation due to the release of primitive stresses imposed and stored in the intact rock.[13] In this model, a net of many mixed and contracted spiral springs, tightly contained in a non-deformable glass ampoule filled with highly viscous resin – *e.g., solidified Araldite* – simulates the primitive state of the intact rock at more profound depths. Breaking the ampoule simulates the stress relaxation due to the gradual weathering and erosion of the rock mass, allowing the slow release of the potential elastic energy stored in the contracted springs, and gradually breaking the resin's envelope unable to adapt itself to the new stress and deformation gradients.

The above observations confirm general rules about the effectiveness of cement grouting of dam foundation but considering that particular circumstances may invalidate them (excluding alluvial or residual soil foundation). Indeed, almost all regressions $K = f(P, R)$ show higher equivalent hydraulic conductivity K near the bedrock top, implying that the cement grouting must be more cautious and severe near the foundation crown.

2.3.6.4 Exploring integral sampling: 1968 to 1978

During the 1968–1979-decade, it was feasible to assess the bedrock's anisotropic hydraulic conductivity based on its litho-structural features measured on oriented integral NH core samples, subsequently adjusted by water pressure tests. Before extracting an integral NX sample, the rock mass was previously reinforced by an iron bar inserted and glued to the walls of an earlier coaxially EX borehole to assure the sample integrity (see Figure 2.43).[14] A theoretical [k] tensor was then modeled according to the principles summarized in Section 2.3.6.1, and their theoretical eigenvalues were finally revised according to the results of the water-pressure tests made after extracting the integral sample.

A computer routine evaluates the permeability tensor's eigenvalues based on the intercepted fractures' attitudes and openings. These data are measured in the integral rock mass cores, assuming similar fracture characteristics near the borehole. Deviations from that assumption are revised from the results of pressure tests, as shown in Example 2.3.

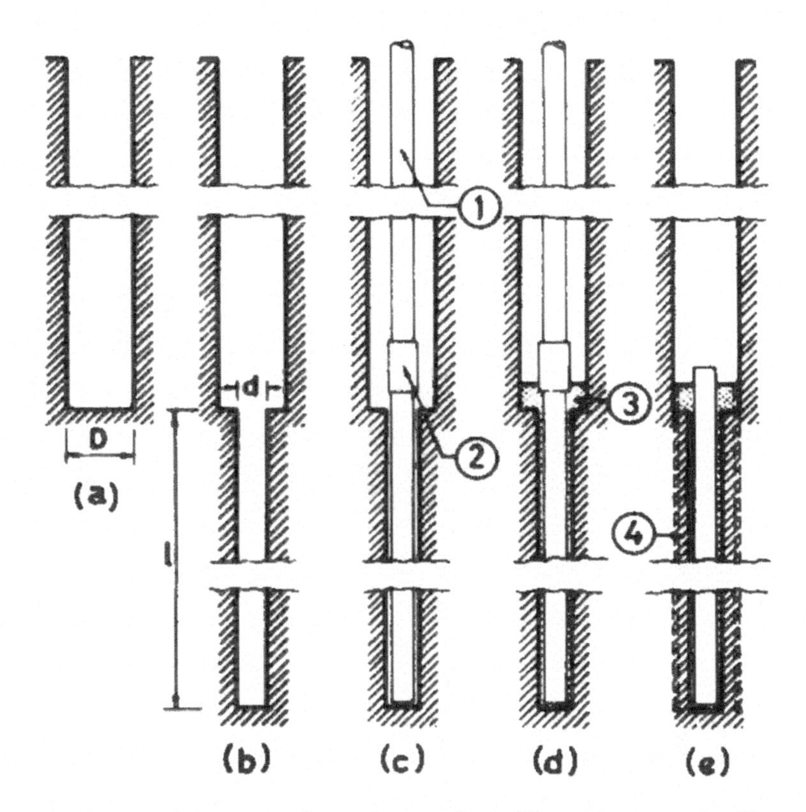

Figure 2.43 Stages of the integral sampling method: (1) positioning rods, (2) connecting element, (3) binder, (4) integral sample.

Example 2.3

The foundation of Agua Vermelha Dam, at the State of São Paulo, Brazil, was investigated by 13 geo-referred integral rock samples and 128 in situ water pressure tests, whose some results are depicted in Table 2.3 and Figure 2.44.

For each test length L, the hydraulic conductivity tensor is evaluated, considering all open and supposedly connected fractures inside and around the volume L^3. To tentatively minimize the sampling bias (Terzaghi, 1965), all fractures above the top and below the integral core, but satisfying the condition $(L+\delta L)\cdot\cos(\theta) < L/2$, where θ is the fracture dip, were sampled (see Figure 2.45).

Six symmetrical cylindrical coordinates define the intersection plane of each fracture. Then, the polar coordinates (λ, θ) of each fracture i result from 3D analytical geometry formulas, and the theoretical transmissivity $\mathbf{T_i}$ of each fracture was estimated by the formulas given in Section 2.3.5. Finally, its equivalent principal hydraulic conductivity k_i referred to a length L was evaluated as:

$$k_i = \frac{T_i}{L}$$

Table 2.3 Average eigenvalues and anisotropy of the hydraulic conductivity

Compact basalt	*Eigenvalue (m/s)*	K_{max}/K_{min}	*Vesicular amygdaloidal basalt*	*Eigenvalue (m/s)*	K_{max}/K_{min}	*Basaltic breccia*	*Eigenvalue (m/s)*	K_{max}/K_{min}
Group CB1	2.95E−05	16	Group VA1	3.89E−06	6	Group BB1	3.31E−06	12
eigenvalues	2.81E−05		eigenvalues	3.31E−06		eigenvalues	3.16E−06	
	1.81E−06			6.03E−07			2.69E−07	
Group CB2	8.51E−07	7	Group VA2	6.60E−06	55	Group BB2	1.00E−07	2
eigenvalues	7.76E−07		eigenvalues	5.62E−06		eigenvalues	5.25E−07	
	1.28E−07			1.20E−07			5.25E−08	
Group CB3	4.78E−07	6	Group VA3	1.69E−06	1,070	Group BB3	2.13E−07	18
eigenvalues	4.26E−07		eigenvalues	1.69E−06		eigenvalues	2.04E−07	
	7.41E−08			1.58E−09			1.20E−08	
Group CB4	3.16E−07	27	Group VA4	4.78E−07	24	Group BB4	Impervious	
eigenvalues	3.01E−07		eigenvalues	4.57E−07				
	1.19E−08			1.97E−08				
Group CB5	6.30E−06	675	Group VA5	Impervious				
eigenvalues	6.19E−06							
	9.33E−09							
Group CB6 eigenvalues	Impervious							

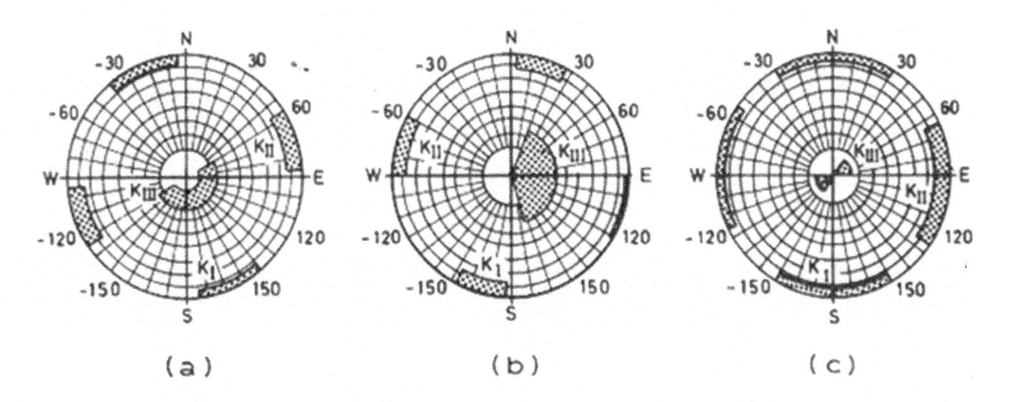

Figure 2.44 Spread [$\mu \pm \sigma$] of the eigenvectors of the compact basalt group (a), vesicular-amygdaloidal basalt group (b), and basaltic breccias group (c) of the anisotropic hydraulic conductivities determined for three different lithologies in the bedrock foundation of the Agua Vermelha Dam, southern Brazil.

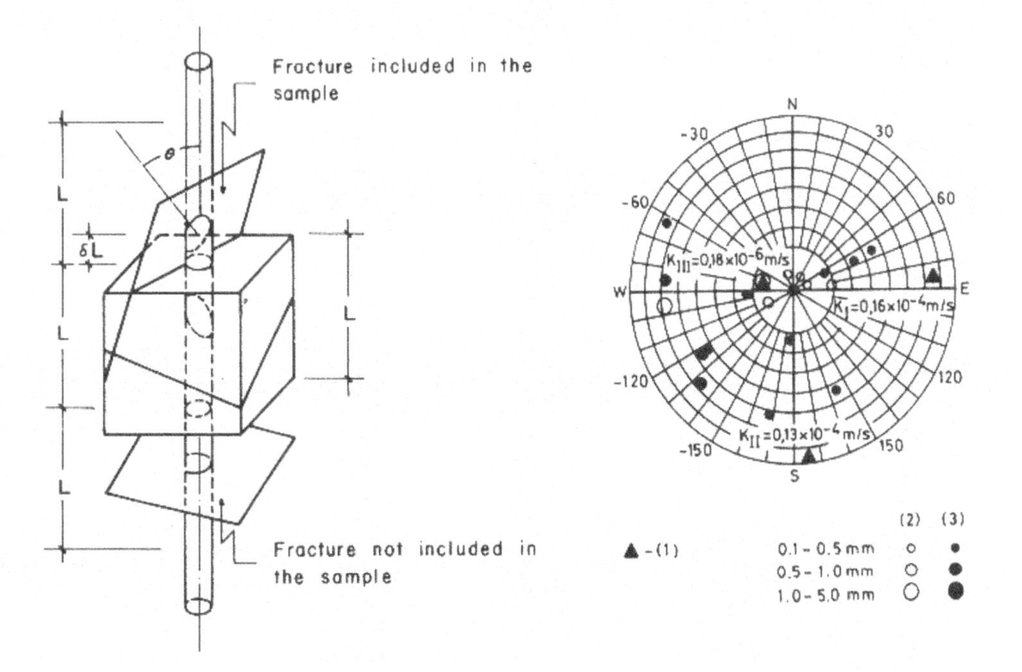

Figure 2.45 A subsystem of volume L^3 is considered for the evaluation of the hydraulic conductivity tensor. To minimize biased data, all litho-structural features above or below this subsystem by a length increment of $\left[(L/2)+\delta L\right]\cdot\cos(\theta)\leq L/2$ was included in the analysis.

As a 3D matrix, the corresponding tensor $|k_i|$ was defined by:

$$|k_i| = k_i \cdot \begin{pmatrix} \sin(\lambda_i)^2 + \cos(\theta_i)^2 \cdot \cos(\lambda_i)^2 & -\sin(\theta_i)^2 \cdot \cos(\lambda_i) \cdot \sin(\lambda_i) & -\sin(\theta_i) \cdot \cos(\theta_i) \cdot \cos(\lambda_i) \\ -\sin(\theta_i)^2 \cdot \cos(\lambda_i) \cdot \sin(\lambda_i) & \sin(\theta_i)^2 \cdot \cos(\lambda_i)^2 + \cos(\lambda_i)^2 & -\sin(\theta_i) \cdot \cos(\theta_i) \cdot \sin(\lambda_i) \\ -\sin(\theta_i) \cdot \cos(\theta_i) \cdot \cos(\lambda_i) & -\sin(\theta_i) \cdot \cos(\theta_i) \cdot \sin(\lambda_i) & \sin(\theta_i)^2 \end{pmatrix}$$

Finally, the resulting theoretical tensor $|K|$ was obtained summing up the contribution of the fractures a, b, c … n (see Section 2.3.5.5):

$$|k| = |k_a| + |k_b| + |k_c| + \ldots + |k_n|$$

To adjust these values as to conform to the water pressure tests, a homogeneous transformation factor F, converting the results of these tests on anisotropic pervious media to equivalent isotropic results, is approximately calculated by[15]:

$$F = \left[\left(\prod_{k=1}^{3} K_k \right)^{1/3} \cdot \sum_{k=1}^{3} \frac{(\alpha_{zk})^3}{K_k} \right]^{(1/2)}$$

In the above formula, K_k is the computed k-theoretical eigenvalue ($k = 1$, 2 and 3), and α_k is its direction cosine for the z-axis. The geometric mean K_{gmean} of the equivalent hydraulic conductivity of the rock mass is estimated by Babushkine and Guirinsky (see Section 2.3.6.3):

$$K_{gmean} = \frac{Q}{2\pi H \cdot F \cdot L} \cdot \ln\left(0.66 \frac{F \cdot L}{r} \right)$$

In this formula, **r** is the non-transformed NX borehole radius (in fact, **r** agrees to the geometric mean of the semi-axis **a** and **b** of the transformed elliptical cross-section of the lateral borehole surface).

A correcting factor ρ was defined to improve the estimated eigenvalues K_k as:

$$\rho = \frac{K_{gmean}}{\left(\prod_{k=1}^{3} K_k \right)^{1/3}}$$

Finally, the best estimated for the eigenvalues K_k were calculated by:

$$(K_k)_{revised} = \rho \cdot (K_k)$$

These correcting factors ρ may occasionally depart more than one order from unity. Thus, to get consistent results, the first estimate of each fracture's transmissivity observed in an integral sample must be carefully made but in relative terms, that is, as compared to the transmissivity of the most prominent observed fracture in that sample.

Important note: The reader must be aware that the rough data collected via rock mass integral samples during the 1968–1978-decade are now rapidly obtained via continuous and oriented logs of high-resolution digital color images or similar outputs defined by acoustic signals, as exemplified below (see Figure 2.46). However, for preparing test samples for laboratory experiments, as shear tests, for example, integral sampling is recommended.

2.3.6.5 *Exploring inverse solution methods: 1978 to 1988*

During the 1978–1988-decade, traditional pumping tests have been adapted to measure the anisotropic character of pervious media as, for instance, recommended by Hantush (1956). For a confined volume of a fractured rock mass, taken as quasi-homogeneous

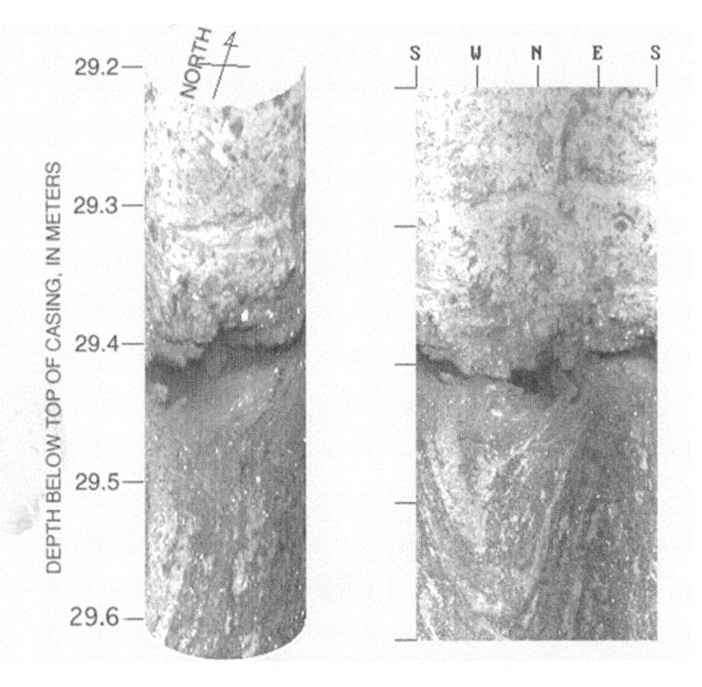

Figure 2.46 Example of a virtual core wrapped (left) and unwrapped (right) images of a bedrock fracture at a depth of 29.4 m collected with a digital television camera. The images indicate that the fracture is at the contact between pegmatite and gneiss. (USGS; public domain image.)

and quasi-anisotropic, an elegant inverse solution to evaluate its hydraulic conductivity and specific storage was devised by Hsieh et al.[16] The following example exemplifies this methodology (abridged from Appendix 5.B, "Using a multiple-borehole test to determine the hydraulic conductivity tensor of a rock mass," in "Rock Fractures and Fluid Flow: Contemporary Understanding and Applications," National Academic Press, Washington, 1996).

Example 2.4

The hydraulic conductivity of a sample volume of the Oracle granite, north of Tucson, Arizona, USA, was estimated with the help of several cycles of pumping tests made on different depths of three boreholes H-2, H-3, and H-6, drilled in a triangular pattern and separated by distances between 7 and 11 m (see Figure 2.47).

To minimize the boundary conditions' influence, as a rule, located far from the test domain, only short transient pumping (or injection) tests are performed. Water is pumped from a packer-bounded interval of one borehole in each test, and corresponding head drops are measured on other packer-bounded intervals of the remaining boreholes (see Figure 2.48).

Six or more than six short transient tests gave enough observation data to write a mathematical equation system involving the six hydraulic parameters sought. Between any pumping point i and any observation point j prevail a relationship involving the distance R_{ij} between these two points, the time t, the pumping rate Q, the head drop Δh_{ij}, the directional conductivity K_{ij}, the determinant of the hydraulic conductivity tensor $|K|$, and specific storage S_S:

$$\Delta h_{ij} = \frac{Q\left(K_{ij}\right)^{1/2}}{4\pi R_{ij}\left(|K|\right)^{1/2}} \cdot \operatorname{erfc}\left[\left[\frac{\left(R_{ij}\right)^2 \cdot S_S}{4K_{ij} \cdot t}\right]^{1/2}\right] \tag{2.20}$$

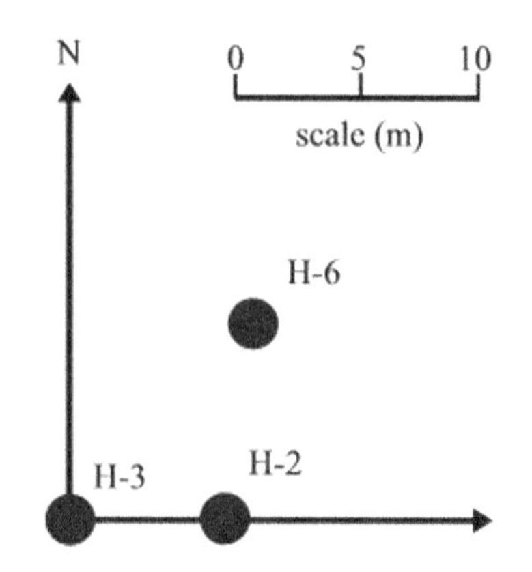

Figure 2.47 Location of the boreholes drilled to perform the 3D tests.

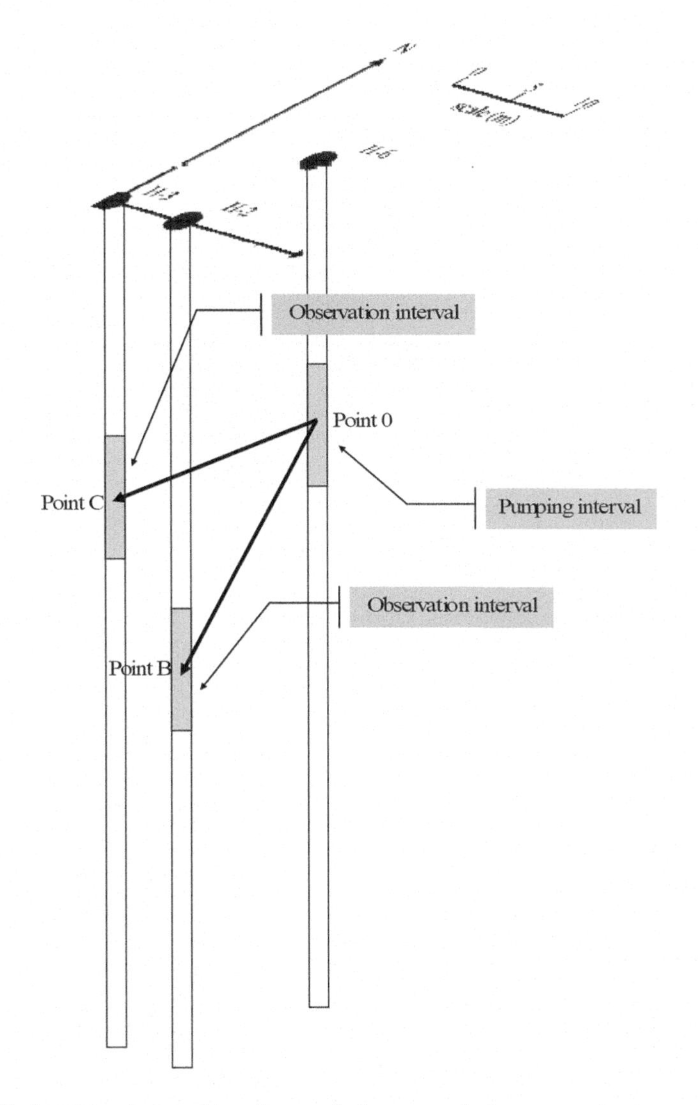

Figure 2.48 Spatial relationship of one of the several short transient pumping tests performed. Water is pumped from the packer-bounded interval O of borehole H-6, and corresponding head drops are measured in the packer-bounded intervals B and C of boreholes H-2 and H-3.

Solving this equation system (usually by a weighted least square method) gives the six components of the second-order symmetric and positive-definite tensor that best describes the anisotropic conductivity $|K|$ of the tested volume and the specific storage S_S.

If the tested rock mass volume can be taken as a statistically homogeneous and anisotropic pervious medium, a 3D plot of the square root of the directional hydraulic diffusivity versus direction should depict an ellipsoid, as shown in Figure 2.49.

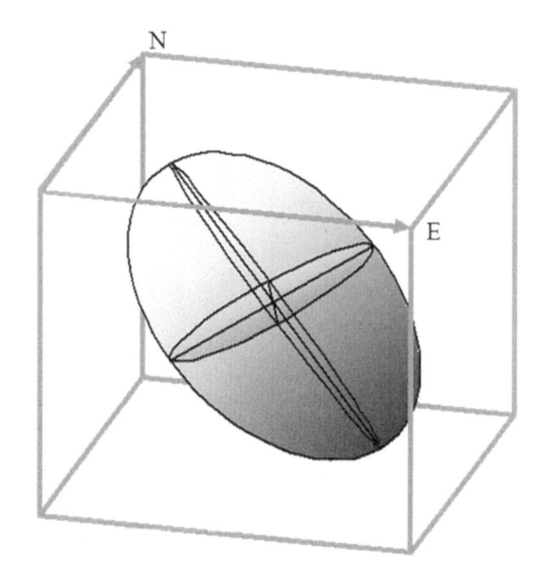

Figure 2.49 Best fitted ellipsoid to the square root of the directional hydraulic diffusivities estimated from transient pumping tests results.

The eigenvalues and corresponding eigenvectors of the best-fit K are shown in Figure 2.50.

The methodology suggested by Hsieh and Neuman may be adapted to measure the average hydraulic conductivity of rock volumes within larger and almost homogeneous units of a flow domain, but having in mind the limitations of extending its applicability to other scales and flow regimes. These results may be extrapolated to the whole domain by appropriate geostatistic models. One of the main advantages of that methodology is that they consider, implicit and collectively, all involved and 3D-varying fractures characteristics that are hard to identify and parameterize only by field geology and rock core inspection as, for example, shape, persistence, transmissivity, and interconnections.

Example 2.5 resumes one variant of that methodology.

Example 2.5

Two significant structures traverse a 60° dip tabular ore body (see Figure 2.51). Direction and dip of neighboring discontinuities of the host rock mass are closely associated with the ore body's attitudes and these two structures.

The footwall of the tabular ore body and the two primary structures are highly permeable. Their intercrossing behaves as very efficient linear drains able to induce the local gravitational lowering of the original lightly inclined water table (see Figure 2.52).

Assuming an infinite and anisotropic pervious medium, Hsieh and Neuman gave a closed steady-state solution to determine the head drop at an arbitrary point k induced

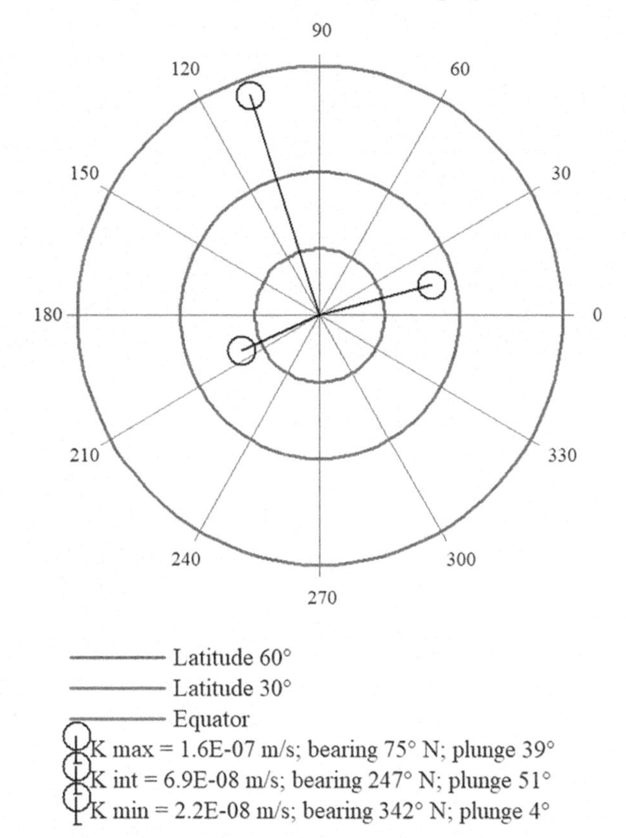

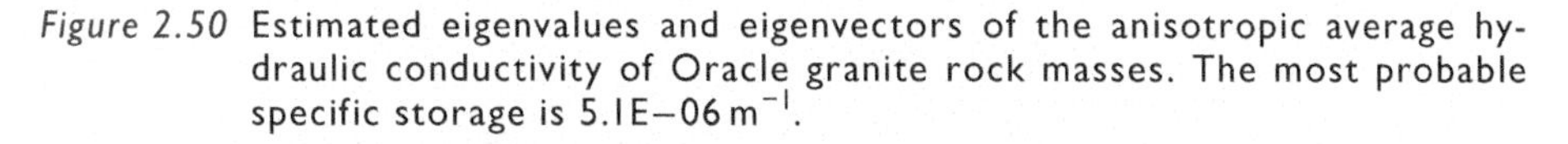

Figure 2.50 Estimated eigenvalues and eigenvectors of the anisotropic average hydraulic conductivity of Oracle granite rock masses. The most probable specific storage is 5.1E−06 m^{-1}.

by a line source or line drain of length 2L, having its mid-point at the origin i of a Cartesian coordinate system (see Figure 2.53).

For more than one tubular drain L_i placed at i different locations (i = 1, 2, 3 ...), the resultant head drop ΔH_k at an arbitrary point k may be calculated by:

$$\Delta H_k = \sum_i \frac{Q_i \cdot \ln\left[\dfrac{\left(\dfrac{\Delta R_{ik}^T \cdot A \cdot \Delta R_{ik}}{L_i^T \cdot A \cdot L_i} + 2 \cdot \dfrac{\Delta R_{ik}^T \cdot A \cdot L_i}{L_i^T \cdot A \cdot L_i} + 1\right)^{1/2} + \dfrac{\Delta R_{ik}^T \cdot A \cdot L_i}{L_i^T \cdot A \cdot L_i} + 1}{\left(\dfrac{\Delta R_{ik}^T \cdot A \cdot \Delta R_{ik}}{L_i^T \cdot A \cdot L_i} - 2 \cdot \dfrac{\Delta R_{ik}^T \cdot A \cdot L_i}{L_i^T \cdot A \cdot L_i} + 1\right)^{1/2} + \dfrac{\Delta R_{ik}^T \cdot A \cdot L_i}{L_i^T \cdot A \cdot L_i} - 1}\right]}{8\pi \cdot \left(L_i^T \cdot A \cdot L_i\right)^{1/2}}$$

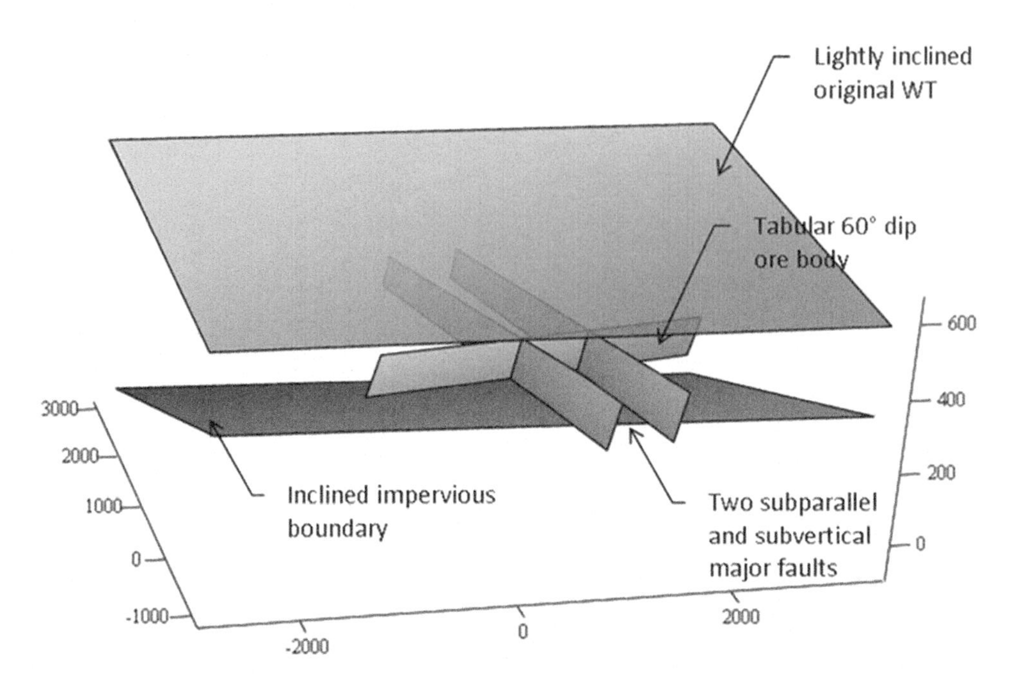

Figure 2.51 Perspective of the tabular ore body and the two significant structures. The impervious bottom, together with the slightly inclined water table, defines the main boundaries of the natural flow domain.

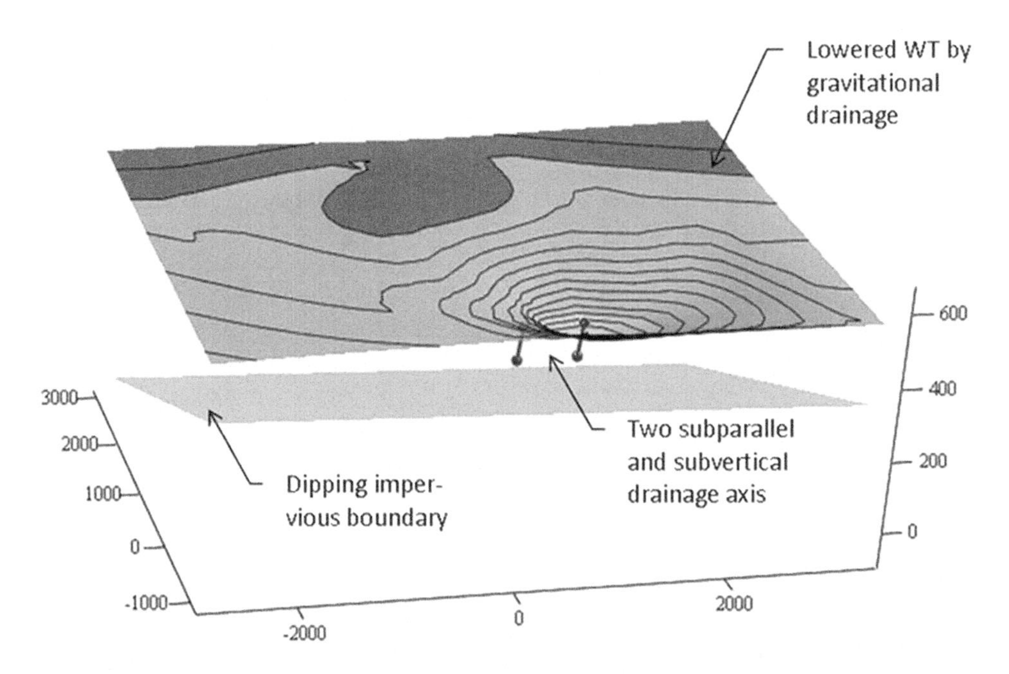

Figure 2.52 Contour map of the lowered water table based on the readings of 16 piezometers.

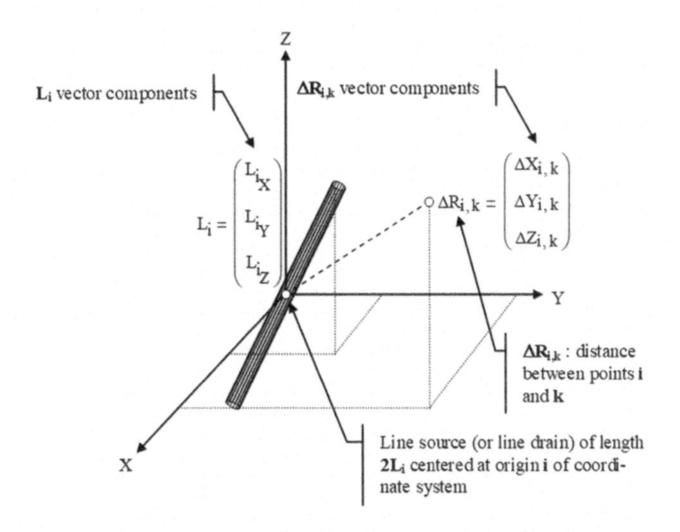

Figure 2.53 Coordinate system referred to a line drain of length $2L_i$.

The remaining variables of the above formula have the following meanings:

ΔR_{IK}: distance between the i tubular drain center and an arbitrary point k described by its Cartesian components (ΔX_{ik}, ΔY_{ik}, ΔZ_{ik},).
Q_I: the total drainage rate percolating to the tubular drain i.
A: the adjoint matrix A corresponding to $|K|K^{-1}$, where $|K|$ is the determinant of the average hydraulic conductivity tensor **K.**
L_I: the Cartesian components of the half-length L_i of the i tubular drain.

For the situation depicted in Figure 2.51, the existing field data: 16 piezometers readings and 2 almost steady-state discharge measurements at the bottom of the two-line drains (0.99 and 0.48 m^3/s), provided enough data to best-fit an approximation to the average hydraulic conductivity tensor **K** of the surrounding rock mass.

This "closed" solution, as deduced for an unlimited domain, can be worked by the image method. However, only the two most influencing image drains were added to the problem: one for the WT boundary and one for the impervious bottom boundary.

K's first guess was defined based on the geometry and the known hydraulic characteristics of the tabular ore body and other significant structures. Then, applying a Levenberg-Marquardt algorithm, it was possible to minimize the quadratic error sum concerning the difference between the calculated and the measured head drops, giving the following result:

$$\mathbf{K} = \begin{pmatrix} 2.855\times10^{-5} & -1.330\times10^{-5} & -2.970\times10^{-6} \\ -1.330\times10^{-5} & 1.280\times10^{-5} & 2.440\times10^{-6} \\ -2.970\times10^{-6} & 2.440\times10^{-6} & 1.442\times10^{-6} \end{pmatrix} \text{m/s}$$

Figure 2.54 shows the corresponding eigenvalues and eigenvectors.

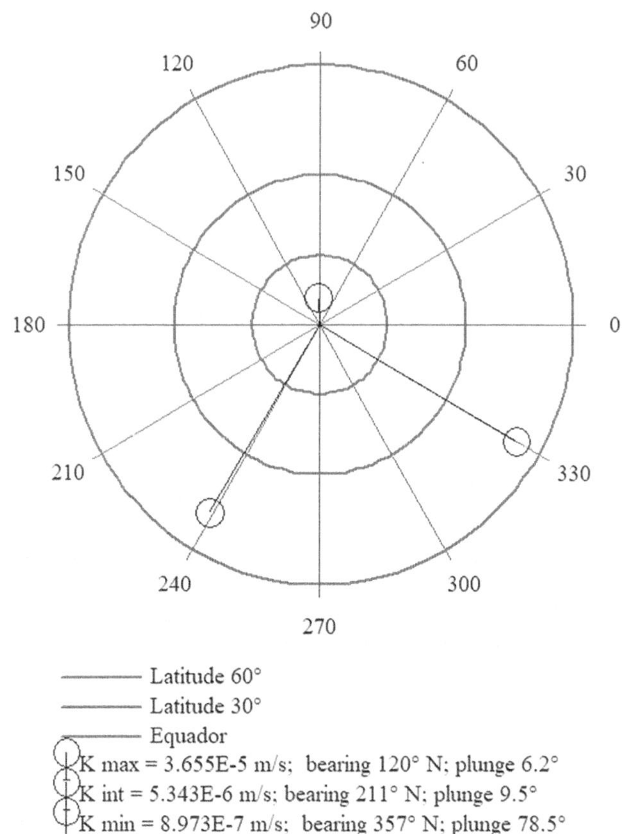

Figure 2.54 Eigenvalues and eigenvectors of K.

2.3.7 Governing equations of groundwater flow

2.3.7.1 Preliminaries

Within a *space domain* V, including its *boundaries* S, the *scalar-field* of the hydraulic potentials H and the *vector-field* of the flow rates **q** is described at any time t either by "closed" mathematical solutions or by "numerical" solutions of a system of differential equations deduced from the *energy* and *mass* conservation principles. However, unique solutions for particular problems depend on additional information, either prescribing H and **q** values inside the solution domain at a specific time t, called *initial conditions*, or prescribing stimuli from *outside* of the solution domain, called *boundary conditions*, that is, values of H and **q** on the boundaries of the flow domain.

Besides satisfying the governing equations and being continuously dependent on the subsidiary information, solutions of *well-posed* problems must be unique. Unfortunately, only for simple cases, closed mathematical solutions can be found. Conversely, numerical approximations usually fit all practical applications.

2.3.7.2 *Energy conservation principle: Darcy's law*

The empirical relationship between the specific discharge **q**, the hydraulic gradient **J**, and the hydraulic conductivity [k] derives from the *energy conservation principle*. If the components J_i of the hydraulic gradient **J** are described as partial derivatives of the hydraulic head H, the corresponding specific discharge components q_i define a system of three partial differential equations:

$$q_i = -\sum_{j=1}^{3}\left(\rho k_{ij}\frac{\partial H}{\partial x_i}\right) \tag{2.21}$$

Now, considering Newton's Second Law, described by the energy conservation principle:

$$[\text{mass}]\times[\text{acceleration}]-\sum[\text{internal forces}+\text{external forces}]=\text{zero}$$

Applying that equation to a control volume inside of a percolating pseudo-fluid, the product of its mass by its acceleration equates the resultant of the distributed internal forces within its mass, plus the external forces applied to its boundaries. For a pervious pseudo-continuous subsystem in a state of dynamic equilibrium, it is possible to distinguish four kinds of external and internal forces (see Figure 2.55):

1. The external driving forces resulting from the differential water pressures acting on its faces.
2. The internal driving forces resulting from the gravity pull.
3. The internal resisting forces, called drag forces, generated by the viscosity of the moving water.
4. Finally, a small inertial force (compared to the drag forces), mobilized only when the moving fluid accelerates or slows down (see Section 2.3.8.1).

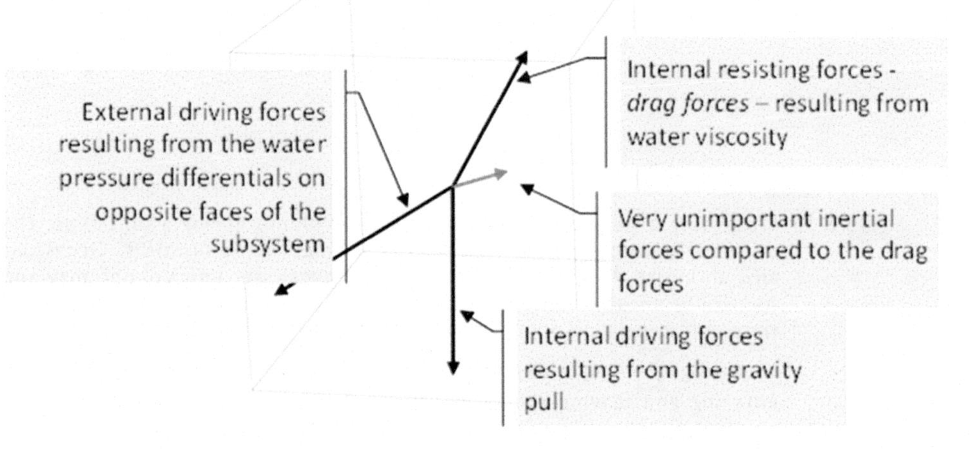

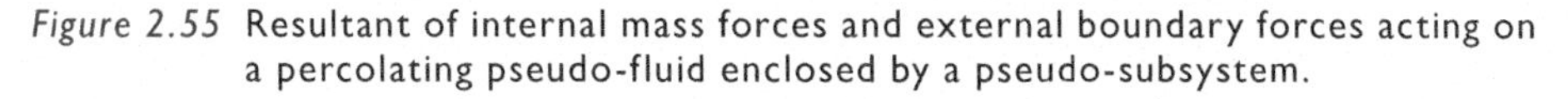
Figure 2.55 Resultant of internal mass forces and external boundary forces acting on a percolating pseudo-fluid enclosed by a pseudo-subsystem.

The resulting driving force per unit volume, called *percolation force*, corresponds to the groundwater unit-weight ρg multiplied by the hydraulic gradient **J**, ρg·**J**. In steady-state, that specific force ρg**J** balances the specific drag **D**. For laminar flow (Darcy's Law), the coefficient **D** and the empirical hydraulic conductivity [k] are related by $\mathbf{D} = \rho g \mathbf{q} \cdot [k]^{-1}$, where the inverse matrix $[k]^{-1}$ is the *hydraulic resistivity*. An insignificant specific inertial force $\rho g(d\mathbf{q}/dt)$ partially counteract the unbalanced percolation and drag forces (both per unit volume).

2.3.7.3 *Mass conservation principle: continuity equation*

2.3.7.3.1 GENERAL EQUATION

During a unit time interval, water percolates through a unit volume of a saturated and slightly compressible pervious subsystem. In that interim, the mass *conservation principle* implies that the differential amount of water entering and leaving the faces of this subsystem matches the differential amount of water removed (or added) from this subsystem due to a contraction (or expansion) of its open discontinuities and also due to the compressibility of water itself (see Figure 2.56). Additionally, an occasional interior source (or sink) may add (or drain) another differential amount of water ρ·Q, where Q denotes a source strength (or sink) expressed as a discharge per unit volume

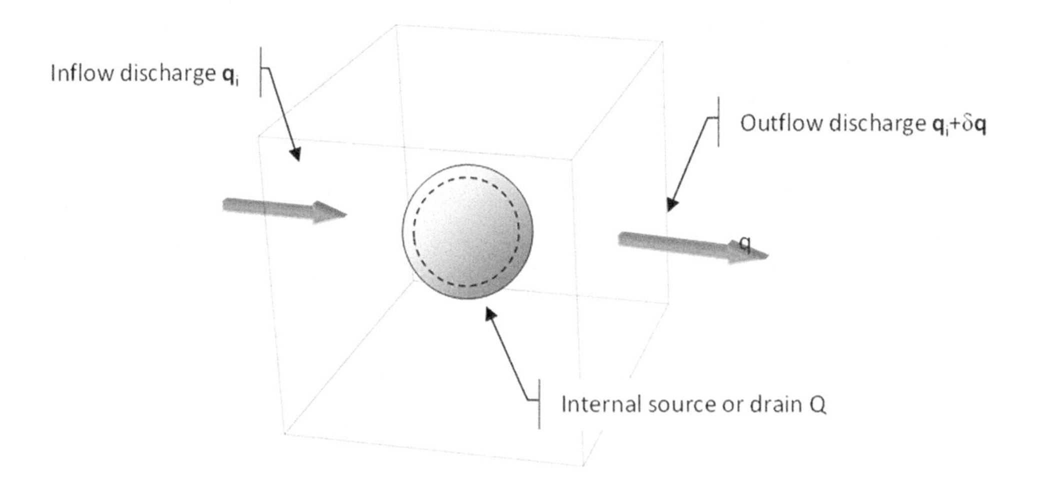

Figure 2.56 Observing a pervious subsystem of unit volume during a unit time, the differential amount of its water content mass ρn is $\partial(\rho n)/\partial t$. Occasionally, an internal source (or sink) of Q strength per unit volume may add (or drain) another differential amount of water mass ρQ. At the same time, differential amounts of water enter and leave its 2 × 3 pairing i-faces totalizing $\sum \partial(\rho q_i)/\partial x_i$. Balancing these differential amounts of water entering and leaving the subsystem and substituting q_i for the Darcy's Law, that is $q_i = -k_{ij}\, \partial H/\partial x_i$, yields the classical continuity equation: $\sum \partial\left(\rho \sum -k_{ij}\, \partial H/\partial x_i\right)/\partial x_i + \partial(\rho n)/\partial t = \rho Q$.

(positive for a *source* or negative for a *sink*). Q units are $L^3/T/L^3$. Then, the formal differential statement of the mass conservation principle, written in $M/L^3/T$ units, is (see Section 2.3.8.2):

$$\sum_{i=1}^{3}\frac{\partial}{\partial x_i}\sum_{j=1}^{3}\left(\rho k_{ij}\left(\frac{\partial H}{\partial x_i}\right)\right)=\sum\nolimits_V\frac{\partial H}{\partial t}-\rho \tag{2.22}$$

In this equation, the parameter $\sum_V$, called the *specific volumetric storage*, is defined by:

$$\sum\nolimits_V \rho g\left(n\rho_0\beta+\frac{\rho}{K_p}\right) \tag{2.23}$$

In the above expression, ρ_0 (M/L^3) denotes the *water density*, equal to 997.1 kg/m^3 at 25°C under atmospheric pressure; β ($M^{-1}LT^2$) denotes the isothermal water *compressibility*, equal to 4.8×10^{-10} Pa^{-1} at 25°C and K_p ($M/L^1/T^2$) denotes the discontinuous subsystem *bulk modulus*, varying around 10^7 Pa for loose sands to 3×10^9 Pa for hard rocks.

The strict units of the term $\sum_V$ is M/L^4. However, considering the water density ρ invariant and normalizing the general continuity equation by ρ, the parameter $\sum_V$, now symbolized by S_V is:

$$S_V=\rho g\left(n\beta+\frac{1}{K_p}\right) \tag{2.24}$$

In this case, the units of S_V reduce to L^{-1}, as commonly quoted in technical literature. Then, the continuity equation, also written in L^{-1} units, is:

$$\sum_{i=1}^{3}\frac{\partial}{\partial x_i}\sum_{j=1}^{3}\left(k_{ij}\left(\frac{\partial H}{\partial x_i}\right)\right)=S_V\frac{\partial H}{\partial t}-C \tag{2.25}$$

The parameter S_V measures the amount of water expelled from a unit volume subsystem due to water expansion and a porosity contraction, resulting from a pressure unit decrease (assuming effective stress variation is always balanced by neutral stress variation). For short-range head variations, S_V can be taken as a constant property to be empirically estimated.

2.3.7.3.2 DUPUIT'S APPROXIMATION

Groundwater flow through quasi-horizontal and thin tabular groundwater systems, having planar dimensions much more significant than its thickness, is more easily described if referred to ENZ's zenithal frame. Depending on how high is the permeability contrast between these strata, groundwater flows practically under virtual confinement or not. When unconfined, characterized by previous strata with extremely low permeability contrasts, groundwater may freely flow over an impervious base. In this case, its upper boundary, the *water table WT*, moves up and down without restraint (see Figure 2.57). When confined at the top, groundwater percolates "sandwiched" between the impervious bottom and the top contact (see Figure 2.58). In both cases, confined and unconfined, small horizontal hydraulic gradients drive the groundwater mass. Then, except when too close to natural water sources or pumping wells, the vertical Z-component $\mathbf{J}_Z$ of the hydraulic gradient $\mathbf{J}$ at any point in the flow domain usually is exceedingly small and can be neglected. That flow condition is translated by $\partial H/\partial z=0$, and that assumption is the starting point of the simplified hydraulic model suggested by Dupuit (1863). In these cases, all equipotential surfaces H are vertical cylindrical surfaces described by the argument (E, N, t) instead of (E, N, Z, t).

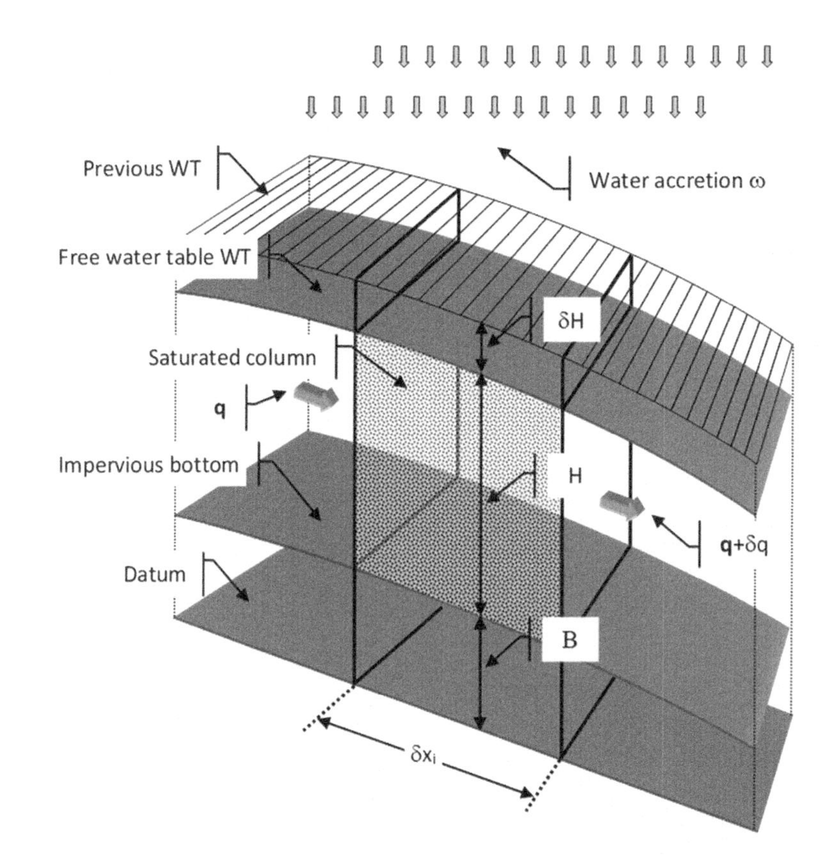

Figure 2.57 In this i-vertical cross-section trough a vertical prismatic cell of an unconfined aquifer, B is the average height of the horizontal impervious bottom (elevation referred to an arbitrary horizontal datum), H is the average saturated thickness (neglecting the capillary fringe), and B + H is the total hydraulic head. The specific water accretion is ω (in $L^3/L^2/T = L/T$ units). The column area base is δS (in L^2 units) and referred to a zenithal frame $\delta S = \delta E \cdot \delta N$. Due to infiltration (or evaporation), the differential height associated to the rise (or fall) of the water table in a time interval δt is δH (in L units).

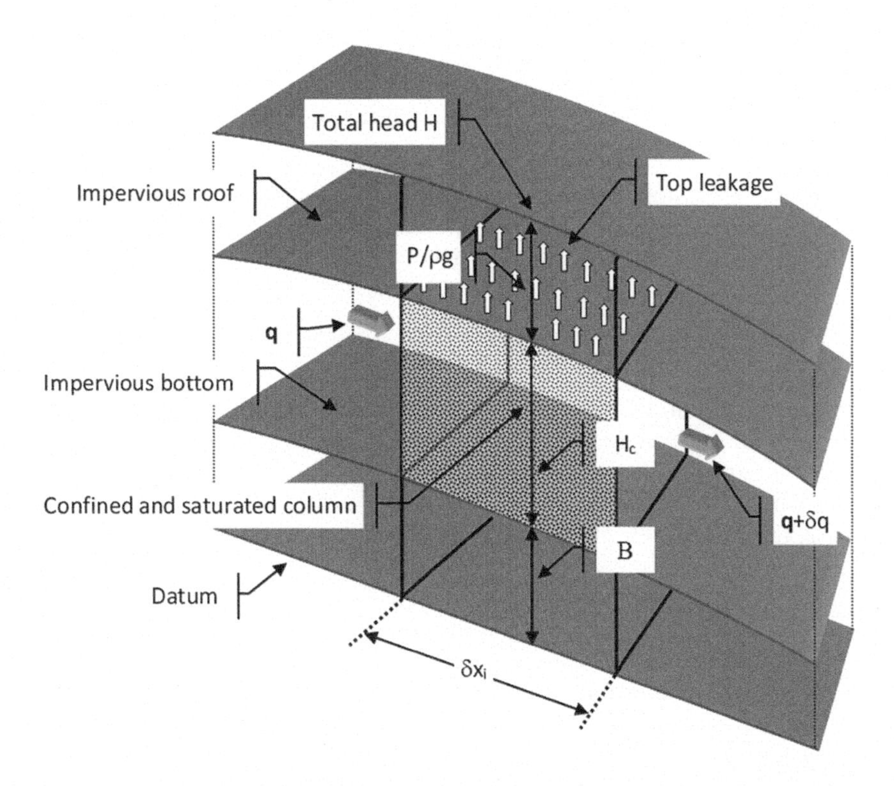

Figure 2.58 In this i-vertical cross-section trough a columnar cell of a confined aquifer, B is the average impervious bottom height (elevation referred to an arbitrary horizontal datum), HC is the average saturated thickness of the confined aquifer and P is the pressure head at the aquifer's (somewhat horizontal) roof. The sum $(B + HC + P/\rho g)$ is the total head, and the specific leakage (or forced accretion) is ω (in $L^3/T/L^2 = L/T$ units). The area of the column base is δS (in L^2 units). In a zenithal frame $\delta S = \delta E \cdot \delta N$.

2.3.7.3.2.1 Unconfined groundwater

Consider an unconfined groundwater columnar cell, where B is the average height of the impervious base, and H is the average height of the saturated water column. Then, the sum (B+H) approximates the average hydraulic head (see Figure 2.57). Over the EN domain, the variable height B only depends on the 3D geometry of the top of the impervious bottom. The variable height H at each point (E, N) depends on the aquifer hydraulic properties and the flow domain's initial and boundary conditions. As equipotential surfaces are cylindrical surfaces, the specific discharge **q** is always a horizontal vector, whose components, described in a zenithal frame, are:

$$\begin{pmatrix} q_E \\ q_N \end{pmatrix} = -\begin{pmatrix} k_{EE} & k_{EN} \\ k_{NE} & k_{NN} \end{pmatrix} \begin{bmatrix} \dfrac{\partial}{\partial E}(B+H) \\ \dfrac{\partial}{\partial N}(B+H) \end{bmatrix} \tag{2.26}$$

During a time-interval δt, the differential amount of water δV_w entering and leaving the saturated columnar cell is:

$$\delta V_W = -\left[\frac{\partial}{\partial E}(\rho q_E H \delta N)\delta E + \frac{\partial}{\partial N}(\rho q_N H \delta E)\delta N\right]\delta t$$

In the above expression, H denotes the variable height of the saturated columnar cell, δE and δN denote the lengths of the orthogonal sides of the base of the prismatic cell, and $\delta E \cdot \delta N$ measures the base area δS. Then:

$$\delta V_W = -\left[\frac{\partial}{\partial E}(\rho q_E H) + \frac{\partial}{\partial N}(\rho q_N H)\right]\delta S\, \delta t$$

Remembering that n_e denotes the effective porosity, the liquid phase mass within the saturated columnar cell measures $n_e \rho H \delta S$. During a time-interval δt and disregarding groundwater and aquifer deformations induced by field stress variations, the amount of mass within the prismatic cell can only change if the water table rises or drops. In this case, the mass variation is:

$$\frac{\partial}{\partial t}(n_e \rho H)\delta S\, \delta t$$

An infiltration or evaporation at the water table causes the gradual rise or fall of the columnar cell height. The water accretion or removal rate is the specific strength $\omega = \left[\partial V_w / \partial t\right] / \delta S$ (in $L^3/T/L^2 = L/T$ units). This rate is *positive* for infiltration and *negative* for evaporation, adding or removing the columnar cell an amount of water mass equal to $\rho \cdot \omega \cdot \delta S \cdot \delta t$ during a time interval δt.

As a result, during a time interval δt, the differential amount of water mass entering and leaving the columnar cell matches the differential amount of water mass due to a rise or a decline of the water table plus the equivalent height of the infiltrated or evaporated water mass. Then, combining the previous equations yields the continuity equation:

$$\left[\begin{array}{l}\dfrac{\partial}{\partial E}\left[\rho H\left[k_{EE}\dfrac{\partial}{\partial E}(B+H) + k_{EN}\dfrac{\partial}{\partial N}(B+H)\right]\right]\ldots \\ +\left[\dfrac{\partial}{\partial N}\left[\rho H\left[k_{NE}\dfrac{\partial}{\partial E}(B+H) + k_{NN}\dfrac{\partial}{\partial N}(B+H)\right]\right]\right]\end{array}\right] = \frac{\partial}{\partial t}(n_e \rho H) - \rho\omega \qquad (2.27)$$

The above partial differential equation is of second-order and nonlinear. Numerical algorithms can be modified to avoid the difficulties implied by this nonlinearity.

For common aquifers, effective porosity n_e may attain 5%–15%. However, for fractured hard rock masses, this value usually is lower than 2%.

2.3.7.3.2.2 Confined groundwater

The average height of a saturated confined columnar cell now only depends on the aquifer 3D geometry. The saturated height denoted by H_C must not be confused with the unconfined height H (see Figure 2.58). The meaning of the variable B remains unchanged and still denotes the height of the impervious base. The pressure head P at the aquifer's roof, the total head, now corresponds to the sum $(B+H_C+P)/\rho g$. However, if P_m symbolizes the pressure head measured in the middle of the confined thickness H_C, instead of the aquifer's roof, the total head corresponds to $(B+H_C/2+P_m)/\rho g$. By Dupuit's assumption, $\partial H/\partial z=0$, then $P_m=H_C/2+P$. However, one must keep in mind that Dupuit's simplified modeling for some situations cannot replace a true 3D unconfined or confined modeling.

The continuity equation for confined systems referred to a zenithal frame, derived as in the preceding section, has the following expression:

$$\left[\frac{\partial}{\partial E}\left[\rho H_C\left[\begin{array}{l} k_{EE}\dfrac{\partial}{\partial E}\left(B+H_C+\dfrac{P}{\rho g}\right)\ldots \\ +k_{EN}\dfrac{\partial}{\partial N}\left(B+H_C+\dfrac{P}{\rho g}\right)\end{array}\right]\right]\ldots \atop +\left[\frac{\partial}{\partial N}\left[\rho H_C\left[\begin{array}{l} k_{NE}\dfrac{\partial}{\partial E}\left(B+H_C+\dfrac{P}{\rho g}\right)\ldots \\ +k_{NN}\dfrac{\partial}{\partial N}\left(B+H_C+\dfrac{P}{\rho g}\right)\end{array}\right]\right]\right]\right] = \frac{\partial}{\partial t}\left(n_e\rho H_C\right)-\rho\omega \qquad (2.28)$$

During a time-interval δt, the liquid phase's mass within a confined saturated columnar cell measures $\left[\partial(n_e\rho H_C)/\partial t\right]\cdot\delta S\cdot\delta t$. In a confined aquifer, that mass can only change by stress-induced deformations. Applying the differentiation chain rule to $\partial(n_e\rho H_C)/\partial t$ and recalling that B and H_C are time-independent variables:

$$\frac{\partial}{\partial t}\left(n_e\rho H_C\right)=\frac{\partial}{\partial P}\left(n_e\rho H_C\right)=\rho g\frac{\partial}{\partial P}\left(n_e\rho H_C\right)\frac{\partial}{\partial t}\left(B+H_C+\frac{P}{\rho g}\right)$$

The first derivative in the right-hand product, $\rho\cdot g\cdot\partial(n_e\rho H_C)/\partial P$, called *storativity* or *columnar storage*, is here denoted by $\sum_C$. Assuming that for usual groundwater pressure variations of the height H_C remains practically invariant, similar reasoning as previously done for the continuity equation yields:

$$\sum\nolimits_C=\rho g H_C\left(n_e\rho_0\beta+\rho\frac{\partial}{\partial P}n_e\right) \qquad (2.29)$$

The first term inside brackets contemplates the compressibility of the water. If the total stresses, defined in the sense of Soil Mechanics, remain invariant, any water pressure variation induces an opposing but similar "effective" stress variation. For short-range pressure changes, $\sum_C$ is almost constant. It may be estimated for deformable rock masses by advanced field pumping tests (Hshieh et al., 1985).

The dimensional units of $\sum_C$ are M/L^3. However, for constant water density ρ, the continuity equation can be divided by ρ, and parameter $\sum_C$ takes a more straightforward expression symbolized by S_C that has no units:

$$S_C = \rho g H_C \left(n_e \beta + \frac{\partial}{\partial P} n_e \right) \tag{2.30}$$

2.3.7.4 Boundary conditions

The Darcy's Law and the Continuity Equation, often combined into a single equation, link the hydraulic head H to the specific discharge **q**. However, unique descriptions for H under unique flow circumstances are only feasible if supplementary ones are adequately stated in addition to these two equations. For each problem, solving strategies depend on the character of these supplementary equations. Typical boundary conditions are defined when:

- Free water body infiltrates or emerges from pervious media.
- Groundwater hits impervious boundaries.
- Groundwater traverses dissimilar pervious media.
- Groundwater emerges into the open air or change from a saturated to an unsaturated state.

Often it is challenging to construct a model surrounded by well-defined boundaries. To reduce its size to practical limits and at the same time retain its significance where this is relevant, heads and flow data at arbitrary boundary conditions must be based on reliable field data (see Figure 2.59).

2.3.7.4.1 SUBMERGED BOUNDARIES

Submerged boundaries typify interfaces where free water infiltrate (or emerge) into (or from) the pervious media (see Figure 2.59: interfaces 1–2–3 and 5–6–7–8). At these boundaries, hydraulic heads H take constant values, and if expressed in L units, they are numerically equivalent to the geometrically associated water level associated with each boundary, but all referred to the same arbitrarily data. Then, for any point at an elevation Z referred to a zenithal frame, but located on a submerged interface, its boundary condition is:

$$Z + \frac{P}{\rho g} = H_{\text{constant}} \tag{2.31}$$

For example, considering again Figure 2.59, the hydraulic head H_{123} at any point located on the interface 1–2–3 remains unchanged for unchanging water level, that is, H_{123}=constant. The same observation applies to the interface 5–6–7–8. However, normally H_{123} and H_{5678} take two different values, that is, $H_{123} \neq H_{5678}$.

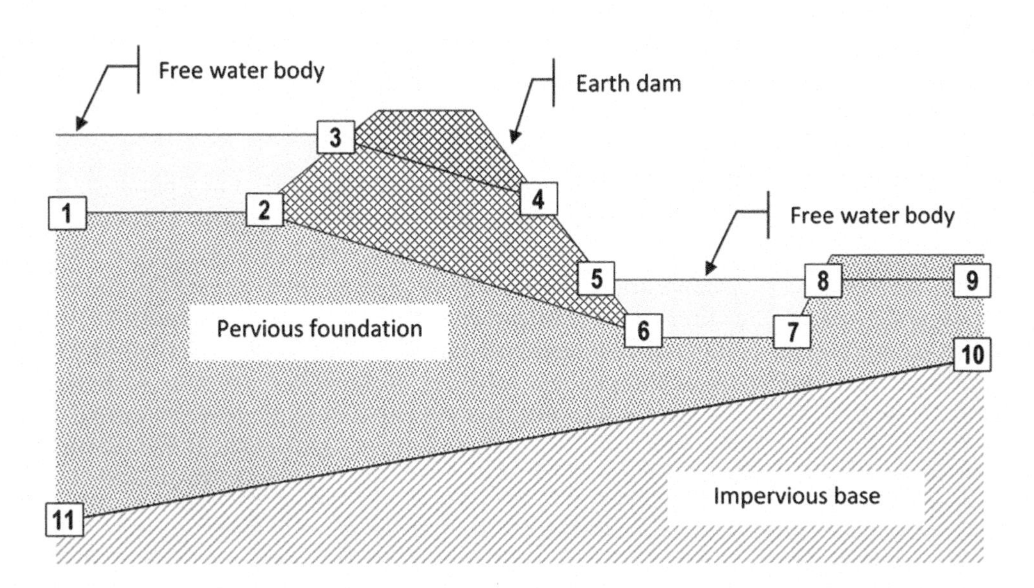

Figure 2.59 Cross-section of a schematic reservoir of water impounded by an earth-dam. Water percolates through the earth dam and the pervious foundation. Boundary conditions at are: 1–2–3 and 5–6–7–8, free water – pervious media interface; 3–4, unconfined groundwater – air interface; 4–5, seepage boundary; 10–11, impervious boundary; 2–6, different pervious media interface. To restrain the model size, heads or flows at boundaries 1–11 and 9–10 must be based on reliable field data.

2.3.7.4.2 IMPERVIOUS BOUNDARIES

A geological contact between a more and a less pervious media may be qualified as *relatively impervious*, from a practical point of view, if the permeability of the less pervious is 10 to 30 times smaller than the permeability of the more pervious (see Figure 2.59: interface 10–11).

To translate in mathematical terms an impervious interface condition, it suffices to equate to *zero* the orthogonal projection of the specific discharge **q** on the outward-normal **n** of this interface element δ**S**. In other words, to annul the *dot product* **q**·**n**, that is, **q**·**n** = 0. For isotropic media, **q**·**n** = 0 imply **J**·**n** = 0. However, for anisotropic media, **q**·**n** = 0 does not imply **J**·**n** = 0 but **J**·**n**≠0 in most cases.

Defining **n** by its direction cosines (α_x, α_y, α_z), that boundary condition for a pervious media described by an anisotropic hydraulic conductivity [k] is:

$$\mathbf{q}\cdot\mathbf{n} = \begin{pmatrix} q_x \\ q_y \\ q_z \end{pmatrix}\begin{pmatrix} \alpha_x \\ \alpha_y \\ \alpha_z \end{pmatrix} = \begin{pmatrix} k_{xx} & k_{xy} & k_{xz} \\ k_{yx} & k_{yy} & k_{yz} \\ k_{zx} & k_{zy} & k_{zz} \end{pmatrix}\begin{pmatrix} J_x \\ J_y \\ J_z \end{pmatrix}\begin{pmatrix} \alpha_x \\ \alpha_y \\ \alpha_z \end{pmatrix} = 0 \qquad (2.32)$$

2.3.7.4.3 SEEPAGE BOUNDARIES

Seepage boundaries occur where groundwater emerges from a pervious media into the open air (see Figure 2.59: interface 4–5). Neglecting capillary heads, the pressure p at these interfaces corresponds to the atmospheric pressure P_{atm}, previously defined as the *zero-origin* of the pressure scale adopted.

As a result, referred to a zenithal frame, the hydraulic head H at any point (E, N, Z) on an immobile seepage boundary match their elevation Z because the pressure term p of the hydraulic head is zero everywhere, that is, $H(E,N,Z)_{\text{at seepage boundary}} = Z$. Also, assuming a constant atmospheric pressure along the seepage boundary, $p = P_{atm} = \textit{invariant}$, the directional gradient $\partial p/\partial \mathbf{r}$ must forcibly be parallel to the outward-normal **n** of the *immobile* seepage boundary. Due to that parallelism, the dot product $\mathbf{n} \cdot \partial p/\partial \mathbf{r}$ equals the gradient modulus $|\partial p/\partial \mathbf{r}|$. Then, the boundary condition is $\mathbf{n} \cdot \partial p/\partial \mathbf{r} - |\partial p/\partial \mathbf{r}| = 0$. On the other hand, the pressure p may be expressed as $p/\rho g = H - Z$. Then, defining **n** by its direction cosines (α_x, α_y, α_z), the full expression of that boundary condition is:

$$\mathbf{n}\left(\frac{\partial}{\partial \mathbf{r}}p\right) - \left|\frac{\partial}{\partial \mathbf{r}}p\right| = \begin{pmatrix} \alpha_E \\ \alpha_N \\ \alpha_Z \end{pmatrix} \begin{bmatrix} \rho g \frac{\partial}{\partial x} H \\ \rho g \frac{\partial}{\partial N} H \\ \rho g \left(\frac{\partial}{\partial Z} H - 1\right) \end{bmatrix} - \left| \begin{bmatrix} \rho g \frac{\partial}{\partial x} H \\ \rho g \frac{\partial}{\partial N} H \\ \rho g \left(\frac{\partial}{\partial Z} H - 1\right) \end{bmatrix} \right| = 0 \quad (2.33)$$

Expressed in terms of the hydraulic gradient **J** components:

$$\mathbf{n}\left(\frac{\partial}{\partial \mathbf{r}}p\right) - \left|\frac{\partial}{\partial \mathbf{r}}p\right| = \begin{pmatrix} \alpha_E \\ \alpha_N \\ \alpha_Z \end{pmatrix} \begin{bmatrix} \rho g J_E \\ \rho g J_N \\ \rho g (J_Z - 1) \end{bmatrix} - \left| \begin{bmatrix} \rho g J_E \\ \rho g J_N \\ \rho g (J_Z - 1) \end{bmatrix} \right| = 0 \quad (2.34)$$

2.3.7.4.4 UNCONFINED GROUNDWATER-AIR INTERFACE

If the small capillary fringe effects are neglected, a water table defines a steady or unsteady groundwater-air interface, depending on the problem's supplementary equations under analysis (see Figure 2.59: interface 3–4).

The pressure p on the water table corresponds to the atmospheric pressure P_{atm} equal to zero in the preceding item. As the pressure at any point (r, t) on a mobile or immobile water table is zero everywhere, that is, $P_{\text{water table}} = 0$. Referred to a zenithal frame, the hydraulic head H at any point of the water table WT has the same value (in L units) as its elevation Z:

$$H_{WT} = Z \quad (2.35)$$

Furthermore, if the atmospheric pressure remains constant, that is, $P_{atm} = \text{invariant}$, the hydrodynamic derivative DP/dt is zero, that is $DP/dt = \partial P/\partial t + \partial P/\partial r = 0$.

On the other hand, the equality $H = Z + P_{atm}/\rho g$ holds for all points restricted to the water table; this implies $\partial P/\partial t = \partial H/\partial t$, $\partial P/\partial x = \rho g \cdot \partial H/\partial x$, $\partial P/\partial y = \rho g \cdot \partial H/\partial y$, and $\partial P/\partial z = \rho g \cdot \partial H/\partial z - 1$. Then, the water table boundary condition is:

$$\frac{dP}{dt} = \frac{\partial}{\partial t}P + \frac{1}{n_e}\begin{pmatrix} q_E \\ q_N \\ q_Z \end{pmatrix} = \begin{pmatrix} \frac{\partial}{\partial E}P \\ \frac{\partial}{\partial N}P \\ \frac{\partial}{\partial Z}P \end{pmatrix} = 0$$

Written with explicit matrix components:

$$\rho g \frac{\partial}{\partial t}H - \frac{\rho g}{n_e}\begin{pmatrix} k_{EE} & k_{EN} & k_{EZ} \\ k_{NE} & k_{NN} & k_{NZ} \\ k_{ZE} & k_{ZN} & k_{ZZ} \end{pmatrix}\begin{pmatrix} \frac{\partial}{\partial E}H \\ \frac{\partial}{\partial N}H \\ \frac{\partial}{\partial Z}H \end{pmatrix}\begin{bmatrix} \frac{\partial}{\partial E}H \\ \frac{\partial}{\partial N}H \\ \left(\frac{\partial}{\partial Z}H - 1\right) \end{bmatrix} = 0 \qquad (2.36)$$

That boundary condition can also be applied to a moving water table (see Figure 2.60), even when acquiring or losing an apparent vertical mass accretion rate $\rho g \cdot \omega$ (ω in L/T).

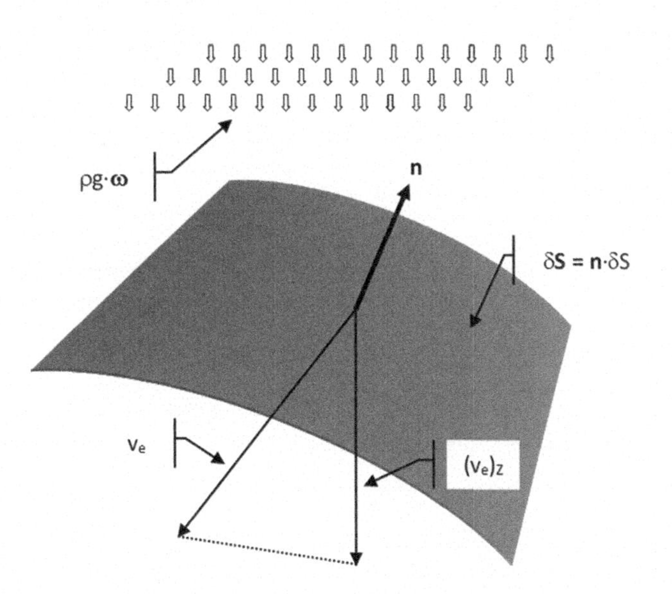

Figure 2.60 Moving water table with vertical accretion. Its downward or upward movement slows down or speeds up depending on whether the accretion rate $\rho g \cdot \omega / n_e$ is positive (infiltration) or negative (evaporation).

In this case, to the vertical component $(\mathbf{v}_e)_Z$ of the "effective" velocity vector $\mathbf{v}_e$ of the moving water table must be added the "effective" mass accretion vector defined by $\rho g \cdot \omega / n_e$. The resulting downward or upward movement of the water table *slows down* or *speeds up* depending on whether the effective mass accretion rate $\rho g \cdot \omega / n_e$ is positive (infiltration) or negative (evaporation). Then, the complete boundary condition is:

$$\frac{dP}{dt} = \frac{\partial}{\partial t}P + \frac{1}{n_e}\left[\begin{pmatrix} q_E \\ q_N \\ q_Z \end{pmatrix} - \begin{pmatrix} 0 \\ 0 \\ \omega_Z \end{pmatrix}\right]\begin{pmatrix} \frac{\partial}{\partial E}P \\ \frac{\partial}{\partial N}P \\ \frac{\partial}{\partial Z}P \end{pmatrix} = 0$$

Written with explicit matrix components considering the former equalities:

$$\rho g \frac{\partial}{\partial t}H - \frac{\rho g}{n_e}\left[\begin{pmatrix} k_{EE} & k_{EN} & k_{EZ} \\ k_{NE} & k_{NN} & k_{NZ} \\ k_{ZE} & k_{ZN} & k_{ZZ} \end{pmatrix}\begin{pmatrix} \frac{\partial}{\partial E}H \\ \frac{\partial}{\partial N}H \\ \frac{\partial}{\partial Z}H \end{pmatrix} - \begin{pmatrix} 0 \\ 0 \\ \omega_Z \end{pmatrix}\right]\left[\begin{matrix} \frac{\partial}{\partial E}H \\ \frac{\partial}{\partial N}H \\ \left(\frac{\partial}{\partial Z}H - 1\right) \end{matrix}\right] = 0 \tag{2.37}$$

For steady-state flows, the term $\rho\, g \cdot \partial H / \partial t$ disappears in both cases.

2.3.8 Addenda

2.3.8.1 Effective velocity and specific discharge

According to Bear, the groundwater "effective" velocity $\mathbf{v}_e$ ascribed to the center of a pseudo-continuous subsystem is a vector-average of all water particles velocities $\mathbf{v}_P$ percolating throughout its "effective" voids during an observation-time interval δt (Bear et al., 1968):

$$\mathbf{v}_e = \frac{1}{\delta V_e\, \delta t}\int_0^{\delta t}\left(\int_0^{\delta V_e} \mathbf{v}_p\, dV\right)dt \tag{2.38}$$

δVe stands for the void "effective" volume contained by the total volume δV_t of the pseudo-continuous subsystem in the above expression. The effective porosity is $n_e = \delta V_e / \delta V_t$.

The specific discharge $\mathbf{q}$ corresponds to the product of the "effective" velocity $\mathbf{v}_e$ by the "effective" porosity n_e and preserves the units of a velocity L/T, that is, $\mathbf{q} = n_e \mathbf{v}_e$. As

$n_e < 1$, the specific discharge $\mathbf{q}$ is always smaller than $\mathbf{v}_e$. Then, a water particle's travel time on a streamline of length L is $t = L/\mathbf{v}_e = L_x/\mathbf{v}_{ex} = L_y/\mathbf{v}_{ey} = L_z/\mathbf{v}_{ez}$.

Recall that a "specific" property $P_{specific}$ applied to subsystems result from the division of a measurable property P by the length L, area L^2, volume L^3 of this subsystem as P/L, P/L^2, P/L^3, e.g., weight W and specific weight W/L^3.

2.3.8.2 *Hydrodynamic gradient*

Darcy's assumption allows a pervious soil or rock mass, made up of a mixture of solid, liquid, and gaseous phases, to be substituted and modeled as a fictitious pseudo-continuous fluid having the same unit weight of the mixed liquid-gaseous phase. Then, considering a transient head field $H(\mathbf{r}, t)$, the value of the consumption of the stored potential energy δH associated with a percolating water particle during a time-interval δt is:

$$\delta H = \left(\frac{\partial H}{\partial t} + \frac{\partial H}{\partial x}\frac{\partial x}{\partial t} + \frac{\partial H}{\partial y}\frac{\partial x}{\partial t} + \frac{\partial H}{\partial z}\frac{\partial x}{\partial t} \right) \tag{2.39}$$

The sum inside the brackets defines the labeled hydrodynamic gradient. It measures the head decay per unit time of a moving water particle. In other words, its temporal head decay-rate. Bearing in mind that $\mathbf{q} = n_e \cdot \mathbf{v}_e$, the hydrodynamic gradient:

$$\frac{dH}{dt} = \frac{\partial H}{\partial t} + \mathbf{q}\frac{\partial H}{\partial r} = \frac{\partial H}{\partial t} + \frac{\mathbf{v}_e}{n_e}\frac{\partial H}{\partial \mathbf{r}} \tag{2.40}$$

The first partial derivative $\partial H/\partial t$ measures the head decay per unit time ($\delta t = 1$ T) and only appears in transient flows. Now, to understand the meaning of the symbolic product $(\partial H/\partial \mathbf{r}) \cdot \mathbf{q}$ it is required to recall that the symbolic differential ratio $\partial \mathbb{P}/\partial \mathbb{Q}$ between two tensors $\mathbb{P}$ and $\mathbb{Q}$ yields a third tensor $\mathbb{R}$ whose rank equals the sum of the ranks of $\mathbb{P}$ and $\mathbb{Q}$. Therefore, $\partial H/\partial \mathbf{r}$ is a vector and the scalar product $(\partial H/\partial \mathbf{r}) \cdot \mathbf{q}$ measures the head spatial decay as the water particle moves during a unit time. Then, for $\delta t = 1$ T, $\mathbf{q} = \mathbf{1}$, and $\partial \mathbf{r} = \mathbf{1}$ L, the hydraulic gradient $\mathbf{J} = -\partial H/\partial \mathbf{r}$ measures the consumption of the potential energy stored in a percolating water particle after traveling a unit length during a unit time with a unit apparent velocity.

Alternatively, Equation 2.40, written in explicit matrix components, is:

$$\frac{dH}{dt} = \frac{\partial H}{\partial t} + \left(\begin{matrix} q_x \\ q_y \\ q_z \end{matrix} \right) \left(\begin{matrix} \frac{\partial H}{\partial x} \\ \frac{\partial H}{\partial y} \\ \frac{\partial H}{\partial z} \end{matrix} \right) = \frac{\partial H}{\partial t} + \frac{1}{n_e} \left[\begin{matrix} (\mathbf{v}_e)_x \\ (\mathbf{v}_e)_y \\ (\mathbf{v}_e)_z \end{matrix} \right] \left(\begin{matrix} \frac{\partial H}{\partial x} \\ \frac{\partial H}{\partial y} \\ \frac{\partial H}{\partial z} \end{matrix} \right) \tag{2.41}$$

In the technical literature, the hydrodynamic gradient is also called *convective* or *advective* derivative, *substantial* or *material* derivative, *Lagrangian* or *Stokes* derivative, and *derivative following the motion*.

2.3.8.3 *Hydraulic conductivity for a group of random fractures*

If the "modeler" neglects minor hydraulic head losses at discontinuities intersections and that the accuracy and details of the results of the pseudo-continuous model suit his needs at the selected scale of observation, a reasonable approach for the hydraulic conductivity of randomly fractured subsystems stems from the linearity of Darcy's Law. In this case, the following assumptions prevail:

i. Impervious cakes do not coat the interface between the intact rock and the fracture walls.
ii. An average vector-field **J** substitutes the almost similar hydraulic gradient vector-field within the pseudo-subsystem.

Then, for a pseudo-system hydraulically equivalent to one single fracture, the discharge component q_i crossing a control-section S_i is:

$$\mathbf{Q}_i = q_i S_i = \left[\sum_{j=1}^{3} \left(k_{ij} \mathbf{J}_i \right) \right] S_i$$

Now, considering many random fractures a, b, c... the resulting discharge $\mathbf{Q}_i$ is:

$$\mathbf{Q}_i = \left[\sum_{j=1}^{3} \left[\left(k^a\right)_{ij} \left(\mathbf{J}^a\right)_i \right] \right] S_i + \left[\sum_{j=1}^{3} \left[\left(k^b\right)_{ij} \left(\mathbf{J}^b\right)_i \right] \right] S_i + \left[\sum_{j=1}^{3} \left[\left(k^c\right)_{ij} \left(\mathbf{J}^c\right)_i \right] \right] S_i + \ldots$$

As assumed, $\mathbf{J}^a \approx \mathbf{J}^b \approx \mathbf{J}^c \approx \ldots \approx \mathbf{J}$, then the superposition principle can be applied:

$$\mathbf{Q}_i = \left[\sum_{j=1}^{3} \left(k^a\right)_{ij} + \sum_{j=1}^{3} \left(k^b\right)_{ij} + \sum_{j=1}^{3} \left(k^c\right)_{ij} + \ldots \right] \mathbf{J}_i S$$

Substituting the summation in brackets for a single tensor [k]:

$$|k| = |k^a| + |k^b| + |k^c| + \ldots \tag{2.42}$$

In full matrix lettering:

$$|k| = \begin{bmatrix} \left[(k_{11})^a + (k_{11})^b + (k_{11})^c + \ldots\right] & \left[(k_{12})^a + \ldots\right] & \left[(k_{13})^a + \ldots\right] \\ \left[(k_{21})^a + (k_{21})^b + (k_{21})^c + \ldots\right] & \left[(k_{22})^a + \ldots\right] & \left[(k_{23})^a + \ldots\right] \\ \left[(k_{31})^a + (k_{31})^b + (k_{31})^c + \ldots\right] & \left[(k_{32})^a + \ldots\right] & \left[(k_{33})^a + \ldots\right] \end{bmatrix} \tag{2.43}$$

2.3.8.4 Energy conservation principle

Darcy's smart assumption allows modeling a pervious and saturated soil or rock mass as a pseudo-continuous fluid with the liquid-gaseous mixture's unit weight. One of its driving forces is the vector-addition of all hydrodynamic fluid pressures acting on its faces.

The i-component of the driving force acting on the face δS_i is $\left[\left(-\partial p/\partial x_i\right)\cdot \delta x_i\right]\cdot \delta S_i$, where δx_i is the length of the edge of the subsystem i, δS_i is the i-face area normal do the i-axis. That component equals $\left(-\partial p/\partial x_i\right)\cdot \delta V$, where δV is the pseudo-continuous subsystem volume derived from the product $\delta x_i \cdot \delta S_i$. Another driving force is the vertical gravity pull $-\rho \cdot \mathbf{g} \cdot \delta V$, where ρ is the fluid density and **g** is the gravity acceleration, a vector quantity. These two driving forces induce flow against a viscous resisting force whose i-component is $D_i \cdot \delta V$, where **D** is the drag force per unit volume, also a vector quantity. If referred to the solid phase unit volume δV_e, that specific drag is $\mathbf{D}/n_e$, where n_e is the effective porosity.

In steady-state flow, without acceleration, these driving and resisting forces are in a dynamic equilibrium satisfying the following equations, related to the zenithal frame ENZ:

$$\left(-\frac{\partial}{\partial E}p + D_E\right)\delta V = 0.$$

$$\left(-\frac{\partial}{\partial N}p + D_N\right)\delta V = 0.$$

$$\left[\left(-\frac{\partial}{\partial Z}p - \rho g\right) + D_Z\right]\delta V = 0.$$

Eliminating δV from these equations and keeping in mind that the first three terms inside the brackets $-\partial p/\partial E$, $-\partial p/\partial N$, and $\left(-\partial p/\partial N - \rho g\right)$ correspond to the components of the hydraulic gradient J_E, J_E, and J_Z, this system of equations is to the following vector-equation:

$$\rho g \mathbf{J} + \mathbf{D} = 0.$$

Therefore, in steady-state, the specific viscous drag **D** balances the impulse $\rho g\mathbf{J}$. The **D** and $\rho g\mathbf{J}$ are opposing vectors. In transient-state, the resultant of $\rho g\mathbf{J} + \mathbf{D}$ accelerates or slows down the groundwater flow. In other words, an insignificant specific inertial force $\rho g\left(d\mathbf{q}/dt\right)$ counteracts the unbalanced percolation force and drag force (both forces are specific quantities per unit volume). The i-component of the resisting inertial force per unit volume is $\left[\rho g\left(\partial q_i/\partial t\right)\cdot \delta S_i \cdot \delta St\right]/\delta V$. If referred to the solid phase volume, it is $\left[\rho g/n_e \cdot \left(\partial q_i/\partial t\right)\cdot \delta S_i \delta St\right]/\delta V_e$, where n_e is the "effective" porosity. However, the inertial force is unimportant compared to the viscous drag.

Upon substitution of the hydraulic gradient **J** for $\mathbf{D}/\rho g$ in the empirical Darcy's law, the specific drag force **D** is as a function of the specific discharge **q** and the inverse of the hydraulic transmissivity $[k]^{-1}$, called hydraulic resistivity, giving $\mathbf{D} = \rho g \cdot \mathbf{q} \cdot [k]^{-1}$. In addition to size-effects, the hydraulic resistivity also incorporates the influence of the kinetic fluid viscosity. If necessary, write the eigenvectors of the hydraulic conductivity as $k_i = (\rho g/\nu) \cdot K_i$ to distinguish these effects, where K is the intrinsic permeability with dimensions L^2, only depending on the shape and spatial arrangement of pores, fractures, and conduits. Note that the vectors **D** and **q** are only collinear for isotropic and homogeneous media.

2.3.8.5 Mass conservation principle

Consider a saturated and slightly compressible Darcy's pseudo-continuous system. During a unit time interval, the mass conservation principle implies that the differential amount of water mass entering and leaving throughout the surface S of this system matches the differential amount of water mass within its volume V due to the contraction (or expansion) of its voids and to the compressibility water. Occasionally, during the same time-interval, distributed specific sources (or sinks) Q within the system volume V may add (or take) from its voids an extra amount of water. Then, the mathematical statement of the continuity equation is:

$$\int_0^V \frac{\partial(\rho n)}{\partial t}\, dV + \int_0^V \nabla(\rho \mathbf{q})dV - \int_0^V \rho Q\, dV = 0. \tag{2.44}$$

In this expression, the first integral of the left member measures the sum of the differential changes of the amount of water filling the voids within the system volume elements dV, due to variations of its porosity n and water density ρ. The second integral of the left member measures the net amount of the differential inflows and outflows of water $\rho\mathbf{q}$ throughout all the six surface elements $d\mathbf{S}$. By the convention adopted in this book, the outgoing-normal **n** of any surface elements $d\mathbf{S}$ is always positive. In contrast, also by a traditional convention adopted in practice, the scalar product $\rho\mathbf{q} \cdot d\mathbf{S}$ is considered "negative" for outflows and "positive" for inflows. That somewhat arbitrary choice reflects that the gradient $\partial H/\partial \mathbf{n}$ has the opposite direction of the outgoing-normal **n** for outflows and vice-versa. The integral of the right member measures the net amount of extra water added into or removed from the system volume V by all sources and sinks Q within it (positive Q for injection).

Recalling Gauss' theorem, in the above equation, the volume integral substitutes the surface integral:

$$\int_0^S \rho \mathbf{q}\, dS = \int_0^V \nabla(\rho \mathbf{q})dV$$

In this expression, the symbol ∇, called the *del operator*, is:

$$\nabla = \sum_{i=1}^{3} \frac{\partial}{\partial x_i} \mathbf{u}_i \tag{2.45}$$

Then, the continuity statement takes the form already given above.

$$\int_0^V \frac{\partial(\rho n)}{\partial t}\, dV + \int_0^V \nabla(\rho \mathbf{q}) dV - \int_0^V \rho Q\, dV = 0.$$

However, shrinking the volume integration V indefinitely around any point inside this volume, and canceling all volume elements dV and integral signs ∫, yields the differential form of the continuity equation (provided all integrands are continuous):

$$\frac{\partial(\rho n)}{\partial t} + \nabla(\rho \mathbf{q}) - \rho Q = 0 \tag{2.46}$$

Darcy's Law allows writing the specific discharge $\mathbf{q}$ in the above equation in terms of the hydraulic head H:

$$\frac{\partial(\rho n)}{\partial t} + \sum_{i=1}^{3} \frac{\partial}{\partial x_i} \sum_{j=1}^{3} \left[\rho \left(-k_{ij} \left(\frac{\partial H}{\partial x_j} \right) \right) \right] - \rho Q = 0 \tag{2.47}$$

As $H = z + p/\rho g$, it follows that $\partial p/\partial t = \rho g \cdot \partial H/\partial t$. Therefore, applying the "chain rule" to the first term of the left member of the preceding equation:

$$\frac{\partial(\rho n)}{\partial t} = \frac{\partial(\rho n)}{\partial p}\frac{\partial p}{\partial t} = \rho g \frac{\partial(\rho n)}{\partial p}\frac{\partial H}{\partial t} = \rho g \left(n\frac{\partial \rho}{\partial p} + \rho \frac{\partial n}{\partial p} \right) \frac{\partial H}{\partial t} \tag{2.48}$$

The water density ρ at a confined pressure p may derive from $\rho_0[1+\beta p]$, where ρ_0 is the water density at atmospheric pressure and β is the isothermal compressibility coefficient of water. Then, the first differential ratio $\partial\rho/\partial p$ in the brackets of the right member corresponds to $\rho_0\beta$.

Assuming a linear elastic behavior and that any pressure relief (or pressure increase) are fully loaded on (or unloaded from) the discontinuous solid phase, the second differential ratio $\partial n/\partial p$ in the right member's brackets corresponds to the inverse of the subsystem bulk modulus K_p. Supposing that any expansion (or contraction) of its open discontinuities primarily reflects variations of its "effective" hydrostatic stress $\partial\sigma_{hydrostatic}$ (as defined in soil mechanics) and supposing that the total average stress σ remains invariant, any variation ∂p induces an equal but opposed hydrostatic stress variation $-\sigma_{hydrostatic}$. Moreover, ignoring the solid matter's minor deformation, any expansion (or contraction) of the open discontinuities induced by the variation $-\partial\sigma_{hydrostatic}$, mostly follows its void volume change. Then, in this case, $K_p = -\partial\sigma_{hydrostatic}/[\partial(nV)/V] = \partial p/\partial n$.

Considering all previous observations, the constant value $\sum_V$, called *specific volumetric storage*, may replace the expression $\rho g \cdot \left[n \cdot \partial\rho/\partial p + \rho \cdot \partial n/\partial p \right]$:

$$\sum\nolimits_V \rho g \left(n \rho_0 \beta + \frac{\rho}{K_p} \right) \tag{2.49}$$

$\sum_V$ unit is M/L^4, but normalizing the general continuity equation by constant ρ, $\sum_V$ takes its reduced expression, denoted by S_V, quoted in L^{-1} units:

$$S_V = \rho g \left(n \beta + \frac{1}{K_p} \right) \tag{2.50}$$

Then, substituting $\sum_V$ for $\partial(\rho n)/\partial t$, the continuity equation takes the form:

$$\sum_{i=1}^{3} \frac{\partial}{\partial x_i} \sum_{j=1}^{3} \left[\rho k_{ij} \left(\frac{\partial H}{\partial x_j} \right) \right] = \sum_V \frac{\partial H}{\partial t} - \rho Q \tag{2.51}$$

For constant ρ:

$$\sum_{i=1}^{3} \frac{\partial}{\partial x_i} \sum_{j=1}^{3} \left[k_{ij} \left(\frac{\partial H}{\partial x_j} \right) \right] = S_V \frac{\partial H}{\partial t} - Q \tag{2.52}$$

For a more rigorous discussion about the *mass conservation principle,* see Bear's full development (Bear, 1968).

Notes

1 1 cP = 0.01 poise = 0.01 g/cm s = 0.001 Pa s = 1 mP s = 0.001 Ns/m^2.
2 Bear, J., *Dynamics of Fluids in Porous Media*, American Elsevier, New York, 1972.
3 Obviously, all remarks made for 3D systems in this book are applied to 2D or 1D systems but reconsidering what is to be kept or changed in each case.
4 A L^p norm of a vector or a matrix X is defined as $\|X_p\| = \left[X^p \right]^{1/p}$, then, L^2 corresponds to $\|X_2\| = \left[X^2 \right]^{1/2}$.
5 Operational details in Franciss, F. O., *Soil & Rock Hydraulics*, Balkema, Rotterdam, 1985.
6 For details, see Chapter 5 devoted to *Systematic Relations* in Zwikker, C., *Physical Properties of Solid Materials*, Pergamon Press, London, 1954.
7 As already noted, a symbolic partial derivative $\partial P/\partial Q$ may be defined as a tensor R whose rank equals the sum of the ranks of P and Q.
8 Nelson.

9 Hsieh, P. A., Neuman, S. P., "Field determination of the three-dimensional hydraulic conductivity tensor of anisotropic media", *Water Resources Research,* Vol. 21, No. 11, 1085, 1985.
10 In Bogomolov, G. V., *Bases de l'Hydrogéologie, Chap. V,* Editions de la Piax, Moscow, 1955.
11 All information about the 19 investigated dam-sites are found in Franciss, F. O., Grade de Docteur Ingénieur, 1ère Thèse: Contribution à l'Étude du Mouvement de l'Eau à travers les Milieux Fissurés, 2ème Thèse: Géomorphologie et Géologie du Brésil, Faculté des Sciences de l'Université de Grenoble, 1970.
12 Snow, D. T., "Rock Fracture Spacings, Openings and Porosities", *Journal of Soil Mechanics and Foundations Division*, Vol. 94, No. 1, 73–91, 1968, SMI, ASCE.
13 Emery, C. L., "The Strain in Rocks in Relation to Highway Design", Department of Mining Engineering, University of British Columbia, Canada, 1966.
14 Rocha, M., Franciss, F., "Determination of Permeability in Anisotropic Rock-Masses from Integral Samples", *Rock Mechanics*, Vol. 9, Nos. 2–3, 67–93, 1977, Springer-Verlag.
15 See the complete (not simplified) transformation formula in Snow, T. D. "Three-hole Pressure Test for Anisotropic Foundation Permeability", *Rock Mechanics and Engineering Geology*, Vol. IV, No. 4, 298–316, 1961.
16 Hsieh, P. A., Neumann, S. P., Stiles, G. K. and Simpson, E. S., "Field Determination of the Three-Dimensional Hydraulic Conductivity Tensor of Anisotropic Media, Part 1. Theory and Part 2. Methodology and Application to Fractured Rocks; U. S. Geological Survey, Menlo Park, California", *Water Resources Research*, Vol. 21, No. 11, 1985.

References

Bear, J., Zaslavsky, D. and Irmay, S., 1968, *Physical Principles of Water Percolation and Seepage, Paris*, UNESCO.

Darcy, H, 1856, *Les fontaines publiques de la ville de Dijon*, Victor Dalmon, Paris.

Dupuit, J., 1863, *Etudes Théoriques et Pratiques sur le Mouvement des Eaux dans les Canaux Découverts et a travers les Terrains Perméables*, Dunod, Paris, 2nd Ed.

Franciss, F. O., 1970, Contribution a l'Étude du Mouvement de l'Eau a travers les Milieux Fissurés, Thèse, Faculté de Sciences de l'Université de Grenoble.

Hshieh, P. A., Neumann, S. P., Stiles, G. K. and Simpson, E. S., 1985, Field Determination of the Three-Dimensional Hydraulic Conductivity Tensor of Anisotropic Media, *Water Resources Research*, Vol. 21, pp. 1667–1676.

Koerber, G. G., 1962, *Properties of Solids*, Prentice-Hall, Englewood Cliffs, 286 p.

Louis, C., 1969, Flow Phenomena in Jointed Media and their Effect on the Stability of Structures and Slopes in Rock, Imperial College, Rock Mechanics Progress Report.

Nelson, R. A., 2001, *Geologic Analysis of Fractured Reservoirs*, Butterworth-Heinemann, Woburn, Second Ed., 332 p.

Rocha, M. and Franciss, F. O., 1977, Determination of Permeability in Anisotropic Rock Masses from Integral Samples, *Rock Mechanics*, Vol. 9, Nos. 2–3, pp. 97–93.

Quadros, E., 1982, Determinação das Características do Fluxo de Água em Fraturas de Rochas, Universidade de São Paulo, Brazil.

Chapter 3

Approximate solutions

3.1 Overview

The mathematical simulation of an aquifer system's hydraulic behavior requires the analytical or numerical integration of two simultaneous partial differential equations, called *governing equations*, ensuing from the simultaneous application of energy and mass conservation principles to pseudo-continuous subsystems of groundwater systems. These partial differential equations connect one or more partial derivatives of the *dependent* variable head H by the *independent* variables (x_1, x_2, x_3, t). The *order* of a partial differential equation corresponds to the order of its highest derivative. They are *linear* if the dependent variable H and the partial derivatives $\partial H/\partial x_i$ and $\partial H/\partial t$, including their products, are only raised to the first power; otherwise, they are nonlinear. However, they are quasi-linear if their highest derivatives remain linear.

The *continuity equation*, resulting from combining these two *governing equations*, associates the *spatial* and *temporal* coordinates (**r**, t) of all points of the flow domain to their hydraulic heads H, via the hydraulic conductivity second-order tensor [k]. The *integration* of the continuity equation gives the spatial-temporal description of the hydraulic head H in the flow domain as a scalar field. The derivatives of the hydraulic head H give the hydraulic gradients vectors **J**, as a vector field. Finally, [k] and J's empirical relationships yield the specific discharges vectors **q**, as another vector-field.

Domains typified by simple structures and properties encourage analytical solutions, but for most groundwater problems, particularly for fractured rocks, a closed analytical solution does not exist. In that case, numerical techniques may lead to *approximate solutions*.

Table 3.1 lists the essential features of the numerical and analytical solutions, pointing out their merits and disadvantages (Chung-Yau, 1994).

Fractured rock masses are inhomogeneous and anisotropic. Their hydraulic properties may vary enormously over the flow domain. Fortunately, these properties may correlate with appropriate descriptors of the flow domain's geologic features, duly calibrated by field and laboratory tests. However, a consistent knowledge of these properties alone does not lead to a specific correct solution. Information about its history and the external influences on their spatial and temporal boundaries are vital to the modeler. Every solution for the *continuity equation* depends on some *supplementary equations*: the *boundary conditions* and the *initial values*. The problem is then said to be *well-posed* if, for every supplementary equation set, one can find a *unique solution* continuously dependent on this set of supplementary equations. When boundary

Table 3.1 Analytical and numerical methods: pros and cons

Analytical	*Numerical*
Continuous solutions exist at every point	Piecewise solutions only exist at grid points, but interpolated out of them
Solutions are exact or very accurate	Solutions are approximate but allow error control
Solutions parameters usually have a clear physical meaning	Solution parameters usually have not a clear physical meaning
Solutions do not exist for most practical problems	More than one approximate solution may be found for practical problems, depending on the desired accuracy

conditions are partially unknown, it may be possible to estimate the missing data by a plausible hypothesis about the past hydraulic behavior on the borders, top, and bases of a 3D aquifer system.

3.2 Differential operators

Assuming constant density ρ and using a symbolic *differential operator*, that is, operators that perform as ordinary algebraic symbols and comply with the fundamental laws of algebra, the general continuity equation is compactly written as:

$$D_V\, H(\mathbf{r},t) - Q(\mathbf{r}) = 0. \tag{3.1}$$

In this equation, the symbolic operator D_V stands for a second-order linear differential operator and $Q(\mathbf{r})$ for a space-dependent, and sometimes also time-dependent, water source (or drain) per unit volume and per unit time, that is, $Q = \lim\left[\left(\delta V_W/\delta t\right)/\delta V\right]$. In this case, the symbolic expression for D_V is:

$$D_V = -\sum_{i=1}^{3} \frac{\partial}{\partial x_i} \sum_{i=1}^{3} \left[k_{ij} \frac{\partial}{\partial x_j} \text{-} \right] + S_V \left[\frac{\partial}{\partial t} \text{-} \right] \tag{3.2}$$

Here, S_V denotes the specific volumetric storage, as discussed in Chapter 2 (see Section 2.3.7.3).

For homogeneous and anisotropic pervious media and coordinate axes complying with the eigenvectors of [k], all cross products vanish, and the differential operator D_V takes its more traditional form:

$$D_V = -\sum_{i=1}^{3} \frac{\partial}{\partial x_i} \left(k_i \frac{\partial}{\partial x_i} \blacksquare \right) + S_V \left(\frac{\partial}{\partial t} \blacksquare \right) \tag{3.3}$$

If the time-dependent term $S_V \cdot (\partial \blacksquare / \partial t) = 0$ vanishes, the differential equation describes a steady flow. Otherwise, it describes an unsteady flow. In some cases, water sources or sinks $Q(\mathbf{r})$ occur at the flow domain's due points.

Second-order linear partial differential equations may be classified into elliptic, parabolic, and hyperbolic (see the end of this chapter). They are homogeneous if all terms contain the unknown variable. Otherwise, they are inhomogeneous.

Boundary-value problems are associated with the Laplace equation, in which $Q(\mathbf{r})=0$, or to the Poisson, in which $Q(\mathbf{r})\neq 0$, giving time-independent solutions. These equations do not make clear the time-rate of energy consumption to sustain the groundwater flow. However, the hydraulic gradient J's formal definition always implies power-consumption (see Chapter 2, Section 2.3.8.2).

The Laplace equation describes "natural" equilibrium processes. Their maximum and minimum solution values occur at the flow boundaries. The Poisson equation describes equilibrium processes disturbed by steady sources and drains. Thus, their maximum and minimum solution values may also occur within the flow domain.

Parabolic equations describe transient flows tending to equilibrium as t grows up indefinitely. Maximum and minimum solution values gradually shift within the flow domain, decreasing or increasing with time. Their time-dependent nature requires the prescription of *initial values* to particularize solutions. In rare cases, it also requires the prescription of *time-dependent boundary values*.

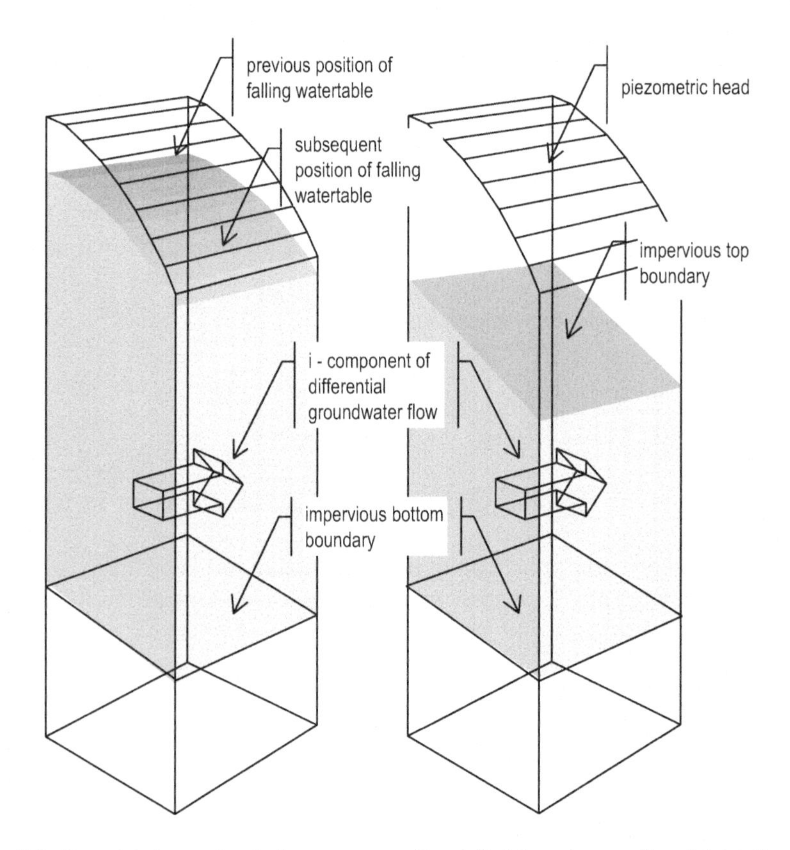

Figure 3.1 Dupuit's hypothesis for an unconfined (left) and a confined (right) aquifer over an impervious base.

If applied to a *confined aquifer*, resting on a non-horizontal impervious bottom, the Dupuit's assumption also yields a linear second-order differential equation (see Figure 3.1):

$$-\left[\begin{array}{l}\frac{\partial}{\partial E}\left[H_C\left[k_{EE}\frac{\partial}{\partial E}\left(B+H_C+\frac{P}{\rho g}\right)+k_{EN}\frac{\partial}{\partial N}\left(B+H_C+\frac{P}{\rho g}\right)\right]\right]\ldots\\+\left[\frac{\partial}{\partial N}\left[H_C\left[k_{NE}\frac{\partial}{\partial E}\left(B+H_C+\frac{P}{\rho g}\right)+k_{NN}\frac{\partial}{\partial N}\left(B+H_C+\frac{P}{\rho g}\right)\right]\right]\right]\end{array}\right]$$

$$+\frac{\partial}{\partial t}(n_e-H_C)-\omega=0. \qquad (3.4)$$

In this equation, the independent variables H_C and B denote the confined saturated height and the base elevation. The dependent variable $P/\rho g$ stands for the pressure head at the aquifer roof, defining the dependent head: $H=(B+H_{C+}P/\rho g)$. The dependent pressure may be referred for in the middle of the confined thickness H_C and denoted by P_m. The differential equation must change accordingly but describe the same process, recalling $\partial H/\partial z=0$, then $P_m=H_C/2+P$.

Collecting the unknown variable H and known variables B and H_C and introducing another differential operator D_C this equation also is written as:

$$D_C H-\left[\begin{array}{l}\frac{\partial}{\partial E}\left[H_C\left(k_{EE}\frac{\partial}{\partial E}B+k_{EN}\frac{\partial}{\partial N}B\right)\right]\ldots\\+\left[\frac{\partial}{\partial N}\left[H_C\left(k_{NE}\frac{\partial}{\partial E}B+k_{NN}\frac{\partial}{\partial N}B\right)\right]\right]\end{array}\right]+\left[\frac{\partial}{\partial t}(n_e-H_C)-\omega\right]=0. \qquad (3.5)$$

Then, the symbolic linear differential operator D_C is:

$$D_C H-\left[\begin{array}{l}\frac{\partial}{\partial E}\left[H_C\left(k_{EE}\frac{\partial}{\partial E}\bullet+k_{EN}\frac{\partial}{\partial N}\bullet\right)\right]\ldots\\+\left[\frac{\partial}{\partial N}\left[H_C\left(k_{NE}\frac{\partial}{\partial E}\bullet+k_{NN}\frac{\partial}{\partial N}\bullet\right)\right]\right]\end{array}\right] \qquad (3.6)$$

All preceding symbolic operators are linear. However, the Dupuit's approximation, when applied to an *unconfined aquifer*, yields a nonlinear second-order differential equation. Whenever possible, a simple change of variable may transform and approximately linearize this equation. Indeed, considering the general continuity equation for unconfined flow and constant ρ:

$$-\left[\begin{array}{l}\frac{\partial}{\partial E}\left[H\left[k_{EE}\left[\frac{\partial}{\partial E}(B+H)\right]+k_{EN}\left[\frac{\partial}{\partial N}(B+H)\right]\right]\right]\ldots\\+\frac{\partial}{\partial N}\left[H\left[k_{NE}\left[\frac{\partial}{\partial E}(B+H)\right]+k_{NN}\left[\frac{\partial}{\partial N}(B+H)\right]\right]\right]\end{array}\right]+\frac{\partial}{\partial t}(n_e H)-\omega=0. \qquad (3.7)$$

By the differentiation rule of a variable raised to a power and for n_e assumed constant: $H\cdot(\partial H/\partial\ldots)=(1/2)\cdot(\partial H^2/\partial\ldots)$ Then, this equation is:

$$-\left[\begin{array}{l}\left[\begin{array}{l}\frac{\partial}{\partial E}\left[H\left[k_{EE}\left(\frac{\partial}{\partial E}B\right)+k_{EN}\left(\frac{\partial}{\partial N}B\right)\right]\right]\ldots\\+\frac{\partial}{\partial N}\left[H\left[k_{NE}\left(\frac{\partial}{\partial E}B\right)+k_{NN}\left(\frac{\partial}{\partial N}B\right)\right]\right]\end{array}\right]\ldots\\+\frac{1}{2}\left[\begin{array}{l}\frac{\partial}{\partial E}\left[k_{EE}\left(\frac{\partial}{\partial E}H^2\right)+k_{EN}\left(\frac{\partial}{\partial N}H^2\right)\right]\ldots\\+\frac{\partial}{\partial N}\left[k_{NE}\left(\frac{\partial}{\partial E}H^2\right)+k_{NN}\left(\frac{\partial}{\partial N}H^2\right)\right]\end{array}\right]\end{array}\right]+\frac{n_e}{2H}\frac{\partial}{\partial t}\left(H^2\right)-\omega=0.$$

For most unconfined aquifers, the Dupuit's hypothesis can be adequately applied, since the areal variations of H are relatively small compared to the aquifer thickness. Then, substituting a new auxiliary variable V for H^2 and considering an average value for $H\approx H_{avr}$.

$$-\left[\begin{array}{l}\left[\begin{array}{l}\frac{\partial}{\partial E}\left[H_{av}\left[k_{EE}\left(\frac{\partial}{\partial E}B\right)+k_{EN}\left(\frac{\partial}{\partial N}B\right)\right]\right]\ldots\\+\frac{\partial}{\partial N}\left[H_{av}\left[k_{NE}\left(\frac{\partial}{\partial E}B\right)+k_{NN}\left(\frac{\partial}{\partial N}B\right)\right]\right]\end{array}\right]\ldots\\+\frac{1}{2}\left[\begin{array}{l}\frac{\partial}{\partial E}\left[k_{EE}\left(\frac{\partial}{\partial E}V\right)+k_{EN}\left(\frac{\partial}{\partial N}V\right)\right]\ldots\\+\frac{\partial}{\partial N}\left[k_{NE}\left(\frac{\partial}{\partial E}V\right)+k_{NN}\left(\frac{\partial}{\partial N}V\right)\right]\end{array}\right]\end{array}\right]+\frac{1}{2H_{av}}\frac{\partial}{\partial t}\left(n_e V\right)-\omega=0.$$

That transformed equation can also be compactly written as a linear:

$$D_U V-\left[\begin{array}{l}\frac{\partial}{\partial E}\left[H_{av}\left[k_{EE}\left(\frac{\partial}{\partial E}B\right)+k_{EN}\left(\frac{\partial}{\partial N}B\right)\right]\right]\ldots\\+\frac{\partial}{\partial N}\left[H_{av}\left[k_{NE}\left(\frac{\partial}{\partial E}B\right)+k_{NN}\left(\frac{\partial}{\partial N}B\right)\right]\right]\end{array}\right]-\omega=0. \tag{3.8}$$

Now, the linear differential operator D_U is:

$$D_U=-\frac{1}{2}\left[\begin{array}{l}\left[\frac{\partial}{\partial E}\left[k_{EE}\left(\frac{\partial}{\partial E}\bullet\right)+k_{EN}\left(\frac{\partial}{\partial N}\bullet\right)\right]\right]\ldots\\+\frac{\partial}{\partial N}\left[k_{NE}\left(\frac{\partial}{\partial E}\bullet\right)+k_{NN}\left(\frac{\partial}{\partial N}\bullet\right)\right]\end{array}\right]+\frac{1}{2H_{av}}\frac{\partial}{\partial t}\left(n_e\bullet\right) \tag{3.9}$$

This variable change implies also changing all specified boundary conditions in terms of V instead of H^2. Finally, the square root of the final solution V(**r**, t) re-establishes the head H(**r**, t).

Boundary conditions can also be written in a compact symbolic form. As explained in Chapter 2, the 3D equation of continuity results from applying the mass conservation principle to an elementary volume δV within a flow domain. By similar reasoning, the 2D Neumann or *natural boundary condition* may result from the application of the same principle to an elementary surface δ**S** on the boundary of the flow domain. If α_{ni} denotes the direction cosines of the outward-normal **n** to a boundary element δS, a specified incoming (or outgoing) perpendicular flow to δS, $\mathbf{q}_n \cdot \delta S$ must be equal to a corresponding Darcy's flow component at the backside of δS, defined by the inner product $\sum\sum k_{ij}\mathbf{J}_j \cdot \alpha_{ni} \cdot \delta S$. Then, equating both expressions and removing δS:

$$\sum_{i=0}^{3}\sum_{j=1}^{3}\left(k_{ij}\,J_j\right)\alpha_{ni} - \mathbf{q}_n = 0$$

With the hydraulic gradient J expressed as a function of the hydraulic head H:

$$-\sum_{i=0}^{3}\sum_{j=1}^{3}\left(k_{ij}\frac{\partial}{\partial x_j}H\right)\alpha_{ni} - \mathbf{q}_n = 0 \qquad (3.10)$$

The above equation translates the Neumann or *natural* boundary condition, compactly written as:

$$D_N\,H(\mathbf{r},t) - \mathbf{q}_n(\mathbf{r}) = 0. \qquad (3.11)$$

The differential operator D_N is:

$$D_N = -\sum_{i=0}^{3}\sum_{j=1}^{3}\left(k_{ij}\frac{\partial}{\partial x_j}\bullet\right)\alpha_{ni} \qquad (3.12)$$

On the other hand, the Dirichlet or *essential* boundary condition prescribes the hydraulic head H values, at least at one boundary point, to guarantee a unique solution.

3.3 Uniqueness of solutions

Two "apparently" different solutions for the same domain and the same *boundary* or/and *initial conditions* must be identical: that is, must be *unique*. Statements and uniqueness theorems discussions are a well-developed subject of higher mathematics. However, their validity can be intuitively perceived. What is essential to keep in mind is that correctly *supplementary equations, that is*, well-posed *boundary* or/and *initial conditions*, are crucial to assure a *unique solution*.

Correct prescriptions for boundary and initial conditions depend on the type of the partial differential equation: *hyperbolic*, *elliptic*, or *parabolic* and on the domain boundaries' character: *open* or *closed*. An open border expands indefinitely in one or more dimensions, usually t, and a closed one surrounds the entire domain.

Dirichlet and Neumann prescriptions are typically the most common *boundary conditions* to guarantee unique groundwater flow solutions in a state of equilibrium: *steady-state solutions*. For anisotropic rock masses, the Neumann condition must consider the non-parallelism between the hydraulic gradient and the specific discharge.

The third kind of boundary condition, called Cauchy or Robbins or *mixed* condition, stipulates Dirichlet and Neumann's linear combinations. If not compatible, these prescriptions may not fit into concrete situations or imply forcing functions; this is a common source of aquifer simulation errors, such as if active pumping wells replace pure gravity-induced draining shafts.

Prescribed values of the hydraulic head and its derivatives over the flow domain at an elected time, called *initial conditions*, are essential to uncover unique non-equilibrium simulations: *unsteady-state solutions*. Unbalanced solutions may tend to equilibrium for stable initial conditions as time goes on, attaining a stable final state. Therefore, fixed boundary conditions describe final states of equilibrium after the gradual evolution of prescribed initial conditions.

The third type of problem, called *initial-boundary* problems, require concomitant prescriptions of initial conditions and well-matched boundary conditions.

Even if a unique solution exists theoretically, it may stay uncovered or hard to get. Particularly for complex systems that do not conform to simple classification schemes, as their erratic properties imply differential equations with varying coefficients. In this case, more than one approximate solution, each one with a different accuracy level, may suit the model user needs; however, even in this option, their boundary conditions must be consistently defined.

Table 3.2 summarizes, without proof, the uniqueness theorems that mostly apply to common types of boundaries, boundary conditions, and differential equations (Band, 1960).

3.4 Approximate errors

An *approximate solution*, shortly called an *approximant*, advantageously replaces a complicated or unknown analytical solution. Simple approximants are *linear polynomials* or *interpolation polynomials*, allowing easy and fast computations by algebraic rules

Table 3.2 Uniqueness theorems

Boundary condition	*Boundary type*	*Hyperbolic equation*	*Elliptic equation*	*Parabolic equation*
Dirichlet or Neumann	Open	Under-specified (therefore many solutions)		Unique solution
Dirichlet or Neumann	Closed	Under-specified (therefore many solutions)	Unique solution	Over-specified (usually no solution possible)
Cauchy	Open	Unique solution	Over specified (usually no solution possible)	
Cauchy	Closed	Over-specified (usually no solution possible)		

under a known accuracy level. Careful inspection of how well dissimilar approximants simulate the same physical system under the same context may reveal dominant trends that may favor discovering more significant facts.

If a linear approximant h(**r**, t) replaces the exact but unknown solution H(**r**, t), the modeler must adequately consider specific *approximant errors* concerning the following:

- The approximate solution within the solution domain: $\varepsilon = h(\mathbf{r}, t) - H(\mathbf{r}, t)$
- The Dirichlet conditions on its 3D boundaries: $\varepsilon_D = h(\mathbf{r}, t) - H_D(\mathbf{r}, t)$
- The continuity condition within the solution domain: $\varepsilon_V = D_V \cdot h(\mathbf{r}, t) - Q(\mathbf{r})$
- The Neumann conditions on its 3D boundaries: $\varepsilon_N = D_N \cdot h(\mathbf{r}, t) - \mathbf{q}_n(\mathbf{r})$

Note: Errors ε, ε_D, and ε_N have the same formal meaning but are distinguished here for where they are applied.

Well-known numerical methods, differing by how they minimize these errors, have their advantages and inconveniences regarding reliability, costs, performance, and practicability. For simple systems, a unique approximant applied to the flow domain may be adequate. A finite set of local approximants, simultaneously applied to the flow domain sub-domains, may be recommended for complex systems.

Linear *polynomial approximants* h(**r**, t), modeled as a sum of (n+1) products of numerical coefficients a_i by convenient elementary analytical functions $h_i(\mathbf{r}, t)$, acceptably describes the hydraulic behavior of an aquifer system

$$h(\mathbf{r},t) = a_0\, h_0(\mathbf{r},t) + a_1\, h_1(\mathbf{r},t) + a_1\, h_3(\mathbf{r},t) + \ldots + a_n \tag{3.13}$$

Written more compactly:

$$h(\mathbf{r},t) = \left[\sum_{i=0}^{n} \left[a_i\, h_i(\mathbf{r},t)\right]\right] \tag{3.14}$$

Adequate elementary functions $h_i(\mathbf{r}, t)$ depend on the type of performance to describe. These functions called *base functions*, share some common analytic attributes. The numerical coefficients a_i differ for each particular solution.

Two properties of the elementary analytical solutions of the continuity equation help to select efficient base functions. The first one, the *superposition principle*, asserts that if elementary analytical solutions $h_i(\mathbf{r}, t)$ solves the homogeneous equation $D \cdot h_i(\mathbf{r}, t) = 0$, then a linear combination of these solutions is also a solution. Indeed, considering the linear character of the operator D, the following equality holds:

$$\sum_{i=0}^{n} \left[D\left[a_i\, h_i(\mathbf{r},t)\right]\right] = D\left[\sum_{i=0}^{n} \left[a_i\, h_i(\mathbf{r},t)\right]\right]. \tag{3.15}$$

The second property asserts that any partial derivative $\partial h_i(\mathbf{r}, t)/\partial x_j$ of an elementary solution $h_i(\mathbf{r}, t)$ is also a solution. Then, the linear character of the partial derivative operator permits the following equality:

$$\sum_{i=0}^{n}\left[D\left[a_i \frac{\partial}{\partial x_j}\left[h_i(\mathbf{r},t)\right]\right]\right]=\frac{\partial}{\partial x_j}\sum_{i=0}^{n}\left[D\left[a_i\, h_i(\mathbf{r},t)\right]\right] \tag{3.16}$$

For a non-homogeneous equation $D_V{\cdot}H(\mathbf{r}, t)-Q(\mathbf{r})=0$, a finite sum of low order terms $g_k(\mathbf{r}, t)$ also approximate the forcing function $Q(\mathbf{r})$:

$$Q(\mathbf{r},t)=\sum_{k=j+1}^{n}\left[a_k\, g_k(\mathbf{r},t)\right] \tag{3.17}$$

If each elementary solutions $g_k(\mathbf{r}, t)$ solves $D_V{\cdot}g_k(\mathbf{r}, t)-Q(\mathbf{r})=0$, then their linear combination also do. As a result, the non-homogeneous equation's approximants can be structured by both homogeneous and non-homogeneous equations.

Continuous integrations of low order elementary analytical solutions generate a natural linear set of base functions.

- $1, x, x^2, x^3, \ldots x^n$
- $1, x, y, x^2, y^2, xy, \ldots x^a y^b \quad (a+b<n)$
- $1, x, y, z, \ldots x^a y^b z^c \quad (a+b+c<n)$

However, these base functions do not suitably fit time-dependent solutions. In this case, one simple alternative id to multiply their sum by a time-dependent solution, that is, $f(t)\cdot\Sigma u_i(\mathbf{r})$.

Note that:

- Trigonometric or exponential base functions or few terms of their linear series expansions also model adequate approximants for periodic processes.
- Efficient base functions must be linearly independent.
- Space-dependent power series are effective approximants, and their accuracy improves by adding higher-order terms.
- Derivatives or integrals of approximants result from the sum of the derivatives or integrals of their terms.
- The numerical evaluation of power series requires simple addition, subtraction, and multiplication operations. However, to restrict computation to a finite number of digits, the number of their terms and corresponding numerical coefficients must be limited.
- Time-dependent trigonometric base-functions may fit the periodic behavior of oscillatory boundary conditions.
- For an open time-boundary, the linear combinations of the low order terms of the infinite series expansion of the exponential function may fit the long run reaction (rise or decay) of a transient system.

- The choice of an adequate set of base functions to fit the expected response of a system to an imposed field is relatively simple, as sometimes it is possible to link its typical reactions to the character of the base function graphs.

Approximants h(**r**, t) describe scalar quantities or vector or high order tensor components; however, an s-order Cartesian tensor referred to a 3D-frame and described by a continuous function may have its 3^s components approximated by 3^s distinct general polynomials.

An approximation polynomial h(**r**, t) may be alternatively written as a polynomial interpolation:

$$h(\mathbf{r},t)=\sum_{i=0}^{n}\left[H_i\cdot f_i(\mathbf{r},t)\right] \tag{3.18}$$

Cast in this form, the interpolating functions $f_i(\mathbf{r}, t)$ are linear combinations of the elementary functions $h_i(\mathbf{r}, t)$, and the numerical coefficients H_i correspond to known solution values at n arbitrarily selected points $(\mathbf{r}_i, t_i)$. As the *general polynomial* and the *interpolation polynomial* express the same approximation, the following identity holds:

$$\sum_{i=0}^{n}\left[a_i\cdot h_i(\mathbf{r},t)\right]=\sum_{i=0}^{n}\left[H_i\cdot f_i(\mathbf{r},t)\right] \tag{3.19}$$

A Taylor series expansion defined in a closed interval can also be written as a general polynomial. Its main advantage is the immediate recognition of the upper limit of the approximation error.

The example below helps to understand the above concepts and shows a Taylor series approximant and the equivalence between general polynomial and an interpolation polynomial.

Example 3.1

A 1D analytical function f(x) defined in a closed interval [a, b] may be approximated by the first three terms of a second-order Taylor expansion:

$$h(x)\approx h(x_0)+\frac{\frac{\partial}{\partial x}\left[h_0(x_0)\right]}{1!}(x-x_0)+\frac{\frac{\partial^2}{\partial x^2}\left[h_0(x_0)\right]}{2!}(x-x_0)$$

As known from differential calculus, the remainder R_n of the Taylor series gives an upper estimate of the approximation error:

$$R_n=\frac{\frac{\partial^n}{\partial x^n}\left[h_0(\xi)\right]}{n!}\cdot(x-x_0)^n$$

As ξ is an unknown value in between a and b, the exact error cannot be estimated, but its upper value can.

The first three terms of the Taylor expansion can also be written as a polynomial approximation (and vice-versa) as:

$$h(x) \approx a_0 + a_1 x + a_2 x$$

By simple algebraic operations, it can be shown that the coefficients a_0, a_1, and a_3 are:

$$a_0 = h(x_0) - \left[\frac{\frac{\partial}{\partial x}[h_0(x_0)]}{1!}\right] \cdot x_0 + \left[\frac{\frac{\partial^2}{\partial x^2}[h_0(x_0)]}{2!}\right] \cdot (x_0)^2$$

$$a_1 = \left[\frac{\frac{\partial}{\partial x}[h_0(x_0)]}{1!}\right] - 2 \cdot \left[\frac{\frac{\partial^2}{\partial x^2}[h_0(x_0)]}{2!}\right] \cdot x_0$$

$$a_2 = \left[\frac{\frac{\partial^2}{\partial x^2}[h_0(x_0)]}{2!}\right]$$

The function f(x) may also be approximated at any point x of a closed interval $[x_0, x_3]$ by a Lagrangian quadratic interpolation polynomial H(x). If H_0, H_1, and H_2 are already known values of f(x) at three non-equidistant points x_0, x_1, and x_2, then:

$$H(x) = \frac{(x - x_1)(x - x_2)}{(x_0 - x_1)(x_0 - x_2)} H_0 + \frac{(x - x_0)(x - x_2)}{(x_1 - x_0)(x_1 - x_2)} H_1 + \frac{(x - x_0) + (x - x_1)}{(x_2 - x_0)(x_2 - x_1)} H_2$$

The polynomial derivatives of this approximant can also estimate the first and second derivatives of the function f(x):

$$\frac{\partial}{\partial x} H(x) = \left[\frac{(x - x_2) + (x - x_1)}{(x_0 - x_1)(x_0 - x_2)}\right] H_0 + \left[\frac{(x - x_2) + (x - x_0)}{(x_1 - x_0)(x_1 - x_2)}\right] H_1 + \left[\frac{(x - x_1) + (x - x_0)}{(x_2 - x_0)(x_2 - x_1)}\right] H_2$$

$$\frac{\partial^2}{\partial x^2} H(x) = \frac{2}{(x_0 - x_1)(x_0 - x_2)} H_0 + \frac{2}{(x_1 - x_0)(x_1 - x_2)} H_1 + \frac{2}{(x_2 - x_0)(x_2 - x_1)} H_2$$

The resulting error is:

$$f(x) - H(x) = \frac{(x - x_0)(x - x_1)(x - x_2)}{2} \frac{\partial^3}{\partial x^3} f(x) \leq \frac{(x - x_0)(x - x_1)(x - x_2)}{2} M$$

The quantity M never exceeds the absolute value of $\partial^3 f(x)/\partial x^3$. Residual estimates for higher-order polynomials derive from generalizations of the Taylor expansion.

Hermitian interpolation polynomials (and one variant known as Padé compact formula) differ from the Lagrangian approach by using at each interpolating point not only the base functions but also their first derivatives together. Generalizations include higher-order derivatives.

A general polynomial can alternatively be written as an interpolation polynomial. For example, consider how to fit as a general polynomial a simple first-order polynomial approximant h(x, y) to an incompletely defined higher-order surface (see Figure 3.2):

$$h(x,y)=\begin{pmatrix} 1 & x & y \end{pmatrix}\begin{pmatrix} a \\ b \\ c \end{pmatrix}$$

Select the interpolating points (x_i, y_i, H_i), (x_j, y_j, H_j), and (x_k, y_k, H_k). Now, express the known values H_i, H_j, and H_k as functions of the coordinates (x_i, y_i), (x_j, y_j), and (x_k, y_k) and of the coefficients of the polynomial approximant a, b, and c:

$$\begin{pmatrix} H_i \\ H_j \\ H_k \end{pmatrix}=\begin{pmatrix} 1 & x_i & y_i \\ 1 & x_j & y_j \\ 1 & x_k & y_k \end{pmatrix}\begin{pmatrix} a \\ b \\ c \end{pmatrix}$$

Then, the coefficients a, b, and c may be calculated as:

$$\begin{pmatrix} a \\ b \\ c \end{pmatrix}=\begin{pmatrix} 1 & x_i & y_i \\ 1 & x_j & y_j \\ 1 & x_k & y_k \end{pmatrix}^{-1}\begin{pmatrix} H_i \\ H_j \\ H_k \end{pmatrix}=\frac{1}{2S}\begin{pmatrix} A_i & A_j & A_k \\ B_i & B_j & B_k \\ C_i & C_j & C_k \end{pmatrix}\begin{pmatrix} H_i \\ H_j \\ H_k \end{pmatrix}$$

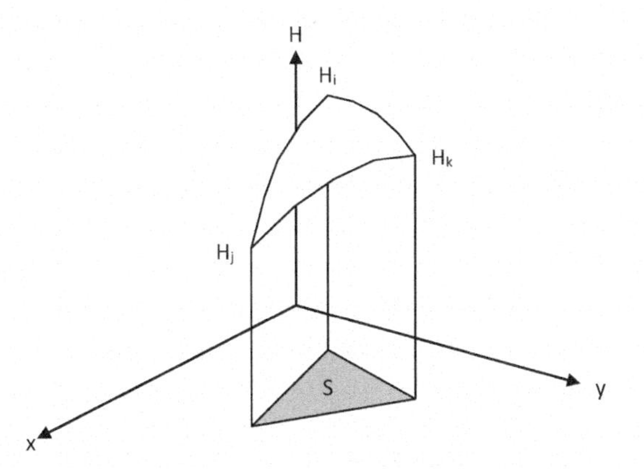

Figure 3.2 Higher-order surface incompletely defined by only three known values H.

In this expression, S denotes the area of the triangle defined by the interpolating points:

$$2S = \left[x_i \left(y_k - y_j \right) + x_j \left(y_i - y_k \right) + x_k \left(y_j - y_i \right) \right].$$

The matrix of coefficients $A_i \ldots C_k$ is:

$$\begin{pmatrix} A_i & A_j & A_k \\ B_i & B_j & B_k \\ C_i & C_j & C_k \end{pmatrix} = \begin{pmatrix} y_j x_k - x_j y_k & x_i y_k - y_i x_k & y_i x_j - x_i y_j \\ y_k - y_j & y_i - y_k & y_j - y_i \\ x_j - x_k & x_k - x_i & x_i - x_j \end{pmatrix}.$$

Substituting the coefficients, a, b, and c for its expressions in the general polynomial yields:

$$h(x, y) = \begin{pmatrix} 1 & x & y \end{pmatrix} \left[\frac{1}{2S} \begin{pmatrix} A_i & A_j & A_k \\ B_i & B_j & B_k \\ C_i & C_j & C_k \end{pmatrix} \begin{pmatrix} H_i \\ H_j \\ H_k \end{pmatrix} \right].$$

3.5 Approximation methods

3.5.1 Preliminaries

Generally, to find an approximate description of an aquifer system's hydraulic behavior, it is enough to know what differential operator complies with the problem at hand and its particular boundary and initial conditions. Then, locally minimize the approximation errors at the entire flow domain's sub-domains – *including its internal and external boundaries* – by one of the main weighted residuals methods: *Collocation, Least-Squares,* and *Galerkin.*[1] For relatively simple flow problems, these methods help to find *global approximants* for the entire flow domain. They help find specific local approximants for contiguous sub-domains for complex groundwater flow domains, as done by some mathematical routines such as *Finite Differences, Finite Element,* or *Boundary Elements.*

The next example helps to understand how a cumbersome analytical solution may be replaced by a simpler one without severe accuracy loss for practical purposes.

Example 3.2

Suppose a continuous and constant excess load is suddenly placed at the top of a fully saturated, deformable, and normally consolidated horizontal soil stratum. In that case, the original aquifer hydrostatic pressure distribution grows up immediately as the soil pore water takes up the induced excess of vertical stress due to that top-loading. By Terzaghi's solution for 1D consolidation, this pressure excess is transferred gradually from the pore water to the soil grains. This gradual pressure transfer is theoretically described by the degree of consolidation as a function of the time elapsed after the top loading. This parameter, denoted by U, customarily expressed as a percentage, is related by another normalized parameter, the time factor T, by the following expressions:

$$U = \left[1 - \sum_{m=0}^{\infty} \frac{\left(2 \int_0^{2H} u_i \sin\left(\frac{M z}{H} \right) dz \right) e^{-M^2 T}}{M \int_0^{2H} u_i \, dz} \right] 100.$$

In the above formula T is calculated by:

$$T = \frac{c_V t}{H^2}$$

In these formulas, u_i stands for the excess initial pore pressure at the center of the soil strata, c_v for the coefficient of vertical consolidation (experimentally determined by laboratory tests), H for the shortest escape path for the pressurized groundwater, and t for the time elapsed since the start of the process.

However, from U values between 0% and 60%, this percentage can be conveniently calculated by the simple approximate formula U(T):

$$U = \sqrt{\frac{4T}{\pi}}$$

3.5.2 *Collocation method*

The Collocation Method may be used when hydraulic heads and its transforms by the differential operators are, respectively, known at m and m_V points within the flow domain. Also, when Dirichlet and Neumann conditions are, respectively, known at m_D and m_N points at boundaries. Then, if the errors ε, ε_V, ε_N, and ε_D for the chosen n-degree approximant are equated to zero, the resulting system of n simultaneous equations gives its numerical coefficients. If $m + m_V + m_N + m_D > n + 1$, this system may be solved by a best fit model. Considering an interpolation polynomial, then the system of simultaneous equations is symbolically written as:

$$\varepsilon = \sum_{i=0}^{n} \left[H f_i(\mathbf{r},t) \right] - H(\mathbf{r},t) = 0 \quad \text{for all m points within the flow domain} \tag{3.20}$$

$$\varepsilon_V = D_V \sum_{i=0}^{n} \left[H f_i(\mathbf{r},t) \right] - Q(\mathbf{r}) = 0 \quad \text{for all } m_V \text{ points within the flow domain} \tag{3.21}$$

$$\varepsilon_N = D_N \sum_{i=0}^{n} \left[H f_i(\mathbf{r},t) \right] - q_n(\mathbf{r}) = 0 \quad \text{for all } m_N \text{ points at the Neumann boundaries} \tag{3.22}$$

$$\varepsilon_D = \sum_{i=0}^{n} \left[H f_i(\mathbf{r},t) \right] - H_D(\mathbf{r}) = 0 \quad \text{for all } m_D \text{ points at the Dirichlet boundaries} \tag{3.23}$$

A similar system may be written for a polynomial approximation.

The next example clarifies the essence of this method.

Example 3.3

The table below assembles data from piezometers PA, PB, PC, and PD installed to measure the water table elevation on a restricted area of an unconfined aquifer seeping over an almost impervious non-horizontal bottom.

Date of observation 2004–10–28	*South coordinate E (m)*	*North coordinate N (m)*	*Aquifer bottom B (m)*	*Hydraulic head H (m)*
Piezometer PA	1.26E+03	3.75E+03	31.331	62.214
Piezometer PB	2.26E+03	3.65E+03	29.719	59.592
Piezometer PC	1.76E+03	2.85E+03	29.569	67.758
Piezometer PD	1.66E+03	3.35E+03	30.581	73.878

The hydraulic heads inside the influence area of these piezometers were previously estimated by a simple linear interpolation (see Figure 3.3):

A polynomial approximation by the Collocation Method may produce the best picture of the head contour map. Based on local geological evidence, the elevation B of the aquifer bottom is approximated by a planar surface defined by the equation:

$$B = b_0 + b_1 E + b_2 N$$

Known elevation levels B of the aquifer bottom at the four piezometers axes give four algebraic equations to give the coefficients b_0, b_1, and b_2:

$$b_0 + 1.3 \times 10^3 b_1 + 3.7 \times 10^3 b_2 - 31.331 = 0$$

$$b_0 + 2.3 \times 10^3 b_1 + 3.6 \times 10^3 b_2 - 29.719 = 0$$

$$b_0 + 1.8 \times 10^3 b_1 + 2.8 \times 10^3 b_2 - 29.569 = 0$$

$$b_0 + 1.7 \times 10^3 b_1 + 3.3 \times 10^3 b_2 - 30.581 = 0$$

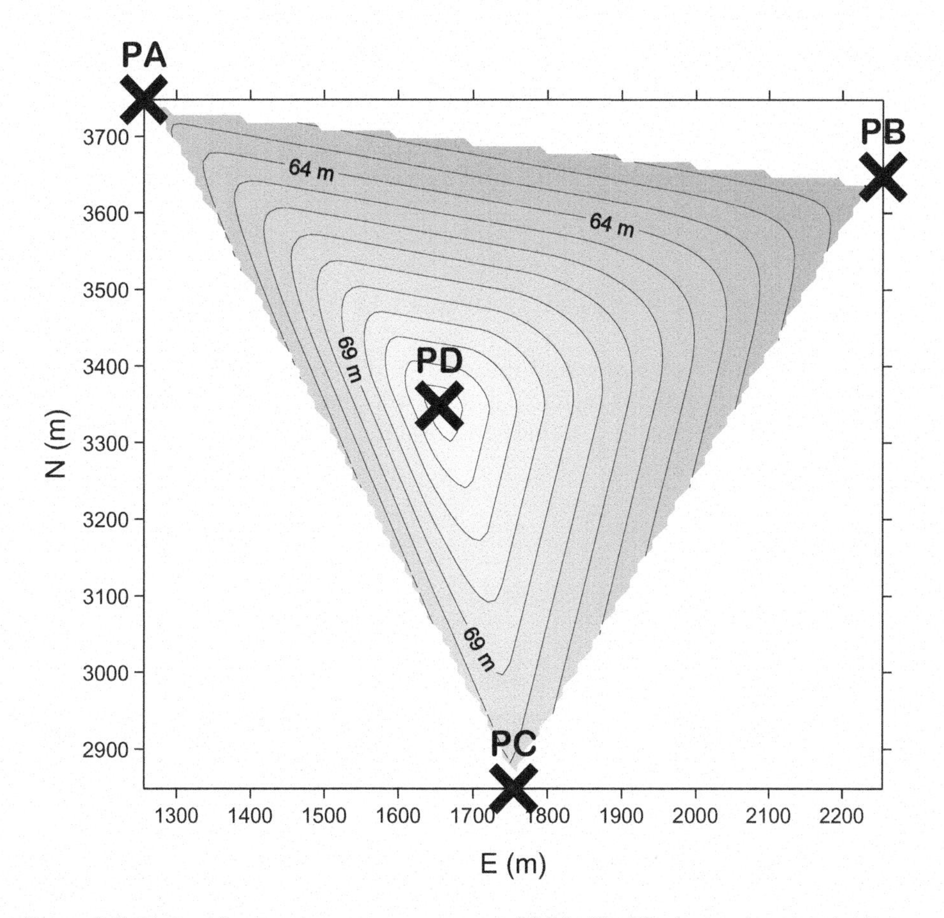

Figure 3.3 Hydraulic head contour map on 2004–10–28 drawn by simple linear interpolation of the monitored head values at the piezometers PA, PB, PC, and PD.

Minimizing the residuals of the b coefficients in these four simultaneous equations by the Least Squares method gives their best estimate:

$$b = \begin{pmatrix} 28.998 \\ -1.5\times 10^{-3} \\ 1.125\times 10^{-3} \end{pmatrix}$$

Next, the variable hydraulic head B+H may be estimate by another polynomial approximation:

$$B + H = a_0 + a_1 E + a_2 N + a_3 E^2 + a_4 N^2 + a_5 E N$$

The known hydraulic heads H at the four piezometers give four equations relating the six unknown coefficients a_0, a_1, a_2, a_3, a_4, and a_5:

$$a_0 + 1.3\times10^3 a_1 + 3.7\times10^3 a_2 + 1.6\times10^6 a_3 + 1.4\times10^7 a_4 + 4.7\times10^6 a_5 - 62.214 = 0.$$

$$a_0 + 2.3\times10^3 a_1 + 3.6\times10^3 a_2 + 5.1\times10^6 a_3 + 1.3\times10^7 a_4 + 8.2\times10^6 a_5 - 59.592 = 0.$$

$$a_0 + 1.8\times10^3 a_1 + 2.8\times10^3 a_2 + 3.1\times10^6 a_3 + 8.1\times10^6 a_4 + 5.0\times10^6 a_5 - 67.758 = 0.$$

$$a_0 + 1.7\times10^3 a_1 + 3.3\times10^3 a_2 + 2.7\times10^6 a_3 + 1.1\times10^7 a_4 + 8.2\times10^6 a_5 - 73.878 = 0.$$

Assuming constant recharge w and invariant and isotropic hydraulic conductivity k and defining the normalized recharge as $R_n = w/k$, the general continuity equation of an unconfined aquifer over a variable non-horizontal impervious bottom is (see Chapter 2, Section 2.3.7.3.2):

$$\left[\begin{array}{l}\dfrac{\partial}{\partial E}\left[H\left[\left[\dfrac{\partial}{\partial E}(B+H)\right]\right]\right]\ldots \\ +\dfrac{\partial}{\partial N}\left[H\left[\left[\dfrac{\partial}{\partial E}(B+H)\right]\right]\right]\end{array}\right] + R_n = 0$$

Substituting B and B+H in the above equation for their corresponding polynomial approximations and performing the derivations yields the residual of the continuity equation:

$$\begin{aligned}\varepsilon_V = &(a_1 + 2a_3 E + a_5 N - b_1)(a_1 + 2a_3 E + a_5 N)\ldots \\ &+2\left(a_0 + a_1 E + a_2 N + a_3 E^2 + a_4 N^2 + a_5 EN - b_0 - b_1 E - b_2 N\right)a_3\ldots \\ &+(a_2 + 2a_4 N + a_5 N - b_2)(a_2 + 2a_4 N + a_5 E)\ldots \\ &+2\left(a_0 + a_1 E + a_2 N + a_3 E^2 + a_4 N^2 + a_5 EN - b_0 - b_1 E - b_2 N\right)a_4\ldots \\ &+\frac{\omega}{k}\end{aligned}$$

Equating ε_V to zero at the locations of the four piezometers give four additional equations:

$$\begin{aligned}&\left(a_1 + 2.6\times10^3 a_3 + 3.7\times10^3 a_5 - 1b_1\right)\left(a_1 + 2.6\times10^3 a_3 + 3.7\times10^3 a_5\right)\ldots = 0 \\ &+\left(2a_0 + 2.6\times10^3 a_1 + 7.4\times10^3 a_2 + 3.4\times10^6 a_3 + 2.8\times10^7 a_4 + 9.6\times10^6\right. \\ &\quad \left. a_5 - 2b_0 - 2.6\times10^3 b_1 - 7.4\times10^3 b_2 a_3\ldots\right) \\ &+\left(a_2 + 7.4\times10^3 a_4 + 1.3\times10^3 a_5 - 1b_2\right)\left(a_2 + 7.4\times10^3 a_4 + 1.3\times10^3 a_5\right)\ldots \\ &+\left(2a_0 + 2.6\times10^3 a_1 + 7.4\times10^3 a_2 + 3.4\times10^6 a_3 + 2.8\times10^7 a_4 + 9.6\times10^6\right. \\ &\quad \left. a_5 - 2b_0 - 2.6\times10^3 b_1 - 7.4\times10^3 b_2 a_4 + R_n\right)\end{aligned}$$

$$\left(a_1+4.6\times10^3 a_3+3.6\times10^3 a_5-1b_1\right)\left(a_1+4.6\times10^3 a_3+3.6\times10^3 a_5\right)\ldots=0$$
$$+\left(2a_0+4.6\times10^3 a_1+7.2\times10^3 a_2+1.1\times10^7 a_3+2.6\times10^7 a_4+1.7\times10^7\right.$$
$$\left.a_5-2b_0-4.6\times10^3 b_1-7.2\times10^3 b_2 a_3\ldots\right)$$
$$+\left(a_2+7.2\times10^3 a_4+2.3\times10^3 a_5-1b_2\right)\left(a_2+7.2\times10^3 a_4+2.3\times10^3 a_5\right)\ldots$$
$$+\left(2a_0+4.6\times10^3 a_1+7.2\times10^3 a_2+1.1\times10^7 a_3+2.6\times10^7 a_4+1.7\times10^7\right.$$
$$\left.a_5-2b_0-4.6\times10^3 b_1-7.2\times10^3 b_2\right)a_4+R_n$$

$$\left(a_1+3.6\times10^3 a_3+2.8\times10^3 a_5-1b_1\right)\left(a_1+3.6\times10^3 a_3+2.8\times10^3 a_5\right)\ldots=0$$
$$+\left(2a_0+3.6\times10^3 a_1+5.6\times10^3 a_2+6.4\times10^6 a_3+1.6\times10^7 a_4+1.0\times10^7\right.$$
$$\left.a_5-2b_0-3.6\times10^3 b_1-5.6\times10^3 b_2\right)a_3\ldots$$
$$+\left(a_2+5.6\times10^3 a_4+1.8\times10^3 a_5-1b_2\right)\left(a_2+5.6\times10^3 a_4+1.8\times10^3 a_5\right)\ldots$$
$$+\left(2a_0+3.6\times10^3 a_1+5.6\times10^3 a_2+6.4\times10^6 a_3+1.6\times10^7 a_4+1.0\times10^7\right.$$
$$\left.a_5-2b_0-3.6\times10^3 b_1-5.6\times10^3 b_2\right)a_4+R_n$$

$$\left(a_1+3.4\times10^3 a_3+3.3\times10^3 a_5-1b_1\right)\left(a_1+3.4\times10^3 a_3+3.3\times10^3 a_5\right)\ldots=0$$
$$+\left(2a_0+3.4\times10^3 a_1+6.6\times10^3 a_2+5.8\times10^6 a_3+2.2\times10^7 a_4+1.1\times10^7\right.$$
$$\left.a_5-2b_0-3.4\times10^3 b_1-6.6\times10^3 b_2\right)a_3\ldots$$
$$+\left(a_2+6.6\times10^3 a_4+1.7\times10^3 a_5-1b_2\right)\left(a_2+6.6\times10^3 a_4+1.7\times10^3 a_5\right)\ldots$$
$$+\left(2a_0+3.4\times10^3 a_1+6.6\times10^3 a_2+5.8\times10^6 a_3+2.2\times10^7 a_4+1.1\times10^7\right.$$
$$\left.a_5-2b_0-3.4\times10^3 b_1-6.6\times10^3 b_2\right)a_4+R_n$$

In these equations, the normalized recharge R_n stands for the ratio w/k. Minimizing the residuals of these eight simultaneous equations by the Least Squares method gives the best estimate of the normalized recharge R_n and of the a_i coefficients:

$$a=\begin{pmatrix}116.944\\-5.172\times10^{-3}\\-0.02\\-3.138\times10^{-6}\\9.865\times10^{-7}\\3.5\times10^{-6}\end{pmatrix}$$

$$R_n=6.832\times10^{-5}$$

Then, the hydraulic heads H (expressed in meters) at the nodes of a 10×10 m equally spaced grid covering the influence area of these piezometers is estimated by the matrix equation:

$$H(E,N) = \begin{pmatrix} a_0 & a_1 & a_2 & a_3 & a_4 & a_5 \end{pmatrix} \begin{pmatrix} 1 \\ E \\ N \\ E^2 \\ N^2 \\ EN \end{pmatrix}$$

In the above expression, the numerical coefficients (a_0, a_1, a_2, a_3, a_4, a_5), as well as the variables (E, N), have compatible units. The following matrix assembles the calculated head values at the nodes of a 100×100 m grid:

70.082	69.223	68.381	67.554	66.743	65.948	65.17	64.407	63.66	62.929	62.214
69.743	68.916	68.104	67.309	66.53	65.766	65.019	64.288	63.572	62.873	62.19
69.341	68.545	67.765	67.002	66.254	65.522	64.806	64.106	63.422	62.754	62.103
68.876	68.112	67.363	66.631	65.915	65.215	64.53	63.862	63.209	62.573	61.953
68.348	67.616	66.899	66.198	65.513	64.844	64.192	63.555	62.934	62.329	61.74
67.758	67.057	66.371	65.702	65.049	64.412	63.79	63.185	62.596	62.022	61.465
67.105	66.435	65.781	65.144	64.522	63.916	63.326	62.752	62.194	61.653	61.127
66.389	65.751	65.129	64.522	63.932	63.358	62.799	62.257	61.731	61.22	60.726
65.61	65.004	64.413	63.838	63.279	62.737	62.21	61.699	61.204	60.725	60.262
64.769	64.194	63.635	63.091	62.564	62.053	61.557	61.078	60.615	60.167	59.736
63.865	63.321	62.793	62.282	61.786	61.306	60.842	60.394	59.962	59.547	59.147

A hydraulic head contour map can be constructed by interpolating between these grid nodes (see Figure 3.4):

3.5.3 *Least squares method*

The Collocation Method is quite simple, but it does not restrict the magnitude of ε, ε_V, ε_N, and ε_D outside the collocation points. A natural extension of this method is to equate to zero the sum of the residuals within sub-domains. However, this approach may conceal quite large residuals and lead to computational difficulties (Linz, 1979). The Least Squares Method overcomes that inconvenience by minimizing the sum of squared errors.

Consider first how to find an approximation for $H(\mathbf{r}, t)$ if some of its values are known at few points. In this case, the number and the type of base functions $h_i(\mathbf{r}, t)$ of the approximant $h(\mathbf{r}, t)$ are previously elected (according to the expected physical behavior to be described and to the desired accuracy). Then, the sum of the squared

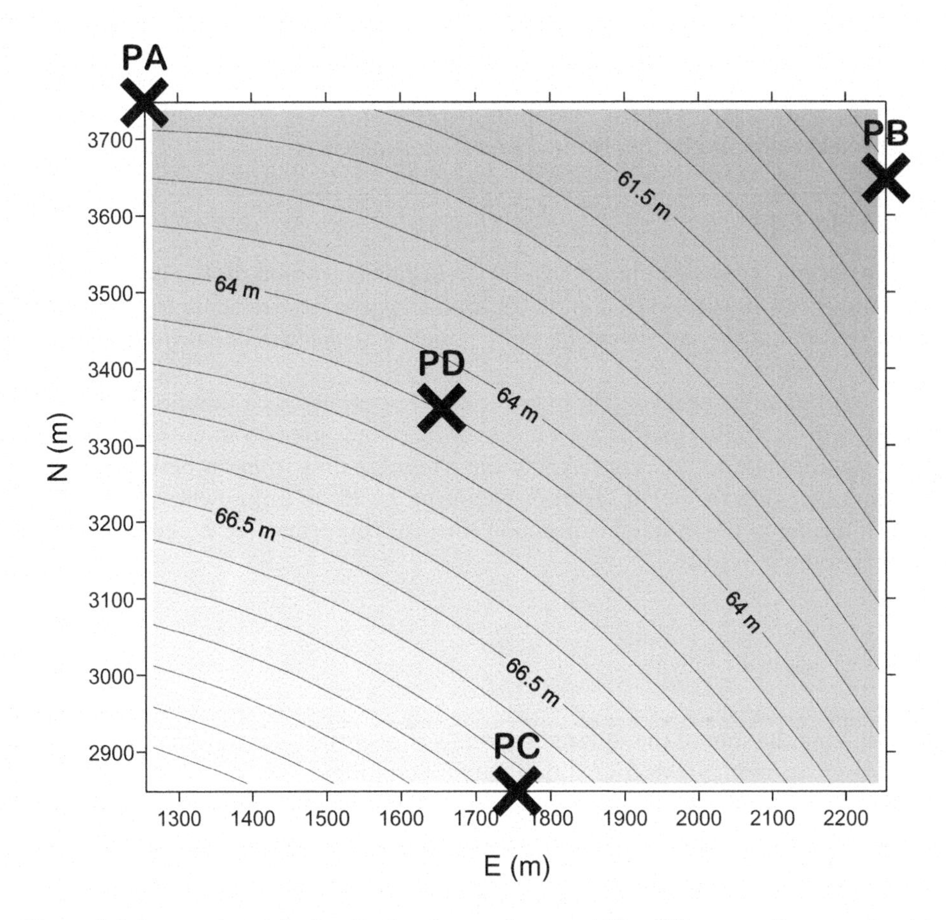

Figure 3.4 Interpolated hydraulic head map from a 100 × 100 m equally spaced grid nodes of calculated head values.

residuals ε^2 only depends on arbitrary variations of its numerical coefficients a_j. It is sufficient to find which set of coefficients a_j minimizes this sum within the flow domain and its boundaries. This set is the solution of the simultaneous equations below (see Section 3.6.1).

$$\int_0^{t_f}\int_0^{V}\left[\sum_{i=0}^{n}\left[a_i\, h_i(\mathbf{r},t)\right]-H(\mathbf{r},t)\right]\left[h_j(\mathbf{r},t)\right]dV\,dt=0 \quad \text{for all j elementary solutions } h_j(\mathbf{r},t) \tag{3.24}$$

$$\int_0^{t_f}\int_0^{S}\left[\sum_{i=0}^{n}\left[a_i\, h_i(\mathbf{r},t)\right]-H(\mathbf{r},t)\right]\left[h_j(\mathbf{r},t)\sum_{k=1}^{3}\alpha_k\right]dS\,dt=0$$

$$\text{for all j elementary solutions } h_j(\mathbf{r},t) \tag{3.25}$$

In the last constraint, the coefficients α_k denote the direction cosines of the outward-normal n to the boundary element dS. It must be emphasized that as the analytic expression for $H(\mathbf{r}, t)$ is unknown, the integrals $\int H(\mathbf{r}, t)$ must be found numerically from the known values of $H(\mathbf{r}, t)$ at monitoring points.

The next example clarifies this method for a simple case.

Example 3.4

The following graph (see Figure 3.5) shows the time-variation of the zinc content, from September 30, 2004 to March 31, 2005, found in groundwater samples collected each week (but not at the same weekday) in a point located 259 m downstream from an industrial process.

Vales of dissolved Zn content in groundwater samples, as a function of the elapsed time t, can be approximated by a cubic polynomial fitted to monitored data by the least square method, but considering the irregular time interval between successive samples (varying from 1 to 7 days). Denoting by P_j its unknown numerical coefficients, a general polynomial expression for j varying from 0 to N_p can be written as (is this example $N_{p=4}$):

$$Zn = \sum_{j=0}^{N_p-1} \left(P_j t^j\right)$$

Minimizing the sum of the squared residuals ε^2 over the total timespan, considering a cubic approximant, yields four simultaneous equations:

$$\int_0^T \left[\sum_{j=0}^{N_p-1} \left(P_j t^j\right)\right] t^0\, dt - \sum_{i=0}^{N-2} \left[\left(\frac{data_i + data_{i+1}}{2}\right)\left(t_{i+1} - t_i\right)\left(\frac{t_{i+1} + t_i}{2}\right)^0\right] = 0$$

$$\int_0^T \left[\sum_{j=0}^{N_p-1} \left(P_j t^j\right)\right] t^1\, dt - \sum_{i=0}^{N-2} \left[\left(\frac{data_i + data_{i+1}}{2}\right)\left(t_{i+1} - t_i\right)\left(\frac{t_{i+1} + t_i}{2}\right)^1\right] = 0$$

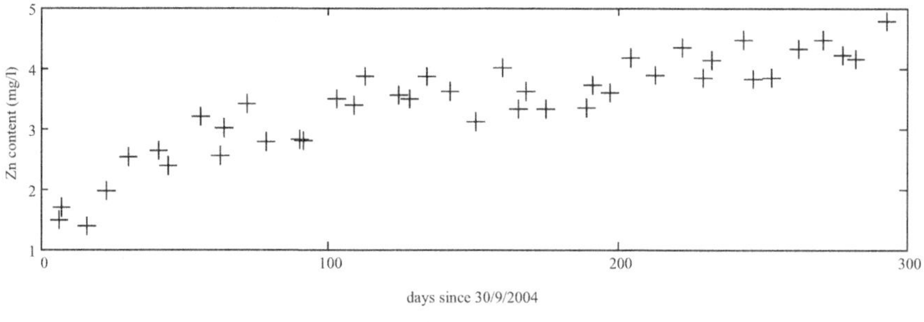

Figure 3.5 Zinc content time-variation monitored in a control point. Faulty sampling, defective analysis, or rain infiltration dilution may explain the disproportionate data spread.

$$\int_0^T \left[\sum_{j=0}^{N_p-1} \left(P_j t^j \right) \right] t^2 dt - \sum_{i=0}^{N-2} \left[\left(\frac{data_i + data_{i+1}}{2} \right) (t_{i+1} - t_i) \left(\frac{t_{i+1} + t_i}{2} \right)^2 \right] = 0$$

$$\int_0^T \left[\sum_{j=0}^{N_p-1} \left(P_j t^j \right) \right] t^3 dt - \sum_{i=0}^{N-2} \left[\left(\frac{data_i + data_{i+1}}{2} \right) (t_{i+1} - t_i) \left(\frac{t_{i+1} + t_i}{2} \right)^3 \right] = 0$$

In the above equations, T denotes the time elapsed since sampling began. The following expression in these equations corresponds to trivial numerical integrations multiplied by the weighting factors t^0, t^1, t^2, and t^3, but considering the irregular time interval δt between consecutive Zn values.

From a structural point of view, these four simultaneous equations have a precise meaning. The first equation expresses the equivalence of the approximant areas and by the known solution (in this case, defined by discrete values, despite its lousy accuracy). The second equation forces the centers of gravity of these areas to be equally distant from the ordinate axis (equivalence or first moments). The third one requires equivalence of inertia moments (second moments) and the fourth, of third moments.

Solving these simultaneous equations gives the unknown numerical coefficients of the cubic approximant:

$$P = \begin{pmatrix} 0.908 \\ 0.314 \\ -0.012 \\ 1.627 \times 10^{-4} \end{pmatrix}$$

Figure 3.6 shows the resulting approximant.

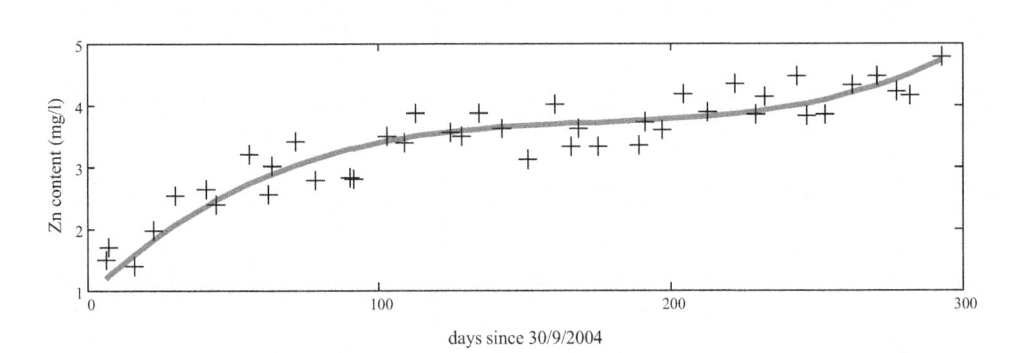

Figure 3.6 Cubic polynomial fitted to the scattered data, taking into account the irregular variation of the time interval δt between successive Zn contents values.

The preceding simultaneous equations can be written using the symbolic continuous inner product notation, and they may also be combined and pondered by weighting coefficients w_V and w_S:

$$w_V\left\langle\left[\sum a_i h_i(\mathbf{r},t)-H(\mathbf{r},t)\right], h_i(\mathbf{r},t)\right\rangle_{V,t} - w_S\left\langle\left[\sum a_i h_i(\mathbf{r},t)-H(\mathbf{r},t)\right], h_i(\mathbf{r},t)\right\rangle_{S,t} = 0$$

These minimum constraints within the flow domain or the minimum constraints at the flow domain boundaries are exactly satisfied by
 the above equations. How well they approximate each location's solution depends on the relative magnitude of the weighting coefficients w_V and w_S.

The Least Squares Method also holds when the differential operators D_V and D_N are applied to the residuals $[h(\mathbf{r}, t)-H(\mathbf{r}, t)]$. In this case, the unknown coefficients a_j after minimizing the sum of the squared residuals $\varepsilon_V{}^2$ and $\varepsilon_N{}^2$ corresponds to the solution of the following simultaneous equations:

$$\int_0^{tf}\int_0^{V}\left[D_V\sum_{i=0}^{n}\left[a_i h_i(\mathbf{r},t)\right]-Q(\mathbf{r},t)\right]\left[h_j(\mathbf{r},t)\right]dV\,dt = 0$$

for all j elementary solutions $h_j(\mathbf{r},t)$ (3.26)

$$\int_0^{tf}\int_0^{S}\left[D_N\sum_{i=0}^{n}\left[a_i h_i(\mathbf{r},t)\right]-q_n(\mathbf{r},t)\right]\left[h_j(\mathbf{r},t)\sum_{k=1}^{3}\alpha_k\right]dS\,dt = 0$$

for all j elementary solutions $h_j(\mathbf{r},t)$ (3.27)

Both requirements can also be combined (occasionally using weighting coefficients). Then, using the convenient inner product notation, these conditions can be written as:

$$\left\langle D_V\sum a_i h_i(\mathbf{r},t)-Q(\mathbf{r},t), h_i(\mathbf{r},t)\right\rangle_{V,t} - \left\langle D_N\left[\sum a_i h_i(\mathbf{r},t)-q_n(\mathbf{r},t)\right], h_i(\mathbf{r},t)\right\rangle_{S,t} = 0 \quad (3.28)$$

The next example clarifies this method.

Example 3.5

Parallel linear drains buried under an embankment must be designed to keep the rising water table caused by the infiltration of 75% of a continuous and extremely exceptional 100 mm rain for four days, below a predetermined maximum level of 0.5 m, corresponding to an average infiltration rate of 18.75 mm/day.

Being T = 4 days, a quadratic polynomial H(x, t, T) may approximate the rise of the water as a product $F_1(x)\cdot F_2(t)$. In this case, x is the distance measured from the drains centerline and t is the elapsed time since the start of the continuous rain:

$$H(x, t, T) = \left(a_0 + a_1 x + a_2 x^2\right)\sin\left(\frac{\pi}{2}\frac{t}{T}\right) + H_L$$

In the above formula, H_L is the vertical distance from the embankment impervious base to the drain base level. Obviously, the units of the numerical coefficients must be compatible, that is, a_0: [L], a_1: [–], a_2: [L^{-1}], H_L: [L] and x: [L].

Flow symmetry implies zero horizontal discharge at the drain's centerlines, that is, $q_n = 0$, at $x = 0$, corresponding to the following Neumann condition at $x = 0$:

$$\left[-k\left[\frac{\partial}{\partial x}\left(H + H_L\right)\right]\right]_{x=0} = 0$$

Substituting H for the selected approximation:

$$\left[-k\left[\frac{\partial}{\partial x}\left(a_0 + a_1 x + a_2 x^2\right)\sin\left(\frac{\pi}{2}\frac{t}{T}\right) + H_L\right]\right]_{x=0} = 0$$

Taking the derivate:

$$\left[-k\left(a_1 + 2a_2 x\right)\sin\left(\frac{1}{2}\pi\frac{t}{T}\right)\right]_{x=0} = 0$$

From which $a_1 = 0$.

The draining capacity is designed to assure zero hydraulic pressure at the drain centers. Then, the following Dirichlet condition at $x = L$ is:

$$\left[\left(a_0 + a_2 L^2\right)\sin\left(\frac{\pi}{2}\frac{t}{T}\right) + H_L\right]_{x=L} = H_L$$

From which $a_2 = -a_0/L^2$.

By replacing a_1 and a_2 for their values, the approximant reduces to:

$$H(x, t, T) = a_0\left(1 - \frac{1}{L^2}x^2\right)\sin\left(\frac{1}{2}\pi\frac{t}{T}\right) + H_L$$

Now, the Least Squares Method can be employed to find a_0. The unconfined Dupuit's continuity equation in (x, t) and the Neumann condition at (L, t) for a 1 m wide strip are, respectively:

$$\frac{\partial}{\partial x}\left[\rho H\left[-k\frac{\partial}{\partial x}\left(H + H_L\right)\right]\right] + \frac{\partial}{\partial t}\left(n\,\rho\,H\right) - \rho\,\omega = 0$$

$$\left[-k\frac{\partial}{\partial x}\left[\left(H + H_L\right)m\right] - q_n\right]_{x=L} = 0.$$

To effectively drain q_n must balance all water accretion ωL, that is, $q_n = \omega L$.

Now, replacing the last expression H(x, t, T) in the Dupuit and Neumann conditions give:

$$-4\rho\frac{(a_0)^2}{L^4}x^2\sin\left(\frac{1}{2}\pi\frac{t}{T}\right)k + 2\rho(a_0)^2\left(1-\frac{1}{L^2}x^2\right)\sin\left(\frac{1}{2}\pi\frac{t}{T}\right)^2\frac{k}{L^2}\ldots = 0.$$
$$+\frac{1}{2}n\,\rho a_0\left(1-\frac{1}{L^2}x^2\right)\cos\left(\frac{1}{2}\pi\frac{t}{T}\right)\frac{\pi}{T} - \rho\omega$$

$$2\rho k\frac{a_0\,m}{L^2}\times\sin\left(\frac{1}{2}\pi\frac{t}{T}\right) - \rho\omega L = 0.$$

Then, minimizing the squared residuals within the flow domain and the drain boundaries, making $h_{j=1}$ and combining these two requisites with equal weighting coefficients:

$$\int_0^T\left[\int_0^L\left[\begin{array}{l}-4\rho\frac{(a_0)^2}{L^4}x^2\sin\left(\frac{1}{2}\pi\frac{t}{T}\right)k + 2\rho(a_0)^2\left(1-\frac{1}{L^2}x^2\right)\sin\left(\frac{1}{2}\pi\frac{t}{T}\right)^2\frac{k}{L^2}\ldots\\ +\frac{1}{2}n\,\rho a_0\left(1-\frac{1}{L^2}x^2\right)\cos\left(\frac{1}{2}\pi\frac{t}{T}\right)\frac{\pi}{T} - \rho\omega\end{array}\right]dx\right]dt\ldots = 0$$
$$+\int_0^T\left(2\rho k\frac{a_0}{L^2}L\sin\left(\frac{1}{2}\pi\frac{t}{T}\right)m - \rho\omega L\right)dt$$

Evaluating these two integrals gives:

$$\frac{2}{3}\rho L n a_0 - 2\rho\omega L T + \frac{4}{\pi}T\rho k\frac{a_0\,m}{L} = 0.$$

Solving for a_0:

$$a_0 = 3\omega L^2 T\frac{\pi}{L^2 n\pi + 6Tkm}$$

Finally, substituting a_0 in the last expression for H(x, t, T):

$$H(x,t,T) = 3\omega L^2 T\frac{\pi}{L^2 n\pi + 6Tkm}\left(1-\frac{1}{L^2}x^2\right)\sin\left(\frac{\pi}{2}\frac{t}{T}\right) + H_L$$

Figure 3.7 shows results for progressive time-periods.

3.5.4 Galerkin's method

3.5.4.1 Orthogonality

A "vector" of an n-dimensional vector space has n components. Similarly, a linear function h(x) can also be regarded as a "vector" of a ∞-dimensional function space, having an infinite number of components h(x) defined for x varying from a to b. This

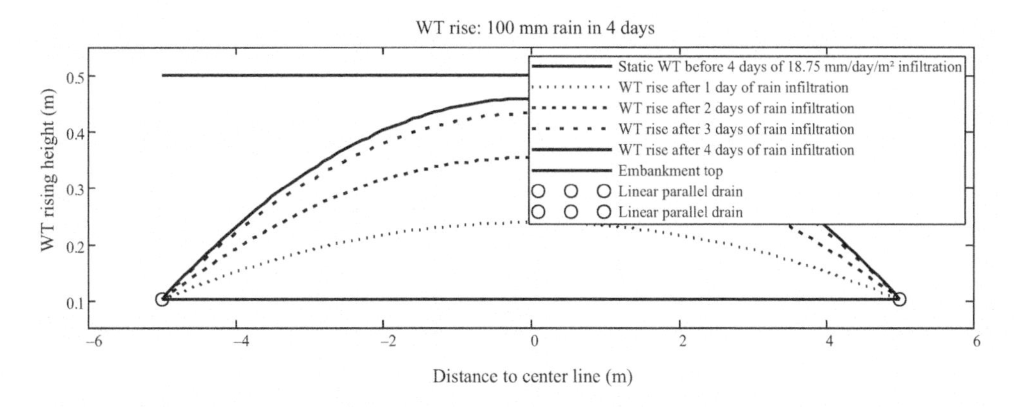

Figure 3.7 Water table rising controlled by parallel linear buried drains.

concept can be easily extended from 1D argument x to the 4D argument (**r**, t). Hence, the solution H(**r**, t) of a linear partial differential equation and its approximant h(r, t) can be considered as "vectors" of "function spaces" in which a "plane" P can be defined by the "vector" h(**r**, t), for randomly changes of its numerical coefficients a_j. If the "vector" H does not belong to this "plane" P, then there exists a unique "vector" h(**r**, t) on P, called the "projection" of H(**r**, t) on P, such as that the " approximant error vector" ε is perpendicular to P and takes its lowest minimum, that is, its infimum (see Figure 3.8). Then, h(**r**, t) is the best approximant to H(**r**, t) in the sense of the Minimum Least Squares method (see Section 3.6.2).

Similar reasoning can be applied for the transformed "error vectors" ε_V, ε_N, and ε_D. However, the Least Squares Method does not guarantee that the "error vectors" ε,

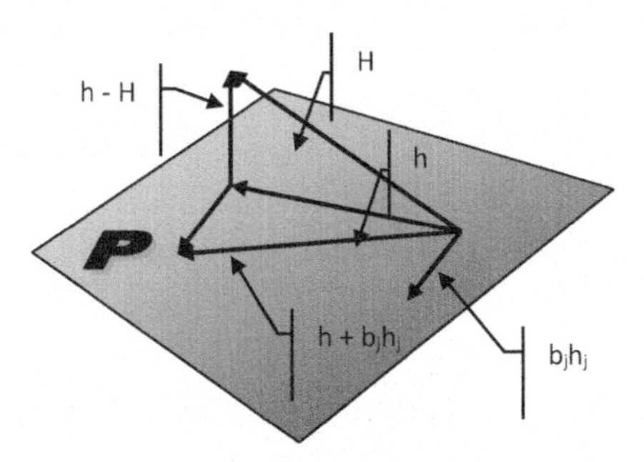

Figure 3.8 "Visualization" of the orthogonality requirement to find the best approximant h(**r**, t) to a function H(**r**, t) in the sense of the Minimum Least Squares Method. Let the "plane" P be defined in the "function space" by the "vectors" h(**r**, t) or, cast as polynomial approximant, by the "vectors" $\sum a_j h_j(\mathbf{r}, t)$ as its numerical coefficients a_j change arbitrarily. For a particular set of values a_j the "vector" h(**r**, t) may match the "orthogonal projection" of H(**r**, t) on P, and in this moment, the "approximant error vector" $\varepsilon = [h(\mathbf{r}, t) - H(\mathbf{r}, t)]$ takes its lowest minimum, *that is, its infimum* value.

ε_V, ε_N, and ε_D are orthogonal to the same "plane" P nor that the "orthogonalization" of one kind of error implies the "orthogonalization" of the others. Additionally, the Least Squares Method requires a continuous approximant up to the second derivative. These insufficiencies are coped by Galerkin's Method, as discussed next.

3.5.4.2 Galerkin's approach

By Galerkin's Method, the coefficients a_j of the best approximant $\sum a_j h_j(\mathbf{r},t)$ are given by a set of simultaneous equations expressing the orthogonality condition for ε, ε_V, ε_N, and ε_D to a "plane" Φ defined by an arbitrary set of linearly independent base "vectors" $\phi_j(\mathbf{r}, t)$:

$$\left\langle \sum a_i h_i(\mathbf{r},t) - H(\mathbf{r},t),\ \phi_j(\mathbf{r},t) \right\rangle_{V,t} - \left\langle \left[\sum a_i h_i(\mathbf{r},t) - -H(\mathbf{r},t) \right],\ \phi_j(\mathbf{r},t) \right\rangle_{S,t} = 0 \quad (3.29)$$

$$\left\langle D_V \sum a_i h_i(\mathbf{r},t) - Q(\mathbf{r},t),\ \phi_j(\mathbf{r},t) \right\rangle_{V,t} - \left\langle D_N \left[\sum a_i h_i(\mathbf{r},t) - q_n(\mathbf{r},t) \right],\ \phi_j(\mathbf{r},t) \right\rangle_{S,t} = 0 \quad (3.30)$$

The "plane" Φ does not necessarily match the "plane" P defined by the approximant "vectors" $\sum a_j h_j(\mathbf{r},t)$ and its transforms by the operators D_V and D_N (see Figure 3.9). However, that "geometric congruence" may be forced if the set of "base vectors" $\{\phi_j(\mathbf{r}, t)\}$ that define the "plane" Φ is constructed by sequentially integrating the "elementary" solutions $h_j(\mathbf{r}, t)$.

Besides, the forcing sinks or sources $Q(\mathbf{r}, t)$ and the boundary inflows or outflows $q(\mathbf{r}, t)_n$ can also be approximated by linear combinations of these solutions:

$$D_V \sum a_i \phi_j(\mathbf{r},t) - Q(\mathbf{r},t) \approx D_V \sum a_i \phi_j(\mathbf{r},t) - \sum b_k \phi_k(\mathbf{r},t) \quad (3.31)$$

$$D_N \sum a_i \phi_j(\mathbf{r},t) - q_n(\mathbf{r},t) \approx D_N \sum a_i \phi_j(\mathbf{r},t) - \sum c_k \phi_k(\mathbf{r},t) \quad (3.32)$$

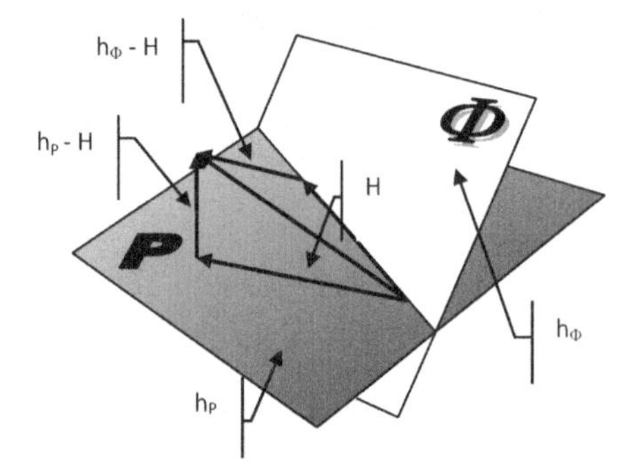

Figure 3.9 "Visualization" of the *orthogonality* requirement to find the best approximant $h(\mathbf{r}, t)$ to a function $H(\mathbf{r}, t)$ by the Galerkin's Method. The "plane" Φ does not necessarily coincide with the "plane" P defined by the approximant "vectors" $h(\mathbf{r}, t)$ and its transforms by the operators D_V and D_N.

In this case, it can be proved that if the orthogonality condition holds for the approximant error ε, then it also holds for the transformed errors ε_V, ε_N, and ε_D.

Next example clarifies the Galerkin's method.

Example 3.6

The variation of the hydraulic head inside a homogenous core of a rockfill dam may be fairly described by a bilinear approximant referred to a non-zenithal frame (see Figure 3.10):

$$h = a_0 x + a_1 z + a_z x$$

The free surface practically matches the x-axis, where the falling head can be approximated by $-x\cos(i)$. Then, a simple orthogonality requirement for the Dirichlet condition may be applied along the free surface:

$$\left\langle \left[(a_0 x + a_1 z + a_2 x) - (-x\cos(i)) \right], 1 \right\rangle_{\text{free surface}} = 0$$

$h = a_0 x + a_1 z + a_z x z$

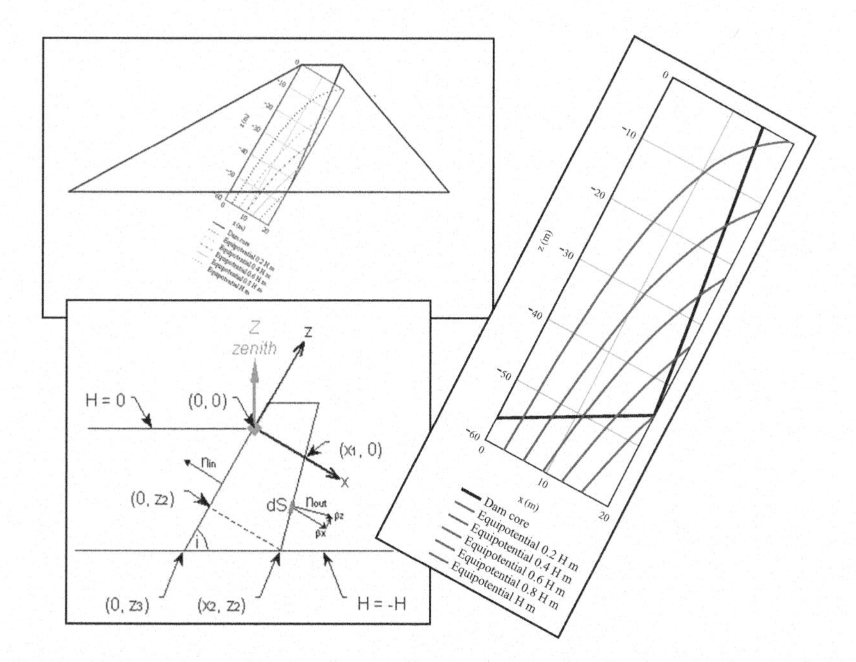

Figure 3.10 Top left sketch: approximated head h contours inside the core. Bottom left sketch: adopted non-zenithal frame xz. The head zero reference level coincides with the water table. The total head difference at the downstream bottom of the core equals $-H$. The points $(0, 0)$, $(x_1, 0)$, $(0, z_3)$, and (x_2, z_2) define the core geometry. The upstream core outward normal equals n_{in}. The outward normal of the seepage face equals n_{out}. The upstream core inclination equals i. The seepage inclination equals β. Right sketch: detail of the head h contours for this example.

Alternatively, written in full:

$$\int_0^{x_1}\left[a_0x + a_1z + a_2xz - x\left(-\cos(i)\right)\right]dx = 0$$

Integrating and considering that $z=0$ at the free surface gives $a_0 = -\cos(i)$. Applying a similar reasoning to the core upstream face where the hydraulic head is zero (equal to the head reference level):

$$\left\langle\left[\left(a_0x + a_1z + a_2xz - 0\right)\right],1\right\rangle_{\text{core upstream face}} = 0$$

Alternatively, written in full:

$$\int_0^{z_3}\left(-\cos(i)x + a_1z + a_2xz\right)dz = 0$$

Integrating and making $x=0$ at the upstream core face gives $a_1=0$. Then, the polynomial approximation reduces to:

$$h = -\cos(i)x + a_2xz$$

In this non-zenithal frame, the total head at any point (x, z) is (See Chapter 2, 2.3.8.2):

$$h = \alpha_x x + \alpha_z z + \frac{P}{\rho g}$$

$$\alpha_x = \cos(Z,x) = \cos(\pi - i) = -\cos(i)$$

$$\alpha_z = \cos(Z,z) = \cos\left(\frac{\pi}{2} - i\right) = \sin(i)$$

At the seepage face, the water seeps out into the free atmosphere. Then, the directional gradient $\partial P/\partial r$ must be parallel to the outward normal n_{out}. Thus, the components of its direction cosines are $\beta_x = \cos(n_{out},x)$ and $\beta_z = \cos(n_{out},z)$:

$$\frac{\frac{\partial}{\partial x}\left[\left(-\cos(i)x + a_2xz - \alpha_x x - \alpha_z z\right)\rho g\right]}{\beta_x} - \frac{\frac{\partial}{\partial z}\left[\left(-\cos(i)x + a_2xz - \alpha_x x - \alpha_z z\right)\rho g\right]}{\beta_z} = 0.$$

Taking the derivatives and simplifying:

$$\left(-\cos(i) + a_2z - \alpha_x\right)\frac{1}{\beta_x} - \left(a_2x - \alpha_z\right)\frac{1}{\beta_z} = 0.$$

The head at the upstream seepage face is z sin (i). Applying a mixed orthogonality requirement at that boundary:

$$\left\langle\left[\left(a_0x+a_1z+a_2xz\right)-\left(z\,\sin(i)\right)\right],1\right\rangle_{\text{seepage face}}\ldots$$
$$+\left\langle\left[\left(-\cos(i)+a_2z+\alpha_x\right)/\beta_x-\left(a_2x+\alpha_z\right)/\beta_z\right],1\right\rangle_{\text{seepage face}}=0$$

Alternatively, written in full:

$$\left[\int_{S_1}^{S_2}\left[\left(-\cos(i)x+a_2xz\right)-\left(z\,\sin(i)\right)\right]dS\right]\ldots$$
$$+\left[\int_{S_1}^{S_2}\left[\left(-\cos(i)+a_2z-\alpha_x\right)\frac{1}{\beta_x}-\left(a_2x-\alpha_z\right)\frac{1}{\beta_z}\right]dS\right]=0.$$

Integrating and solving for a_2 (considering the geometrical relationships between units, levels, point coordinates, direction cosines of the upstream outward normal, face inclinations):

$$a_2=3\left[\frac{(z_3)^2(\beta_z)^4\cos(i)+\left[\begin{array}{l}\left[\left[(\sin(i))(z_3)^2+2z_3\cos(i)x_1\right]\beta_x+2z_2\cos(i)+2z_2\alpha_x\right](\beta_z)^3\ldots\\+\left(2x_1\alpha_z-2x_2\alpha_z\right)(\beta_x)^2\end{array}\right]}{2(z_3)^2(\beta_z)^4+\left[3(z_3)^2x_1\beta_x+3(z_2)^2\right](\beta_z)^3+\left[-3(x_2)^2+3(x_1)^2\right](\beta_x)^2}\right]$$

Substituting a_2 in the approximate head gives an analytic solution. The derivative $\partial h/\partial x$ of the analytic expression gives the exit gradient in the x-direction (see Figure 3.11):

$$h=a_2-\cos(i)x$$

3.5.4.3 "Weak" solutions

In the "finite element method," the flow domain is partitioned into small contiguous sub-domains. For each sub-domain, appropriately called "finite element," corresponds a local approximant of the governing equation's solution, subjected to the finite element natural boundary conditions. To guarantee convergence, as the finite elements' size decreases, the inter-element continuity for their first and second-order derivatives of the local approximants should be preserved. However, full continuity

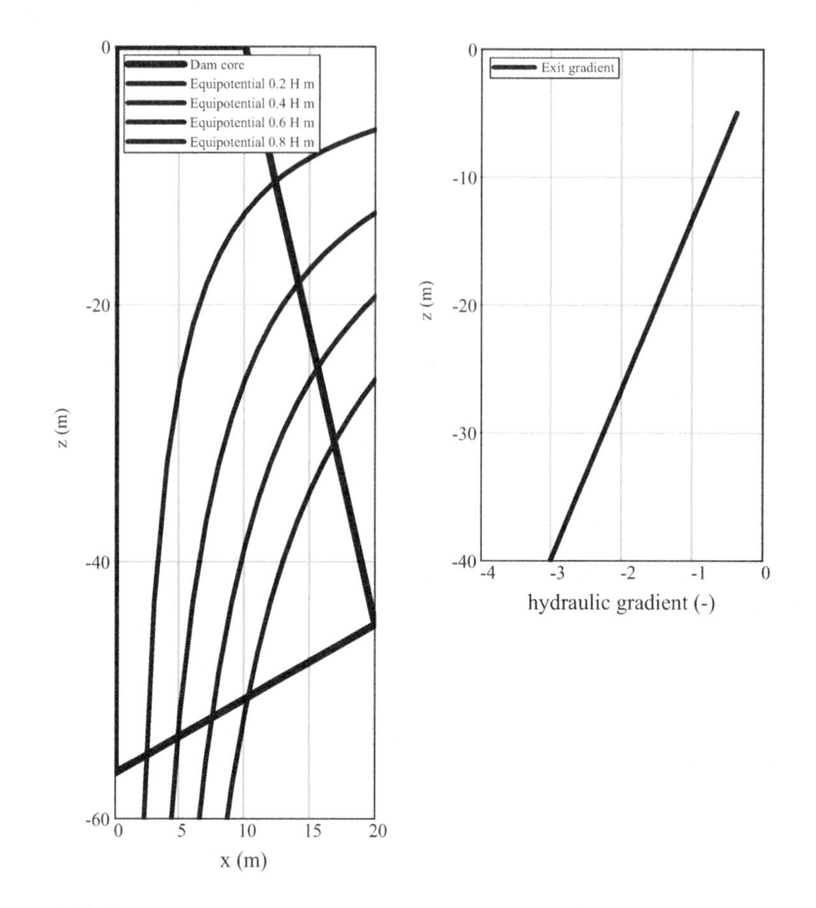

Figure 3.11 Exit gradient in the x-direction and estimated head contours for ($x_I = 15$ m, $z_I = 0$ m), ($x_I = 20$ m, $z_I = -45$ m), and ($x_I = 0$ m, $z_I = -56.55$ m). (Note: Rectangular coordinates xy, as shown in Figure 3.10.)

constraint cannot be applied to approximants that are only continuous up to the first derivative. Then, avoid that difficulty, it is convenient to transform the differential form of that constraint as to eliminate the second-order derivative from the first inner product. To do this, consider the full orthogonality constraint:

$$\left\langle \left[D_V \cdot \sum a_i h_i - Q \right], h_j \right\rangle_{V,t} - \left\langle \left[D_N \cdot \sum a_i h_i - q_n \right], h_j \right\rangle_{S,t} = 0$$

Then, integrating by parts the terms $D_V \cdot \sum a_i h_i$ and $D_N \cdot \sum a_i h_i$ of the above requirement, the resulting transformed orthogonality condition can be applied to approximants that need to be "continuous" only up to the first derivative:

$$\left\langle \sum k_k \cdot \partial h/\partial x_k, \text{grad} \cdot \left(h_j\right) \right\rangle_{V,t} - \left\langle S_V \cdot \partial h/\partial t, h_j \right\rangle_{S,t} + \left\langle Q, h_j \right\rangle_{V,t} - \left\langle q_n, h_j \right\rangle_{S,t} = 0 \quad (3.33)$$

These modified approximants, called "weak solutions," do not satisfy the partial differential equation but only its equivalent integral formulation. Moreover, while a "weak" solution firmly imposes the Dirichlet condition, it only approaches the Neumann condition in the limit, and as an average.

3.5.4.4 Variational notation

If $\delta h(\mathbf{r}, t)$ denotes the *total variation* of an approximant $h(\mathbf{r}, t)$ due to arbitrary *variations* of its unknown coefficients a_i, it follows that:

$$\delta h = \delta a_0 h_0 + \delta a_1 h_1 + \ldots + \delta a_n h_n = \sum_{i=0}^{n} (\delta a_i h_i)$$

On the other hand, the first orthogonality constraint, as shown below, remains unaffected if it is multiplied by an arbitrary variation δa_j of any coefficient a_j:

$$\delta a_j \cdot \left\langle \left[\sum a_i h_i - H \right], h_j \right\rangle = \left\langle \left[\sum a_i h_i - H \right], \delta a_j \cdot h_j \right\rangle = 0$$

Adding up all these j constraints and recalling that $\delta h = \sum \delta a_j \cdot h_j$:

$$\sum \delta a_j \cdot \left\langle \left[\sum a_i h_i - H \right], h_j \right\rangle = \left\langle \left[\sum a_i h_i - H \right], \sum \delta a_j \cdot h_j \right\rangle = \left\langle \left[\sum a_i h_i - H \right], \delta h \right\rangle = 0$$

This sort of integrated notation can be applied to all types of orthogonality requirements. For "weak solutions" it is written usually as follows:

$$\left\langle \sum k_k \cdot \partial h / \partial x_k, \text{grad} \cdot (\partial h) \right\rangle_{V,t} - \langle S_V \cdot \partial h / \partial t, \partial h \rangle_{S,t} + \langle Q, \partial h \rangle_{V,t} - \langle q_n, \partial h \rangle_{S,t} = 0 \quad (3.34)$$

"Variational notation" is employed for orthogonality requirements formulated in terms of maxima and minima of *functionals* (functions of functions), that is, in terms of the principles of the *calculus of variations*, one of the oldest mathematical disciplines and developed almost simultaneous with differential and integral calculus.

The next example clarifies the use of the variational notation.

Example 3.7

The hydraulic behavior of an artesian horizontal leaking aquifer (see Figure 3.12), can be described by a 1D steady flow continuity equation and two Dirichlet boundary conditions:

$$-T \frac{\partial^2}{\partial x^2} h + ch = 0$$

$$0 < x < L$$

$$h_{x=0} = A$$

$$h_{x=L} = B$$

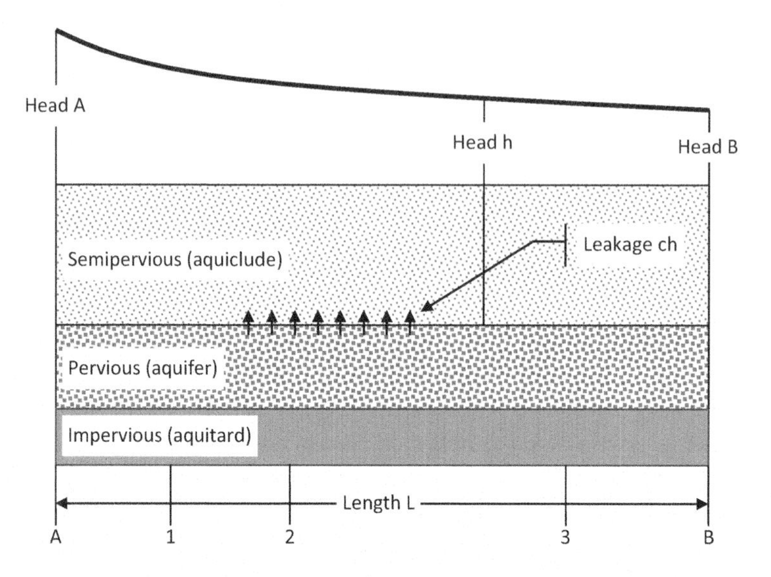

Figure 3.12 Unit width strip of an artesian horizontal leaking aquifer. The upward leakage at the interface aquifer-aquiclude is proportional to the head. Heads at A and B define two *Dirichlet* conditions.

Here, the algebraic symbols' meanings are x: horizontal coordinate, h: hydraulic artesian head, T: aquifer transmissivity, c: leakage proportionality coefficient of the variable upward leakage ch.

Pairs of coordinates x_i and x_j delimit sub-domains $\delta L_{i,\ j}$ (appropriately called "finite elements"). Their head variations can be described locally by "weak solutions," that is, first-order polynomials interpolation instead of twice-differentiable approximants:

$$H = \left[H_i \left[1 - \left(\frac{x - x_i}{x_j - x_i} \right) \right] + H_j \left(\frac{x - x_i}{x_j - x_i} \right) \right].$$

For each finite element, the independent coordinate x can be normalized by the sub-domain length:

$$X = \frac{x - x_i}{x_j - x_i}$$

$$0 \leq X \leq 1$$

Then, the continuity equation takes a simple form:

$$-\frac{\partial^2}{\partial X^2} h + c_{i,j} h = 0$$

Considering the following differentiation rule:

$$dX^2 = \left(\frac{\partial}{\partial X}X\right)^2 = \left(\frac{\partial}{\partial x}\frac{x - x_i}{x_j - x_i}\right)^2 = \frac{1}{\left(x_j - x_i\right)^2}$$

The relationship between the old c and the new constant $C_{i,j}$ is:

$$C_{i,j} = \frac{c}{T}\left(x_j - x_i\right)^2.$$

Similarly, the polynomial interpolation H, the variation δH and their gradients are:

$$H = H_i(1 - X) + H_j X$$

$$\delta H = \delta H_i(1 - X) + \delta H_j X$$

$$\frac{\partial}{\partial X}H = -H_i + H_j$$

$$\frac{\partial}{\partial X}\delta H = -\delta H_i + \delta H_j$$

Thus, applying the Galerkin Method with a variational inner product notation:

$$\left\langle \partial h/\partial X, \partial(\delta H)/\partial X \right\rangle_L - \langle C, \delta H\rangle_V - \langle q_k, \delta H\rangle_{k=i,j} = 0$$

Developing in full:

$$\int_0^1 (-H_1 + H_1)\begin{pmatrix} (-\delta H_i) \\ \delta H_j \end{pmatrix} dX - \int_0^1 C_{i,j}\left[H_1(1 - X) + H_j X\right]\begin{bmatrix} \delta H_i(1 - X) \\ \delta H_j X \end{bmatrix} dX \ldots = 0$$

$$-q_j\begin{bmatrix} \delta H_i(1 - X)_{X=1} \\ \left(\delta H_j\right)_{X=1} \end{bmatrix} - q_j\begin{bmatrix} \delta H_i(1 - X)_{X=0} \\ \left(\delta H_j\right)_{X=0} \end{bmatrix}$$

Integrating the above condition gives a system of two simultaneous equations for each subdomain $i - j$:

$$\begin{bmatrix} \left(1 - \frac{C_{i,j}}{3}\right) & -\left(1 + \frac{C_{i,j}}{6}\right) \\ -\left(1 + \frac{C_{i,j}}{6}\right) & \left(1 - \frac{C_{i,j}}{3}\right) \end{bmatrix}\begin{pmatrix} H_i \\ H_j \end{pmatrix} - \begin{pmatrix} q_i \\ q_j \end{pmatrix} = 0$$

However, as a first-degree approximant gives an average discharge q for each finite element, the terms $-q_i$ and q_j cancel out. Then, applying that result to the four sub-domains A-1, 1–2, 2–3, and 3-B, gives a system of three simultaneous equations for the unknowns H_1, H_2, and H_3, after some matrix algebra manipulation.

Also, instead of imposing two Dirichlet boundary conditions, declaring the same problem for a Dirichlet condition H_A at A and a Neumann condition $\mathbf{q}_B$ at B, solve for the missing Dirichlet condition H_B by the Darcy's Law:

$$q_B = -T\frac{H_B - H_3}{x_B - x_3}$$

This example illustrates the essence of the "finite element method."

3.5.5 *Time-dependent solutions*

Approximate solutions also simulate unsteady groundwater flow. However, as time-boundaries are usually open, except for cyclic behavior, a close functional relationship between time and head variation is not self-evident, except in some circumstances. A common approach is applying a first-order linear approximant for short time intervals and establishing a recursive solution.

The next example illustrates that approach.

Example 3.8

The excess pore pressure dissipation at a horizontal level of the clay core of earth dam can be reasonably modeled by the product of a quadratic space-dependent polynomial by a time-dependent polynomial (see Figure 3.13):

$$u = \left| \begin{array}{l} \left(a_0 + a_1 x + a_2 x^2\right)\left(b_0 + b_1 t\right) \\ -L \le x \le L,\ t \ge 0 \end{array} \right.$$

The starting excess pore pressure u_0 at any point of the centerline of the core base can be estimated by:

$$u_0 = u_{x=0,\ t=0} = \underline{B}\gamma h$$

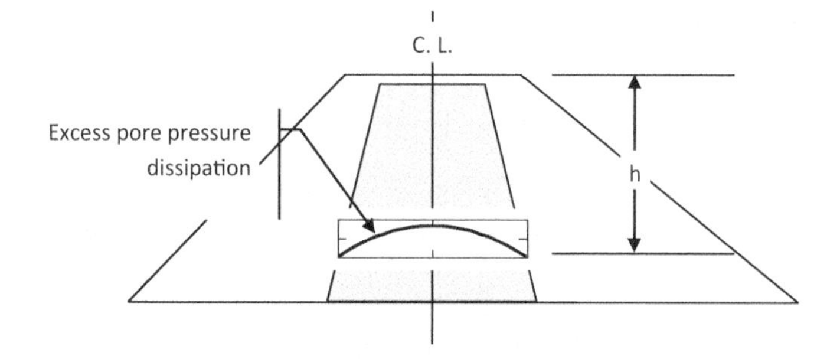

Figure 3.13 Excess pore pressure dissipation within the core of an earth dam following vertical consolidation inducing horizontal drainage.

B (b-bar) denotes an empirical coefficient (ranging from 0.25 to 0.75 for typical core materials), γ the saturated fill unit weight, and h the soil height above the point considered.

The chosen approximant is a second-degree approximant:

$$u = A_0 + A_1x + A_2x^2 + A_3t + A_4xt + A_5x^2t$$

Where the new coefficients A_i are related to the old coefficients a_j and b_k as:

$$\begin{aligned} A_0 &= a_0b_0 \\ A_1 &= a_1b_0 \\ A_2 &= a_2b_0 \\ A_3 &= a_0b_1 \\ A_4 &= a_1b_1 \\ A_5 &= a_2b_1 \end{aligned}$$

Known boundaries and initial conditions help determine four coefficients A_i:

Initial conditions at $x=0$ and $t=0$:

$$\left(A_0 + A_1x + A_2x^2 + A_3t + A_4xt + A_5x^2t\right)_{x=0,\,t=0} = u_0$$

Dirichlet boundary condition at $x=-L$ and $t\geq 0$:

$$\left(A_0 + A_1x + A_2x^2 + A_3t + A_4xt + A_5x^2t\right)_{(x=-L),\,t\geq 0} = 0$$

Dirichlet boundary condition at $x=L$ and $t\geq 0$:

$$\left(A_0 + A_1x + A_2x^2 + A_3t + A_4xt + A_5x^2t\right)_{x=L,\,t\geq 0} = 0$$

Neumann boundary condition at $x=0$ (horizontal slope):

$$\left[\frac{\partial}{\partial x}\left(A_0 + A_1x + A_2x^2 + A_3t + A_4xt + A_5x^2t\right)\right]_{x=0,\,t\geq 0} = 0$$

Solving the preceding equations and substituting the expressions for A_i ($i=0, 1, 2, 3, 4$) in the new approximant yields:

$$u = \left(1 - \frac{x^2}{L^2}\right)\left(u_0 - A_5\frac{t}{L^2}\right)$$

Now, the Galerkin Method may be employed to determine the remaining unknown coefficient A_5.

If c_r symbolizes the experimental radial consolidation coefficient, then the continuity equation for vertical consolidation and horizontal drainage is:

$$-\frac{\partial^2}{\partial x^2}u+\frac{1}{c_r}\left(\frac{\partial}{\partial t}u\right)=0.$$

Substituting u for its approximation in the continuity equation:

$$-\frac{\partial^2}{\partial x^2}\left[\left(1-\frac{x^2}{L^2}\right)\left(u_0-A_5\frac{t}{L^2}\right)\right]+\frac{1}{c_r}\left[\frac{\partial}{\partial t}\left[\left(1-\frac{x^2}{L^2}\right)\left(u_0-A_5\frac{t}{L^2}\right)\right]\right]=0.$$

Deriving yields:

$$\frac{2}{L^2}\left(u_0-A_5\frac{t}{L^2}\right)-\frac{1}{c_r}\left(1-\frac{x^2}{L^2}\right)\frac{A_5}{L^2}=0.$$

Applying a simple orthogonality constraint:

$$\int_0^{t_f}\int_0^{L}\frac{2}{L^2}\left(u_0-A_5\frac{t}{L^2}\right)-\frac{1}{c_r}\left(1-\frac{x^2}{L^2}\right)\frac{A_5}{L^2}dx\ dt=0.$$

Integrating, solving for A_5 and substituting in the new approximant, gives an approximate closed solution for a short interval δt:

$$u=\left(1-\frac{x^2}{L^2}\right)\left[1-\left(\frac{6L^2c_r}{2L^2+3c_rt}\right)\frac{\delta t}{L^2}\right]u_0$$

Applying recursively the above solution yields an expression for $u=u_n/u_0$ after n intervals δt:

$$\frac{u_0}{u_n}=\left(1-\frac{x^2}{L^2}\right)\left[1-\left(\frac{6L^2c_r}{2L^2+3c_rt}\right)\frac{\delta t}{L^2}\right]^n$$

Figure 3.14 shows the graph for the base level of a clay core of an earth dam, showing seven consolidation percentages (%). The radial coefficient of consolidation is $c_r=6\times10^7$ m/s:

Another alternative to cope with unsteady problems is to combine a time-marching "finite difference" algorithm with a "finite element" spatial solution. The next example illustrates the typical technique.

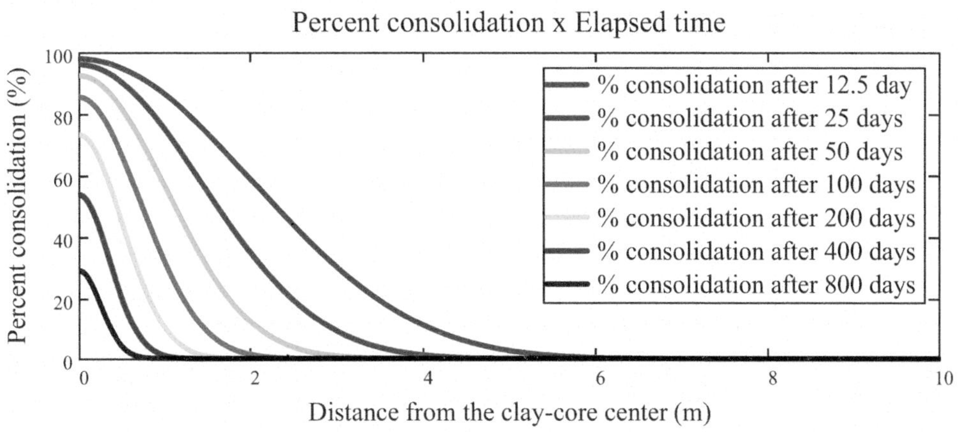

Percent consolidation x Elapsed time

% consolidation after 12.5 day
% consolidation after 25 days
% consolidation after 50 days
% consolidation after 100 days
% consolidation after 200 days
% consolidation after 400 days
% consolidation after 800 days

Percent consolidation (%)

Distance from the clay-core center (m)

Figure 3.14 Percent consolidation at the base of a 20 m wide core of an earth dam, from 12.5 to 800 days (a conservative scenario because consolidation starts during dam construction). Coefficient of consolidation $cv = 6 \times 10^{-7}\,m^2/s$.

Example 3.9

Divide the half-width of the previous example's dam core into four finite elements slabs 0–1, 1–2, 2–3, and 3–4. A "weak" polynomial interpolation may approximate the excess pore pressure dissipation at each finite element slab" as (see Figure 3.15):

$$u_n = \begin{pmatrix} f_j \\ f_{j+1} \end{pmatrix}^T \begin{pmatrix} u_{j,n} \\ u_{j+1,n} \end{pmatrix}$$

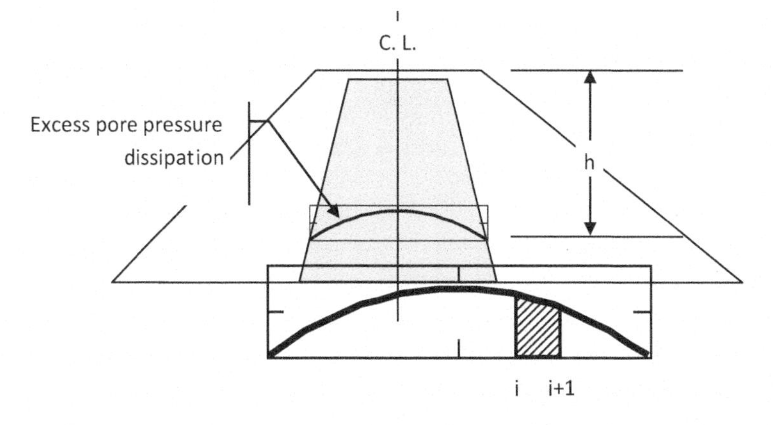

Figure 3.15 Excess pore pressure dissipation within the 20 m wide clay core base of an earth dam. Due to the centerline symmetry, divide the core half-width into four finite elements 0–1, 1–2, 2–3, and 3–4.

In this approximant, the symbols f_j and f_{j+1} denote the interpolating functions at the limits j and j+1 of each finite element:

$$f_j = 1 - \left(\frac{x - x_j}{x_{j+1} - x_j} \right)$$

$$f_{j+1} = \left(\frac{x - x_j}{x_{j+1} - x_j} \right)$$

The capital letters $U_{j,\,n}$, and $U_{j,\,n}$ denote the excess pore pressure dissipation at the points j and j+1 at the moment defined by n·δt where δt is the time interval and n the number of time steps elapsed since the start of the process (initial conditions).

The orthogonality requirement applied to the finite element (j, j+1) gives:

$$\int_{x_j}^{x_{j+1}} \frac{\partial}{\partial x}\left[\begin{pmatrix} f_j \\ f_{j+1} \end{pmatrix}^T \begin{pmatrix} u_{j,\,n} \\ u_{j+1,\,n} \end{pmatrix} \right] \left[\frac{\partial}{\partial x} \begin{pmatrix} f_j \\ f_{j+1} \end{pmatrix} \right]$$

$$dx = \int_{x_j}^{x_{j+1}} \frac{1}{c_r}\frac{\partial}{\partial t}\left[\left[\begin{pmatrix} f_j \\ f_{j+1} \end{pmatrix}^T \begin{pmatrix} u_{j,\,n} \\ u_{j+1,\,n} \end{pmatrix} \right] \begin{pmatrix} f_j \\ f_{j+1} \end{pmatrix} \right] dx \ldots$$

$$+ \left[q_{j+1,\,n} \begin{pmatrix} f_j \\ f_{j+1} \end{pmatrix}_{x_{j+1}} - q_{j+1,\,n} \begin{pmatrix} f_j \\ f_{j+1} \end{pmatrix}_{x_j} \right]$$

As a first-degree approximant only gives an average discharge q for each finite element, the terms $-q_{j,\,n}$ and $q_{j+1,\,n}$ cancel out. Since the two nodal parameters $U_{j,\,n}$, and $U_{j+1,\,n}$ are independent of x they may be out of the integrand. Then, considering the values of f_j and f_{j+1} at $x = x_{j+1}$ and $x = x_j$ and integrating, the above constraint yields:

$$M \begin{pmatrix} u_{j,\,n} \\ u_{j+1,\,n} \end{pmatrix} + N \frac{\partial}{\partial t} \begin{pmatrix} u_{j,\,n} \\ u_{j+1,\,n} \end{pmatrix} = 0$$

Approximating the time-derivative by "finite differences":

$$\frac{\partial}{\partial t} \begin{pmatrix} u_{j,\,n} \\ u_{j+1,\,n} \end{pmatrix} = \frac{1}{\delta t} \begin{pmatrix} u_{j,\,n+1} - u_{j,\,n} \\ u_{j+1,\,n+1} - u_{j+1,\,n} \end{pmatrix}$$

Then, the coefficients M and N are:

$$M = \int_{x_j}^{x_{j+1}} \frac{\partial}{\partial x} \begin{bmatrix} 1 - \left(\dfrac{x - x_j}{x_{j+1} - x_j} \right) \\ \dfrac{x - x_j}{x_{j+1} - x_j} \end{bmatrix}^T \frac{\partial}{\partial x} \begin{bmatrix} 1 - \left(\dfrac{x - x_j}{x_{j+1} - x_j} \right) \\ \dfrac{x - x_j}{x_{j+1} - x_j} \end{bmatrix} dx = \frac{2}{x_{j+1} - x}$$

$$N = \int_{x_j}^{x_{j+1}} \frac{1}{c_r} \begin{bmatrix} 1 - \left(\dfrac{x - x_j}{x_{j+1} - x_j} \right) \\ \dfrac{x - x_j}{x_{j+1} - x_j} \end{bmatrix}^T \begin{bmatrix} 1 - \left(\dfrac{x - x_j}{x_{j+1} - x_j} \right) \\ \dfrac{x - x_j}{x_{j+1} - x_j} \end{bmatrix} dx = \frac{2}{3} \frac{1}{c_r} \left(x_{j+1} - x_j \right)$$

Replacing the time derivative by their finite difference analogs yields the desired time-marching algorithm:

$$\begin{pmatrix} u_{j,\,n+1} \\ u_{j+1,\,n+1} \end{pmatrix} = \left(1 - \frac{M}{N} \delta t \right) \begin{pmatrix} u_{j,\,n} \\ u_{j+1,\,n} \end{pmatrix}.$$

Then, applying that result to the four sub-domains 0–1, 1–2, 2–3 and 3–4, gives a system of three simultaneous time-marching equations for the unknowns H_0, H_1, H_2, and H_3 after some matrix algebra handling.

However, if instead of a string of four finite elements, one chose to apply the time marching algorithm to a single finite element, it is possible to have a more simple preliminary solution. Now, by reasoning as for the previous example, the recurring formula is:

$$u(n) := \left(1 - \frac{3}{L^2} c_r \frac{n\,\delta t}{2} \right)^n$$

Figure 3.16 shows the excess pressure consolidation at the base level and the clay core base.

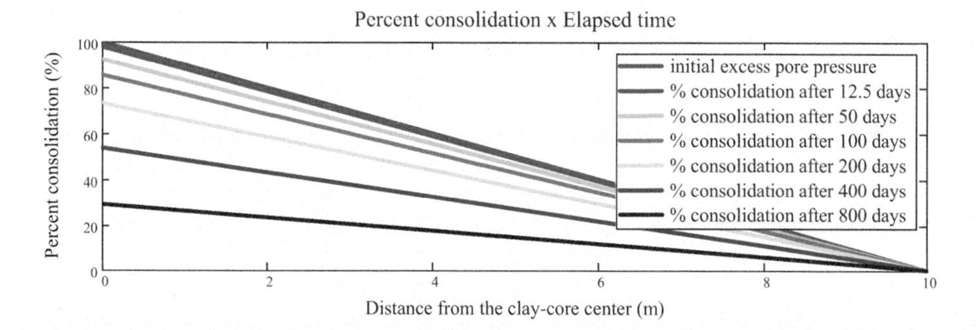

Figure 3.16 The pore pressure decay is almost the same in both examples.

Notice that in this example, the first-degree "weak" finite element approximant, for a coefficient of consolidation $cv = 6 \times 10^{-7} m^2/s$, gave a linear pore pressure graph instead of a curved one, as in the previous example. However, if instead of only one finite element, the modeler solved four simultaneous equations for the four finite elements shown in Figure 3.15, correctly satisfying its boundary conditions, the percent consolidation would correspond to a four-node broken line more comparable to the consolidation line or the previous example.

3.6 Addenda

3.6.1 Classes of second-order partial differential equations

Consider that a dependent variable U is related to n independent coordinates (x_1, $x_2 \ldots x_n$) by a linear second-order partial differential equation. By an appropriate coordinate transform (defined by the solution of an eigenvalue problem), that partial equation loses its mixed derivatives $\partial^2 U/\partial x_i \, \partial x_j \, (i \neq j)$ and gives:

$$\sum_{i=1}^{n}\left(A_i \frac{\partial^2}{\partial x_i^2} U\right) + \sum_{i=1}^{n}\left[B_i\left(\frac{\partial}{\partial x_i} U\right)\right] + CU + D = 0. \quad (3.35)$$

Through any point (x_1, $x_2 \ldots x_n$) of the independent variable space pass one or two peculiar hyper-surfaces, called *characteristics*. For points resting on these hyper-surfaces, the second-order derivatives of U can be mathematically defined but are indeterminate and finite (0/0 indetermination).

For 3D (or 2D) systems, these hyper-surfaces are associated with particular quadric shapes defined by the A_i's coefficients that can be an ellipsoid (or an ellipse), a paraboloid (or a parabola), and a hyperboloid (or a hyperbola). This categorization is generalized for n-dimensional space. These surfaces propagate within the solution domain all information controlled by the boundary conditions. Then, at any moment, these characteristic surfaces separate two distinct but continuously changing subspaces: the *controlled* domains and the *controlling* domains. U's values on the *controlled* domain at any moment depend on the values taken by U on the *controlling* domain at the same moment. Any change of U at one or more points on the influencing domain affect the U values on the conditioned domain. The character of the coefficients A_i of this quadric surface reveals the class of the differential equation:

- Elliptic, if all A_i are non-zero and have the same signal.
- Hyperbolic, if all A_i are non-zero, and have the same signal, but with one exception.
- Parabolic, if one (or more) A_k is (are) zero, but the corresponding B_k is (are) non-zero, and the remaining A_i are non-zero and have the same signal.

As the coefficients, A_i, B_i, C, and D may be functions of the space coordinates x_i and the derivatives $\partial U/\partial x_1$, but not of $\partial^2 U/\partial x_i$, the classification of a linear second-order partial differential equation may also change more than once within the definition domain, mainly for complex systems.

For 3D aquifer flow systems, the dependent variable is the head H, and the independent variables are the time-coordinate t and the space-coordinates (x, y, z). A time-independent solution can be described by an elliptic equation that has no real characteristics and no directional restrictions. In this case, its unique "controlling domain" merges with its unique "controlled domain." As a simple example, consider the steady flow induced by a penetrating well in a confined pervious layer. If this well pumps continuously but preserves a constant drawdown s at the pumping well axis, its pumping rate Q is also constant. However, any change δs in the drawdown s results from a change δQ in the pumping rate and alters the heads H of all points of the confined pervious layer. On the other hand, if s never changes, any head disturbance δH at any point of the confined layer affects all the other heads and pumping rate Q.

For a transient flow regime, described by a parabolic equation, two identical families of real characteristics separate the "controlling domain" from the "controlled domain" by parallel planes. Consider the previous example in an "unsteady" state. Then, at any moment t_n, the "controlled domain" corresponds to points located above the "characteristic" plane (x, y, z, t_n): all future heads for $t>t_n$. On the other hand, the "controlling domain" corresponds to points located below this plane: all past heads for $t<t_n$, defining the history of the system, including its initial values and all precedent boundary values. Therefore, at any moment, t_n, past values "under" the "characteristic" plane (x, y, z, t_n) control all future values "above" this plane.

Hyperbolic 3D equations have two real and distinct characteristics that define two crossing surfaces, separating both influencing and conditioned but confined domains. This type of equation does not occur in groundwater flow.

3.6.2 Minimizing squared residuals

After selecting the most promising approximant, the set of coefficients a_j that minimizes the sum of the squared residuals ε_V^2 within the flow domain satisfies the following condition:

$$\frac{\partial}{\partial a_j}\int_0^{t_f}\int_0^{V}(\varepsilon_V)^2\,dV\,dt = 0 \quad \text{for all } j$$

Or:

$$\int_0^{t_f}\int_0^{V}\varepsilon_V\left(\frac{\partial}{\partial a_j}\varepsilon_V\right)dV\,dt = 0 \quad \text{for all } j \tag{3.36}$$

Substituting, in the last expression, ε_V for $[h(\mathbf{r}, t) - H(\mathbf{r}, t)]$, $\partial\varepsilon_V/\partial a_j$ for $h_j(\mathbf{r}, t)$ and simplifying the resulting equation, one obtains:

$$\int_0^{t_f}\int_0^{V}\left[h(\mathbf{r},t)-H(\mathbf{r},t)\right]\left[h_j(\mathbf{r},t)\right]dV\,dt = 0 \quad \text{for all j elementary solutions } h_j(\mathbf{r},t). \tag{3.37}$$

Writing $h(\mathbf{r}, t)$ as a complete polynomial approximant:

$$\int_0^{tf}\int_0^{V}\left[\sum_{i=0}^{n}\left[a_i h_i(\mathbf{r},t)\right]-H(\mathbf{r},t)\right]\left[h_j(\mathbf{r},t)\right]dV\,dt=0 \quad \text{for all j elementary solutions } h_j(\mathbf{r},t) \tag{3.38}$$

The solution of this system of (n + 1) equations defines the unknown coefficients a_j.

If applied to the flow domain boundaries, this method implies another set of (n + 1) simultaneous equations:

$$\int_0^{tf}\int_0^{S}\left[\sum_{i=0}^{n}\left[a_i h_i(\mathbf{r},t)\right]-H(\mathbf{r},t)\right]\left[h_j(\mathbf{r},t)\sum_{k=1}^{3}\alpha_k\right]dS\,dt=0 \tag{3.39}$$
$$\text{for all j elementary solutions } h_j(\mathbf{r},t)$$

In the last constraint, α_k denotes the direction cosines of the outward-normal n to the boundary element dS.

3.6.3 Minimizing squared residuals changed by differential operators

Transforming the residuals ε_V within the flow domain by the differential operator D_V and the residuals ε_S at its boundaries by differential operator D_S gives:

$$D_V\left[h(\mathbf{r},t)-H(\mathbf{r},t)\right]=D_V\,h(\mathbf{r},t)-Q$$

$$D_N\left[h(\mathbf{r},t)-H(\mathbf{r},t)\right]=D_N\,h(\mathbf{r},t)-q_n$$

Then the set of coefficients a_j minimizing the sum of the squared transformed residuals satisfies two sets of simultaneous equations:

$$\int_0^{tf}\int_0^{V}\left[D_V\sum_{i=0}^{n}\left[a_i h_i(\mathbf{r},t)\right]-Q(\mathbf{r},t)\right]\left[h_j(\mathbf{r},t)\right]dV\,dt=0$$
$$\text{for all j elementary solutions } h_j(\mathbf{r},t)$$

$$\int_0^{tf}\int_0^{S}\left[D_N\sum_{i=0}^{n}\left[a_i h_i(\mathbf{r},t)\right]-q_n(\mathbf{r},t)\right]\left[h_j(\mathbf{r},t)\sum_{k=1}^{3}\alpha_k\right]dS\,dt=0$$
$$\text{for all j elementary solutions } h_j(\mathbf{r},t)$$

3.6.4 The concept of orthogonality

Let the "plane" P be defined in the "function space" by the "vectors" $h(\mathbf{r}, t)$ or, cast as a polynomial approximant, by the "vectors" $\sum a_i h_i(\mathbf{r},t)$ as its numerical coefficients a_j

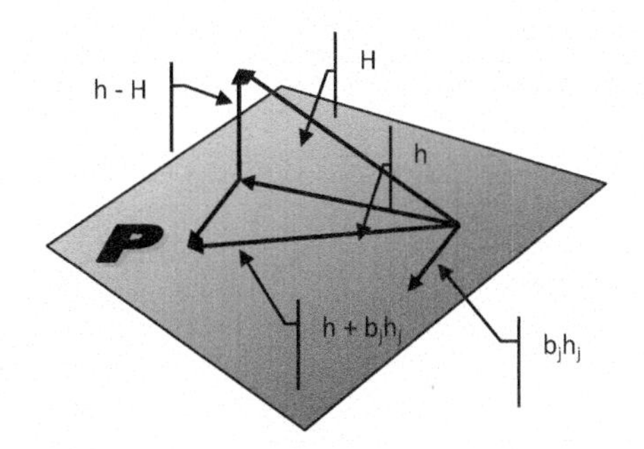

Figure 3.17 For a particular set of values a_j the "vector" $h(\mathbf{r}, t)$ may match the "orthogonal projection" of $H(\mathbf{r}, t)$ on P, and in this moment, the "approximant error vector" $\varepsilon = [h(\mathbf{r}, t) - H(\mathbf{r}, t)]$ takes its lowest minimum, that is, its *infimum* value.

change arbitrarily (see Figure 3.17). Let $(h + b_j h_j)$ be any other "vector" on the "plane" P. If $(h - H)$ is the *infimum*, then the "length" of the "error vector" $(h - H)$ is smaller than the "length" of the vector $[(h + b_j h_j) - H]$ and the following inequality prevails:

$$\left\langle \left[\left(h + b_j h_j\right) - H\right], \left[\left(h + b_j h_j\right) - H\right] \right\rangle - \left\langle (h - H), (h - H) \right\rangle \geq 0 \tag{3.40}$$

Taken into account that $\|x\|^2 = \langle x, x \rangle$ and simplifying the preceding inequality:

$$b_j^2 \left\langle h_j, h_j \right\rangle - 2b_j \left\langle (h - H), h_j \right\rangle \geq 0 \tag{3.41}$$

As that result must hold for arbitrary values of b_j it necessarily implies a formal orthogonality constraint defined by a null inner product:

$$\left\langle (h - H), h_j \right\rangle = 0 \tag{3.42}$$

Note

1 Boris Grigorievich Galerkin, 1871–1945, a Russian mathematician.

References

Chung-Yau, L., 1994, *Applied Numerical Methods for Partial Differential Equations*, Prentice Hall, New York.

Band, W., 1960, *Introduction to Mathematical Physics*, D. Van Nostrand, Princeton.

Linz, P., 1979, *Theoretical Numerical Analysis*, Dover, New York.

Chapter 4

Data analysis

4.1 Introduction

Modeling the spatial distribution of a flow domain's hydraulic properties is not an easy task as it reflects the distribution of 3D geologic features, commonly inferred from insufficient combinations of 2D field mapping and 1D logs. The non-conservative nature of geological features rules out simple conservation laws, such as mass and energy, which could guide the evaluation of this distribution. However, dense distributions of hydraulic heads in hard-rock terrains allow valuable inferences of their hydraulic properties. Furthermore, known eigenvalues and eigenvectors of hydraulic conductivities variation at specific areas correlate with descriptors of lithologic and structural features in other areas, allowing interpolations of their values. However, if the "modeler" bypasses this critical step and replaces the existing flow domain with a homogeneous and isotropic one, his numerical model has little value, despite its computational merits.

Also, as with any other geoscience topic, a hydrogeologic analysis must occasionally be associated with other information types. Hydrologic, geomorphologic, geochemical new data may clarify some obscure questions or bypass some methodologic limitations.

4.2 Analyzing geological data

Geologists traditional tools to describe and evaluate structural features main parameters can be correlated with hydraulic tests. For example, traditional histograms that depict orientation and spacing distributions of several kinds of discontinuities may correlate with observed head contours inferred from monitored data (see Figures 4.1 and 4.2; Franciss, 1994).

4.3 Analyzing hydrologic data

In the Vazante Mine, Minas Gerais St., Brazil, summer rainfall periods and persistent droughts periods alternate each year. However, precipitation volumes seemed to be gradually decreasing. Then it was decided to analyze 40-year regional precipitation data, monitored from 1975 to 2015, to verify what was going on.

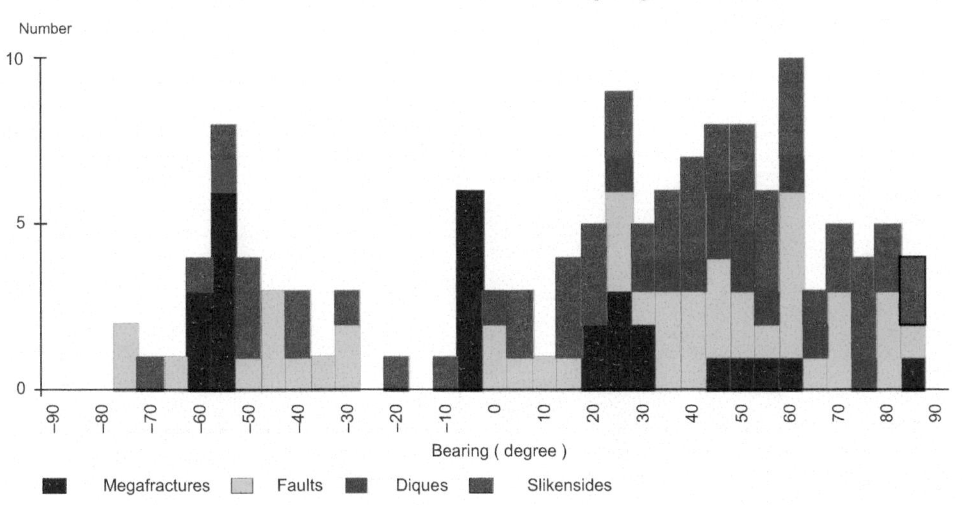

Figure 4.1 Statistical distribution of the orientation of the significant structures affecting the Brazilian Pre-Cambrian Shield (Almirante Barroso Maritime Terminal-TEBAR, São Sebastião, SP, Brazil): sub-vertical diabase dikes traverse a vast gneissic complex; most rock masses conform to typical migmatites; general foliation plunge is N 324°/27° (dip/dip direction); depth of the weathered cover attains 30–40 m; occasionally, deep altered zones, locally related to crossings of megafractures, faults, and dikes, are found below the 100 m depth.

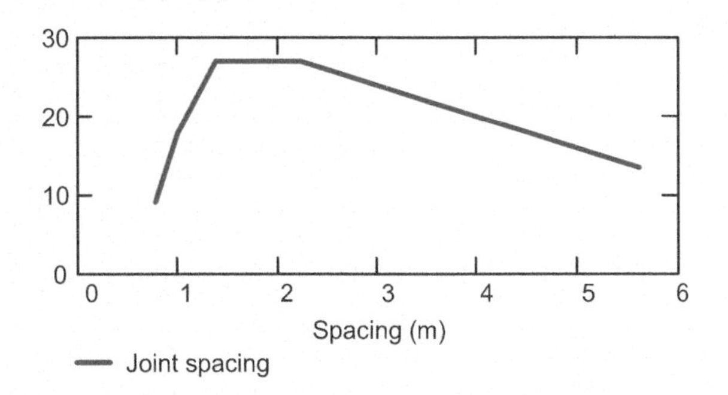

Figure 4.2 Statistical distribution of joint spacing based on scan line counts for the site mentioned in Figure 4.1. Modal value around 1.8 m. Two dominant joint trends, one parallel and other normal do the strike of the foliation, are N 27°/86° and N 136°/82° (dip/dip direction). Secondary joints traverse foliation at low angles.

Data came from six pluviometric stations: Cristalina, Paracatu, Guarda-Mor, Vazante, Coromandel, and Patos de Minas (five in MG St. and two in G St., see Figures 4.3 and 4.4).

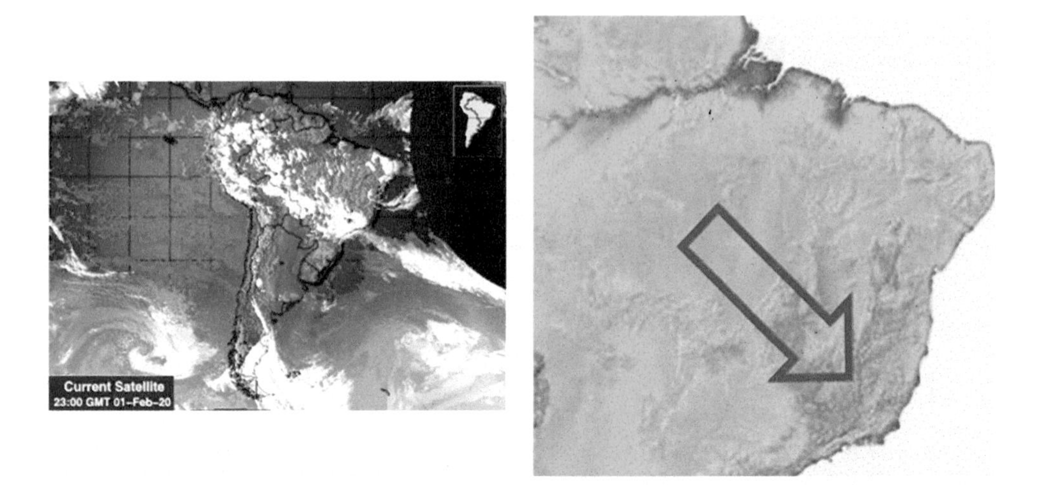

Figure 4.3 Usual pattern of high clouds generated at the Amazon realm moving southeast to the Atlantic coast during the rainy season. Illustration modified from NASA Public Information.

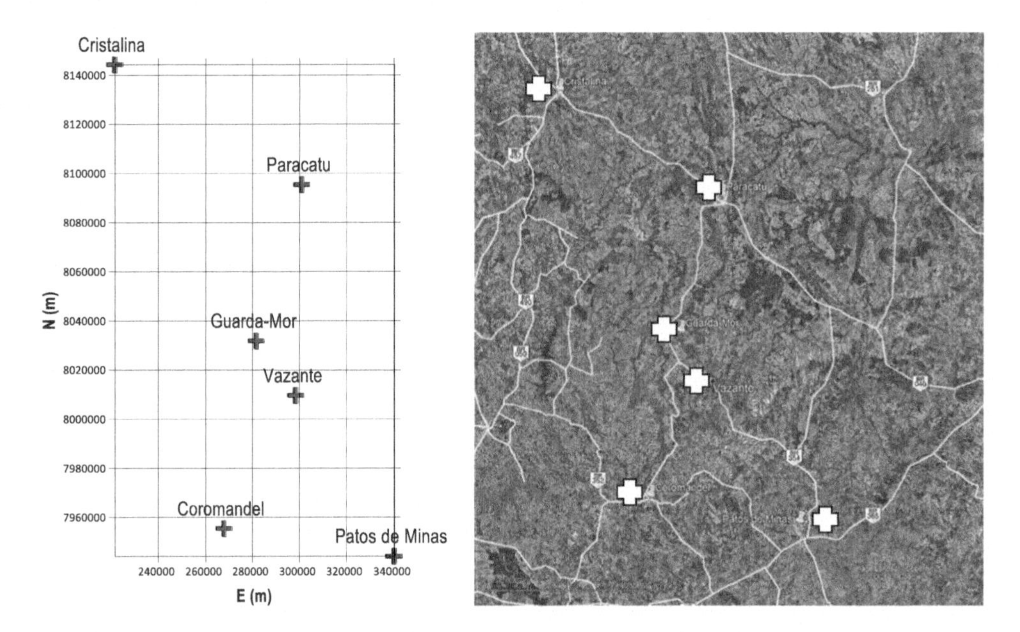

Figure 4.4 Location of the six rainfall gages in Cristalina, Paracatu, Guarda-Mor, Vazante, Coromandel, and Patos de Minas.

However, instead of evaluating based on discrete time series of variable data, as shown in Figure 4.5, a continuous approach was chosen as more significant. Six exponential trend functions replaced the six yearly discrete rainfall data (see Figure 4.6).

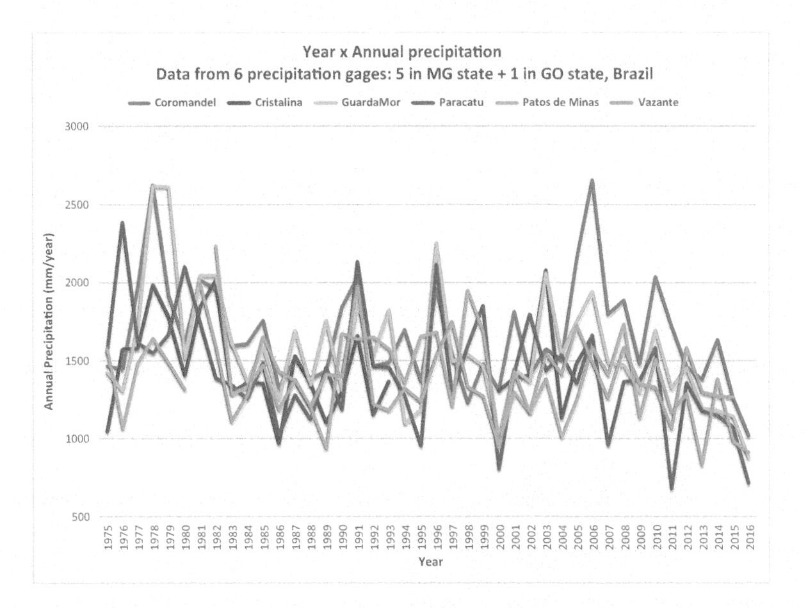

Figure 4.5 Precipitation time-series recorded in the six station-gages from 1975 to 2016. Annual precipitation in mm/year kriged from the recorded 2015 rainfall data.

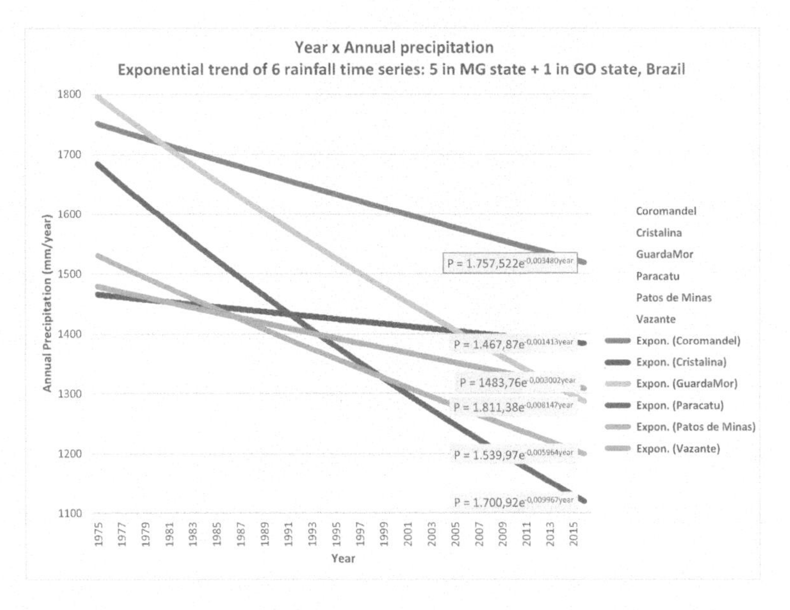

Figure 4.6 Exponential trend of the precipitation time-series recorded in the six station-gages from 1975 to 2015.

The chosen trend functions had the type $P=Ae^{Bt}$, where A and B designate empirical parameters. This model, traditionally used in temporal series analysis of natural events, always gives slightly different results as new data becomes available. It is precisely that property that encourages its use, gradually weighing data closer to the present while reducing past moments' importance.

Thus, it was possible to elaborate 40 yearly precipitation charts, from 1975 to 2015, over the polygon defined by the exterior gage-stations, having a total area of $1{,}07^{10}$ m^2. Then, it was noted that the first and the last charts for the 1975–2015 period (40 years) were quite different (see Figures 4.7 and 4.8).

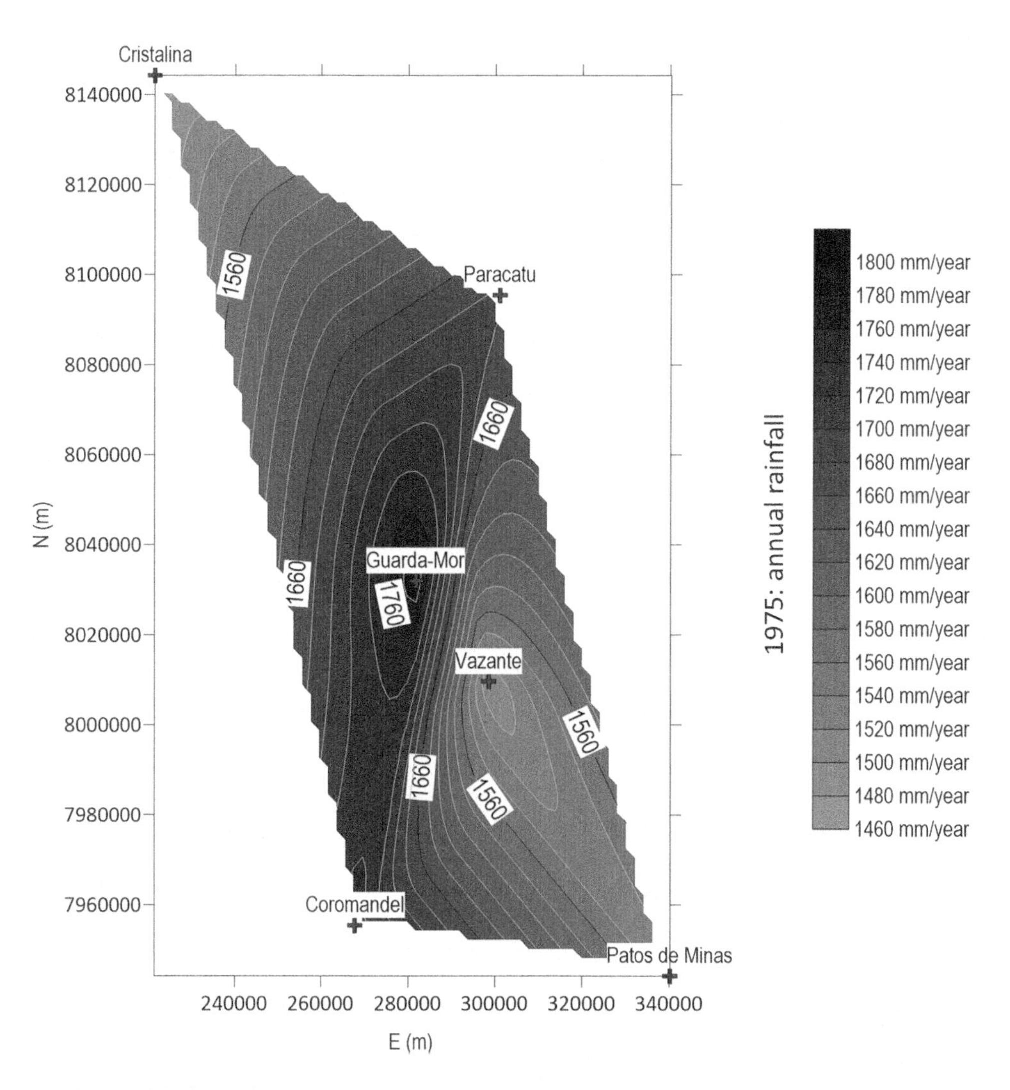

Figure 4.7 Annual precipitation in mm/year kriged from the recorded 1975 rainfall data in the indicated station-gages.

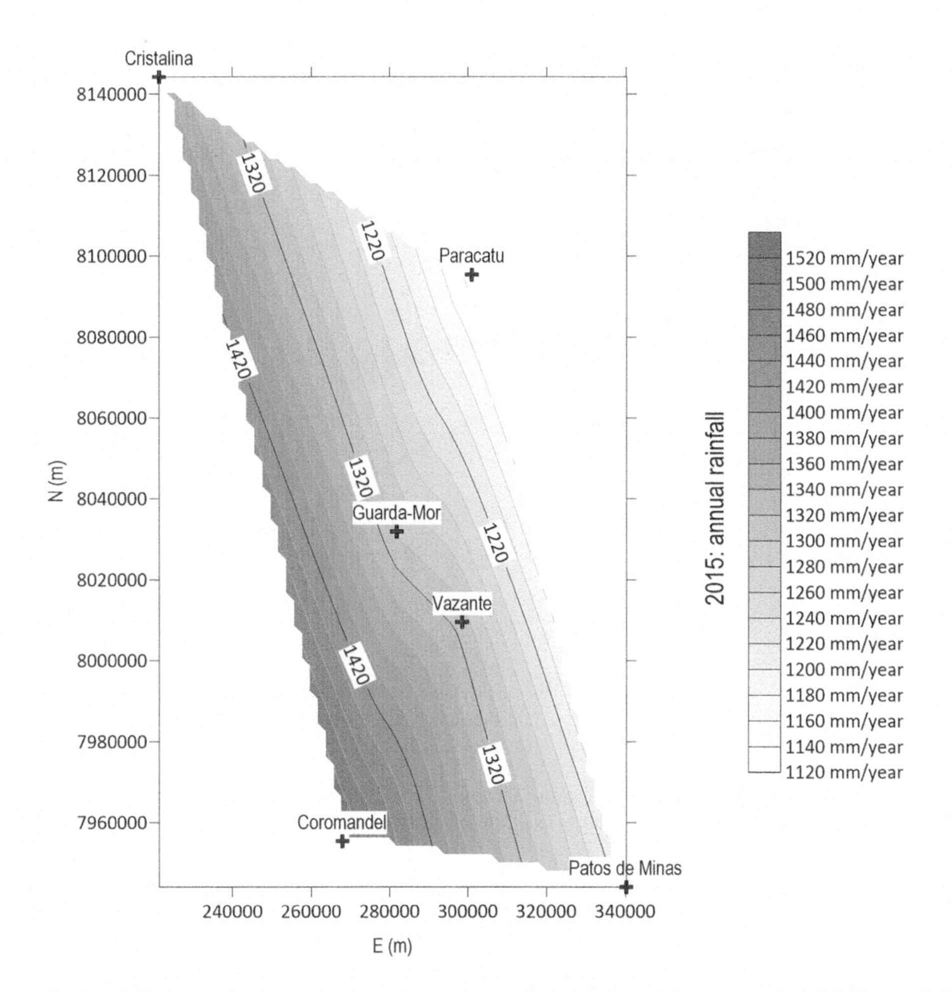

Figure 4.8 Annual precipitation in mm/year kriged from the recorded 2015 rainfall data in the indicated station-gages.

The following table shows small percentual errors between rainfall volumes computed by the original discrete rainfall values and exponential smoothed data, thus validating the selected approach.

Gage-station	*Yearly point rainfall height by discrete data (m^3/m^2)*	*Yearly point rainfall height by continuous data (m^3/m^2)*	*[Discrete]/[continuous] estimation error (%)*
Cristalina	38.10	37.36	1.93%
Paracatu	65.65	59.71	2.72%
Guarda-Mor	65.65	64.17	2.25%
Vazante	48.96	49.80	-1.72%
Coromandel	69.82	68.55	1.81%
Patos de Minas	56.68	57.04	-0.64%

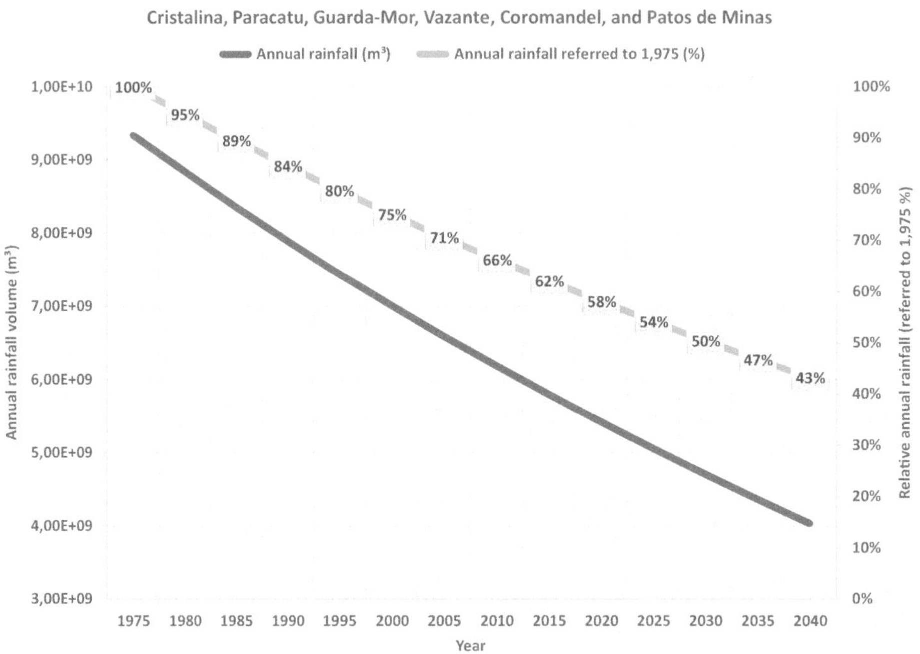

Figure 4.9 Graph showing the year (from 1975 to 2015), the annual rainfall volume, and the yearly point rainfall height for the polygon area defined by the six station-gages.

All 40 annual rainfall volumes defined by kriged iso-precipitation lines in the 40 maps were numerically estimated and then graphed, as shown in Figure 4.9.

The main conclusions of this analysis were as follows:

- Between 1975 and 2015, the total rainfall volume over the concerned area decreased around 38%, corresponding to an average rainfall decline rate in 40 years of about 0.95%/year.
- As the annual groundwater recharge is somewhat proportional to the annual rainfall volume, it is possible to presume that the annual recharge volume over the same area, between 1975 and 2015, suffered a gradual decline more or less proportional to the annual rainfall volumes.

4.4 Analyzing flow rate data

As done for hydraulic head data, flow rate data from springs, watercourses, pumping wells deserve similar treatment. However, it is sometimes possible to draw practical conclusions using hydrology techniques, such as using extreme value distributions. Example 4.1 shows how to adapt hydrologic techniques to analyze flow rate data.

Figure 4.10 Relatively small inrush observed a few days after completing the gallery excavation in dolomite land (Vazante underground mine, State of Minas Gerais, Brazil).

Example 4.1

One of the most annoying accidents during galleries developments for underground mining is sudden water inflows before cementation stabilization.

Figure 4.10 shows a typical but relatively small inflow observed a few weeks after the gallery head crossed an "apparently" sealed fracture in dolomite land.

The coordinates of the significant inrushes, their flow rates, and dynamic hydraulic heads, measured immediately after water irruption, were systematically registered during the development galleries' excavations for Vazante Mine, Minas Gerais St., Brazil, located in dolomite land. As these inflows were turbulent, the measured discharges were normalized by the square root of the hydraulic head: $L^3/T/L^{1/2}$ or $L^{2.5}/T$ (Figure 4.10).

It was possible to fit these data to a Gumbel extreme values distribution defined by:

$$G(A, B) = \sum_{n}\left[F_i - \left[100 \cdot \left(1 - e^{-e^{A \cdot \ln(q_i)+B}}\right)\right]\right]^2 \tag{4.1}$$

In the above equation, F_i denotes relative frequency (%) of I groups of data, and the pair (A, B) is the fitted parameters. Q corresponds to the normalized inrush. Minimization of squared error gave $A = 0.316$ and $B = -0.114$ (see Figure 4.11).

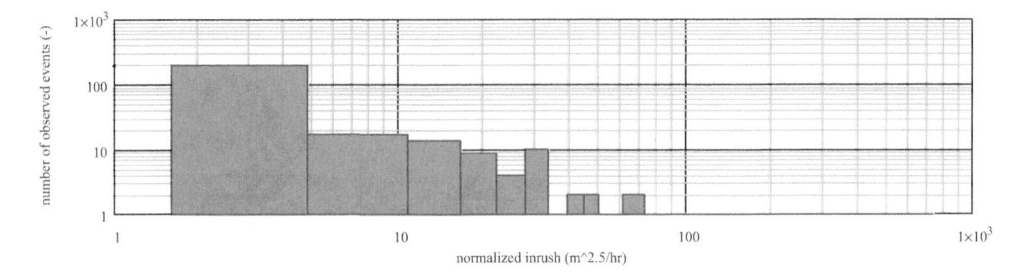

Figure 4.11 Histogram for 266 normalized inflows ($m^{2.5}/h$) systematically registered from December 1982 to December 2002.

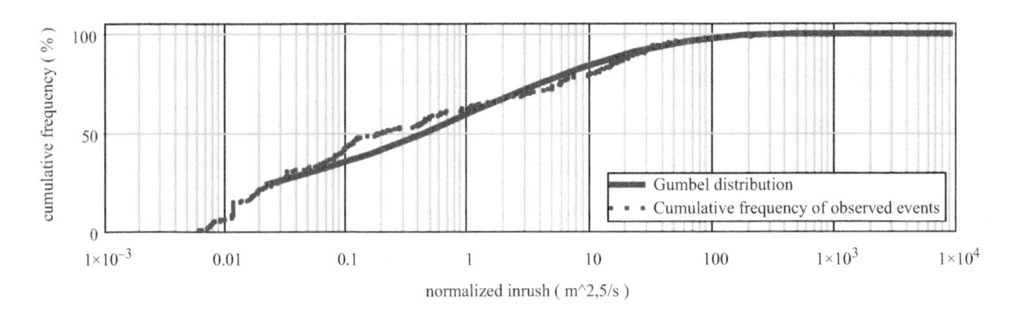

Figure 4.12 Fitted Gumbel distribution of extreme values to the observed events.

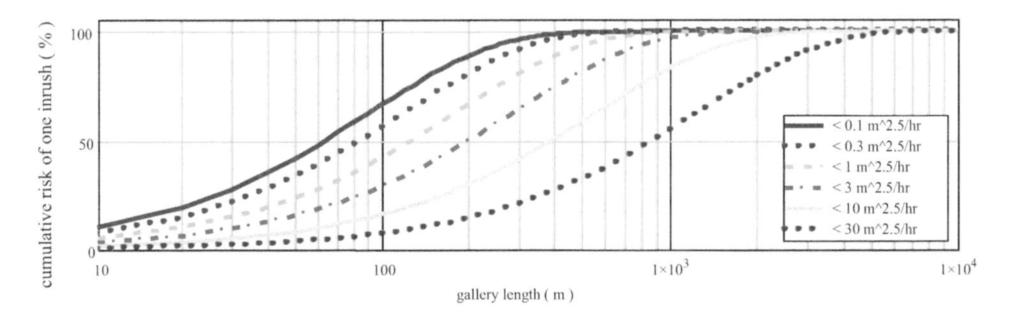

Figure 4.13 Expected risk of at least one inflow for planned excavation of more than 10,000 m of galleries. Each curve corresponds to a normalized inflow from 0.1 to 30 $m^{2.5}/h$.

Consider a gallery segment of length B as the chosen baseline. If P is the probability of a normalized inrush along B less than Q, the probability of an inrush exceeding this value is $1 - P$. Considering the length L of the whole gallery defined by the product kB, the probability of occurrence along the length L of a specific flow lower than Q is $(1 - P)^{kB}$. The complementary probability of occurrence above Q is $1 - (1 - P)^{kB}$. In extreme flows, this "complementary" probability measures the expected risk (see Figure 4.12).

This type of analysis can also be made, considering different lithologies and depths.

4.5 Analyzing hydraulic conductivity data

4.5.1 *Meaning of the hydraulic conductivity*

As formerly discussed, the size of a pseudo-continuous subsystem depends on the observation scale. At the selected scale, a preliminary estimate about the eigenvalues and eigenvectors of the hydraulic conductivity of a subsystem usually is inferred from its lithologic and structural geologic features. In some cases, previous experience from similar places may also allow the first judgments. However, to get useful numerical models results, these parameters must be finally adjusted by solving typical inverse problems, applied either to large-scale field tests or comprehensive monitoring data of groundwater table levels, pumping rates, and gravitational drainage rates.

It is essential to always keep in mind that the hydraulic conductivity typifying a rock mass subsystem is a scale-dependent property applied to a fictitious pseudo-continuous subsystem. Its eigenvalues and eigenvectors range over many orders of magnitude from place to place. Furthermore, it reflects the inherent assumptions concerning its anisotropy and inhomogeneity. When it results from specific field test results, it reflects the boundary conditions and simplifies the analytical model's assumptions. In other words, different mathematical models applied to the same tested volume may yield different values for the hydraulic conductivity, sometimes entirely dissimilar. Moreover, as standard closed solutions usually are applied to ideal homogeneous and anisotropic domains but never encountered in nature, their results lose meaning as the anisotropy, and the test space's heterogeneity increases. Then, given the preceding observations, the modeler must exercise its judgment to appropriately decide what approach best fits his needs from a technical and financial perspective.

The next alerts give the reader an idea of how different the hydraulic behavior of fractured rock masses may be:

(a) No other physical property may have such a vast range of possible values as the hydraulic conductivity, particularly for hard-rocks.

(b) Discontinuities, traces of fractures, faults, and contacts L indicated in a map may range from km to m, and their equivalent hydraulic apertures may range from dm to μ.

(c) The higher the apparent length L, the equivalent hydraulic aperture a, and the continuity parameter κ,[1] the much lower the probability of a fracture occurs in a net. Nevertheless, on the other hand, the higher may be its water transport capacity Q. Indeed, considering two adjacent subsystems (having m^3 volume) within a fractured rock mass, but both saturated under a unit hydraulic gradient **J**, while one of them may be conveying only $10^{-5} m^3/s$, the other, even if contiguous, may be conveying $10^2 m^3/s$, that is, 10^7 times more! (see Figure 4.10).

Moreover, the variation of the eigenvalues and eigenvectors of the hydraulic conductivity [K] within a fractured rock mass domain is always very problematic due to the significant anisotropic heterogeneousness of their discontinuity groups. For example, at a very short distance from a fracture characterized by a very low transmissivity may coexist another one, but more than 1,000 times more transmissible. Moreover, even if their transmissivities have the same order of magnitude, their spatial attitudes may diverge very much from each other.

More important is the identification of the correct dimensions of the rock mass subsystems that allow its description by a statistically quasi-homogeneous whole (as coherently proposed by Bear, 1972); sometimes, this may not be possible given the dimensions of the volume to be analyzed, perhaps definitely intractable from a statistical point of view. In this case, try a more laborious inhomogeneous model, characterized by the space variation of its hydraulic conductivity [K].

Additionally, the modeler must keep in mind that all mathematical models and formulas usually used to compute the eigenvalues and eigenvectors of the anisotropic hydraulic conductivity of rock masses, based on field tests, unavoidably give answers compulsorily biased by the premises of these models, necessarily simplified, despite the elegance and other merits of their derivation logic.

As an example, the monitored data from two long-term field permeability tests made on a $12^{1/4''}$ diameter well, between depths 542 and 722m, in a very thick sequence of Triassic-Jurassic sandstones[2] covered by successive and quasi-continuous Cretaceous basalt-flows, above a stable cratonic setting, gave different [K] values for different modeling approaches (see Figure 4.14).

The numerical criterion to verify the Hantush[3] transient approach's validity gave 0.011, a much lower value than the tolerable limits of 0.1–0.2. The average percentage error between the theoretical and monitored piezometric heads did not exceed 10.2%.

The hydraulic transmissivity T and the hydraulic conductivity K of the Botucatu sandstone evaluated by the Copper-Jacob[4] method, having a non-leaky basalt top

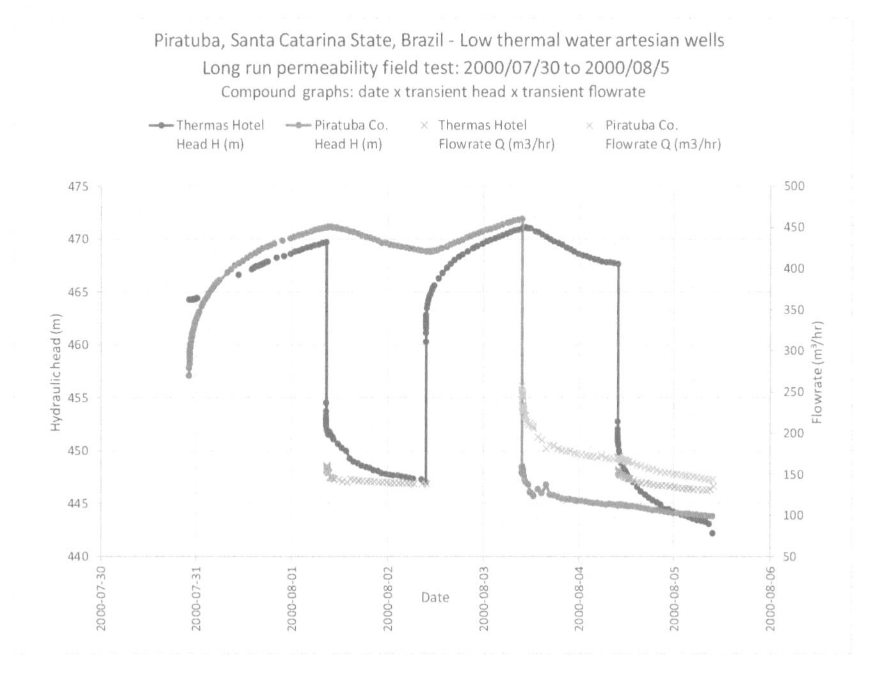

Figure 4.14 Monitored head and flowrate data of two series of long run permeability tests made in two deep hydrothermal wells bored in a Triassic Jurassic sandstones almost horizontal formation. Hydrothermal Piratuba and the Thermas Park Hotel, Piratuba, Santa Catarina St., Brazil.

Table 4.1 Park Hotel results: the more reliable corresponds to the exact Hantush model with leaky basalt cover and unsteady radial flow. The last column of the table shows how much the botucatu permeability is higher than the basalt permeability

Method	*Botucatu average transmissivity T (m^2/s)*	*Botucatu average hydraulic conductivity K (m/s)*	*Botucatu average specific storage S (–)*	*Botucatu specific 180 m of columnar storage S_C (m)*	*Basalt cover Average hydraulic conductivity K (m/s)*	**Quotient $K_{botucatu}/K_{basalt}$**
Copper-Jacob: impervious cover *steady state flow*	1.018E – 3	7.273E – 6	–	–	–	–
Hantush simplified leaky cover: *steady state flow*	3.06E – 4	2.186E – 6	–	–	1.075E – 7	***37.237***
Hantush exact leaky cover: *steady state flow*	3.121E – 4	2.229E – 6	–	–	2.26E – 7	***9.671***
Hantush exact leaky cover: *transient flow*	3.616E – 4	2.583E – 6	1.282E – 4	0.018	1.485E – 7	***17.396***

cover, was necessarily higher than that determined by the exact Hantush method, but correctly admitting a leaky basalt top cover. Such discrepancy does not entirely invalidate its practical use for current applications if discharge predictions, based on Cooper-Jacob parameters, are made by the same Cooper-Jacob methodology. However, hydraulic heads estimates can bear significant errors. Table 4.1 groups the results obtained for the Park Hotel well for four theoretical models.

For the semi-confined sandstone aquifer and from the parameters in Table 4.1, it was possible to estimate the probable decay and the recovery of the hydraulic charge at another point located 500 m apart from the pumping well, but after 6 h of uninterrupted pumping of 100 m^3/h (see Figure 4.15).

Now, the modeler must pay attention to two common issues: one is the best assumptions concerning de-evaluation of the *hydraulic transmissivity* of partially sealed discontinuities, and the other issue concerns the extent of their *hydraulic connectivity* at a chosen observation scale. These two points deserve attention when modeling a theoretical pseudo-continuous network of a fractured rock mass.

Secondary mineralization, chemical weathering, or other kinds of sealings or plugins may partially block a fracture's aperture and hamper its hydraulic transmissivity.[5] A crude appraisal of these obstructions' areal extent may result from detailed fieldwork on bare outcrops or excavated slope faces. Then, the relative areal extent κ of the fracture obstruction noticed on the exposed fracture traces may be estimated

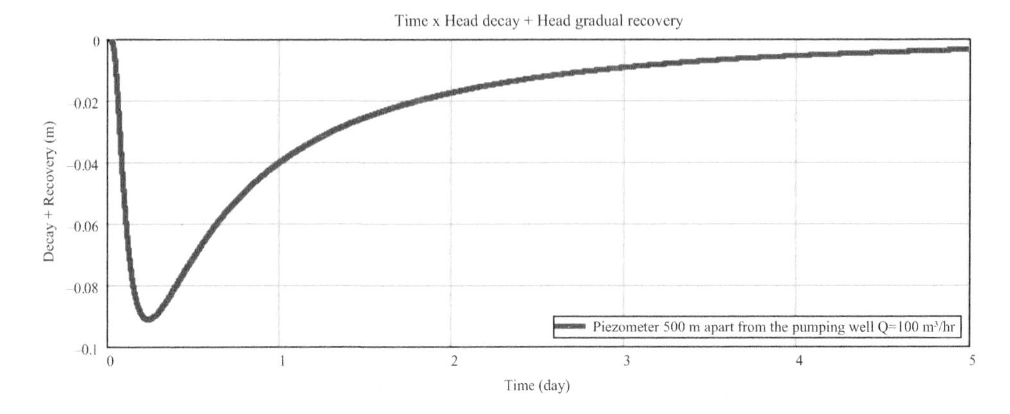

Figure 4.15 For the semiconfined sandstone aquifer (see parameters in Table 4.2), it was possible to estimate the probable decay and recovery of the hydraulic head in a piezometer located 500 m apart the pumping well after 6 h of uninterrupted pumping of 100 m^3/h.

by the sum of the lengths of clogged segments $\sum \delta l_c$ divided by the sum of all trace lengths $\sum \delta l$, that is, $\kappa = \sum \delta l_c \Big/ \sum \delta l$. That practical inference implies that for each group of more or less parallel open fractures noticed on undisturbed borehole core samples, its inferred average spacing e_a is, in fact, the product of the correct average spacing e and the relative areal extent of κ obstructions, that is, $e_a = \kappa \cdot e$ and $e_a < e$.

For partially blocked fractures, that is, for $0 < \kappa < 1$, the Maxwell mixture rule approximates their average theoretical transmissivity T as a mixed product: $T = T_0^{(1-\kappa)} \cdot T_\kappa^{(1-\kappa)}$, where T_0 and T_κ are, respectively, the *non-obstructed* and the *obstructed* fracture transmissivities.

Indeed, a heterogeneous medium, made up of l statistically homogenous subsystems, each defined by a frequency of occurrence p_{lm} and a directional permeability k_{lm} along an m-direction, has a directional permeability defined by[6]:

$$k_m = \left[\sum_{l=1}^{l} p_{lm} \left(k_{lm} \right)^n \right]^{\frac{1}{n}}$$

where $-1 \le n \ne 0 \le 1$ is:

$$n = 1 - \frac{2}{l} \cdot \sum_{l=1}^{l} \alpha_{ml}^2$$

In the formula above, $\alpha_{ml} = \cos(m, l)$ is the direction cosine of the angle between the scanning line mm and the perpendicular line to the lithological contact (see Figure 4.16).

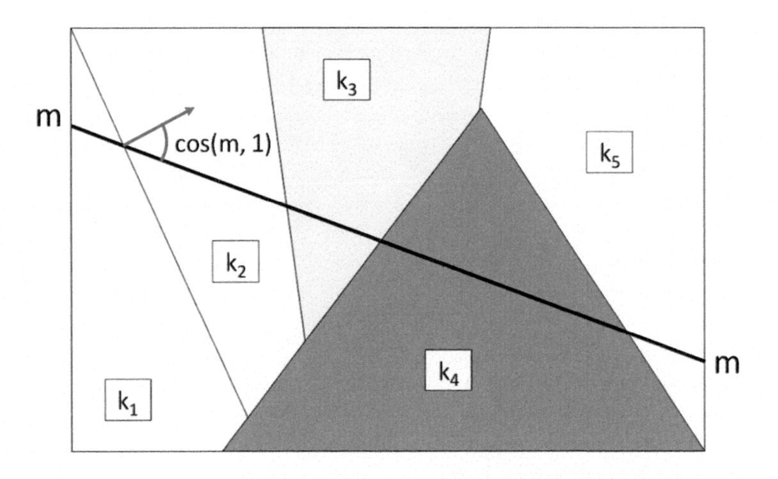

Figure 4.16 Heterogeneous media composed of five different materials of permeability k_l and aleatoric contacts. The directional permeability along the mm direction may be guessed by an empirical formula proposed by Zwikker, where $\alpha_{ml} = \cos(m, l)$ is the direction cosine of the angle between the scanning line mm and the perpendicular line to the lithological contact.

For n = 1 or n = −1, this formula gives the known mathematical expressions for the average horizontal K_H or the average vertical K_V permeability of horizontal strata composed of individual layers defined by k_H and k_V. Then:

$$K_H = \sum_{i=1}^{N} (p_i \cdot k_{iH})$$

$$K_V = \sum_{i=1}^{N} \left(\frac{p_i}{k_V} \right)$$

In the above expressions, **N** is the number of individual layers.

However, for random lithological contacts, where n changes randomly from −1 to 1, n = 0 may be taken as its average value. Then, applying the L'Hospital rule to get the limit, the ratio $[K_m]^n \Big/ \left[\sum p_{lm} (k_m)^n \right] = 1$ when $n \rightarrow 0$ gives the Maxwell mixture rule.

From all the preceding considerations, it follows that it is challenging to appraise the character of fractures' hydraulic transmissivity at a chosen observation scale. To complicate the problem, one must remember that in most cases, seepage along partially opened fractures is, in fact, partially concentrated on anastomosed small preferential conduits within the fracture plane.

Finally, modeling strategies to evaluate intact rock or rock masses' hydraulic conductivity depend on the desired observation scale. Chapter 2 in Sections 2.2.4 and 2.3.6 review some traditional and recent proved methods. The reader is encouraged to read these chapters, given its high interest.

Figure 4.17 Part of Sornin limestone plateau in France. (Google image.)

4.5.2 Mapping of eigenvalues and eigenvectors

According to the principles summarized in Chapter 2, it is always challenging to communicate the meaning of second-order tensor parameters. The methodology adapted from Nabil Al-Ambar's doctoral thesis[7] is a possible way to communicate second-order tensor meanings.

Part of the fractured limestone of Sornin-Vercors, France, is shown in Figure 4.17. The analyzed area amounts to $10^8 km^2$, and the thickness of the limestone strata varies around 250 m. As its thickness/area ratio is about 2.38×10^{-6}, a mathematical 2D model based on Dupuit's assumption is acceptable.

This area was divided into 164 quadrangular 400×400 m subareas. For each area, equivalent 2D-anisotropic hydraulic conductivities were differentiated based on fracture lineaments.[8] Average permeability of 9.68×10^{-8} m/s for the whole area was inferred from data concerning a high runoff volume of $34{,}000 m^3$ ensuing after a short episode of 4 h of intense precipitation. This volume was deduced from discharge measurements at the main local groundwater drainage exit. Then, the mean transmissivity and mean permeability for the Sornin limestone was estimated as $2.5 \times 10^{-5} m^2/s$ and $\approx 10^{-7}$ m/s, respectively. However, the inferred 2D-average hydraulic conductivity of each rectangular area of 400×400 m was inferred as to comply with the average value

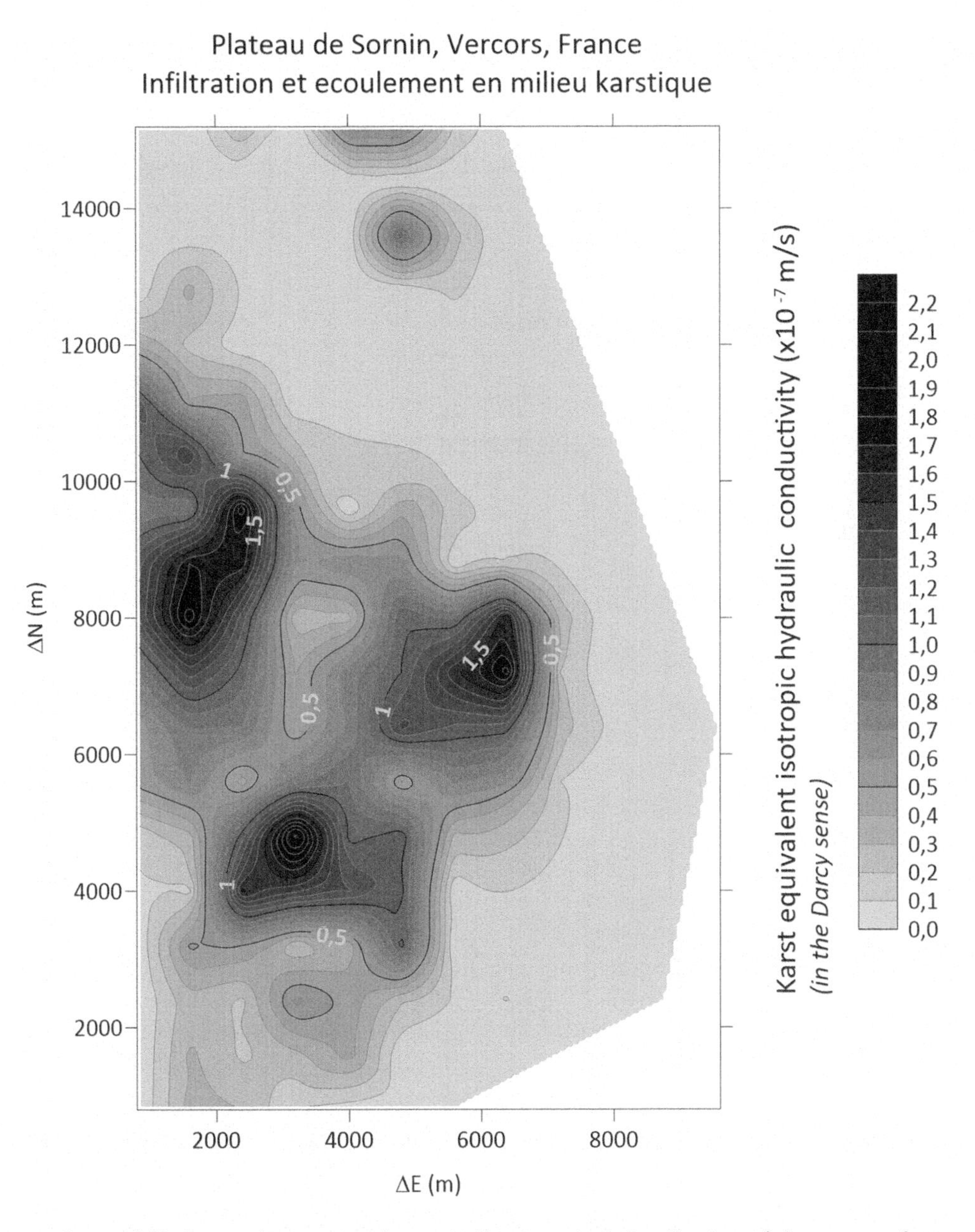

Figure 4.18 Sornin limestone plateau in France: areal distribution of the average fracture hydraulic conductivity complying with the distribution of all fractures and fissures photo interpreted in 1972.

of the whole plateau. Figure 4.18 shows how it was distributed in the whole area, complying with the general average permeability of 10^{-7} m/s.

Figure 4.19 exemplifies how the inferred 2D-anisotropic hydraulic conductivities can be portrayed for typical nine subareas of the plateau.

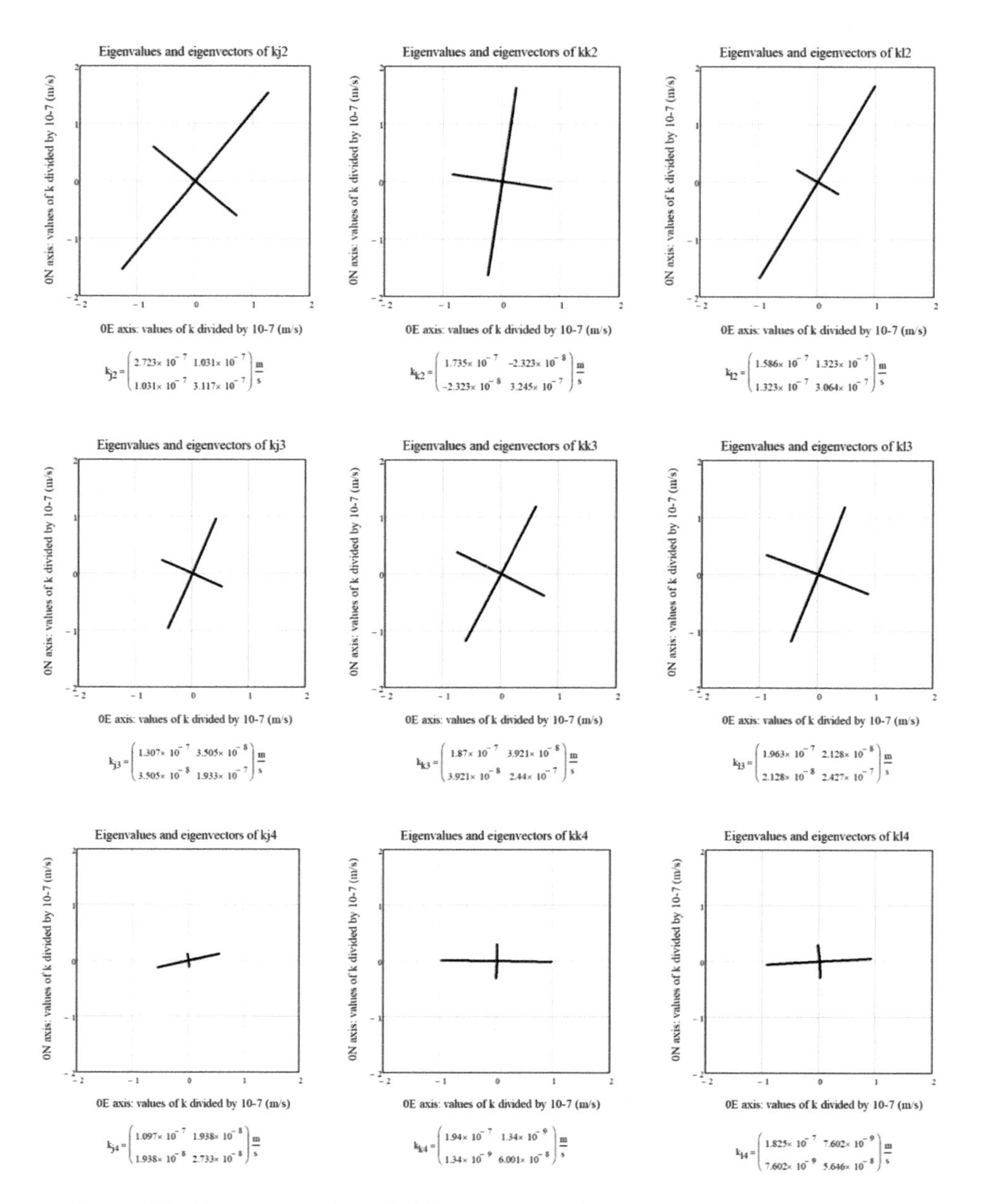

Figure 4.19 How nine inferred 2D-anisotropic hydraulic conductivities could be portrayed for three subareas in the Sornin Plateau. Corresponding K second-order tensors are shown below each graph.

Figure 4.20 shows how all 2D eigenvalues and eigenvectors can be portrayed for the whole area.

None of the above graphical representation of 2D-hydraulic conductivity tensors is entirely satisfying as it is easy to agree. Only 0-order tensors are amenable to impressive XY mapping.

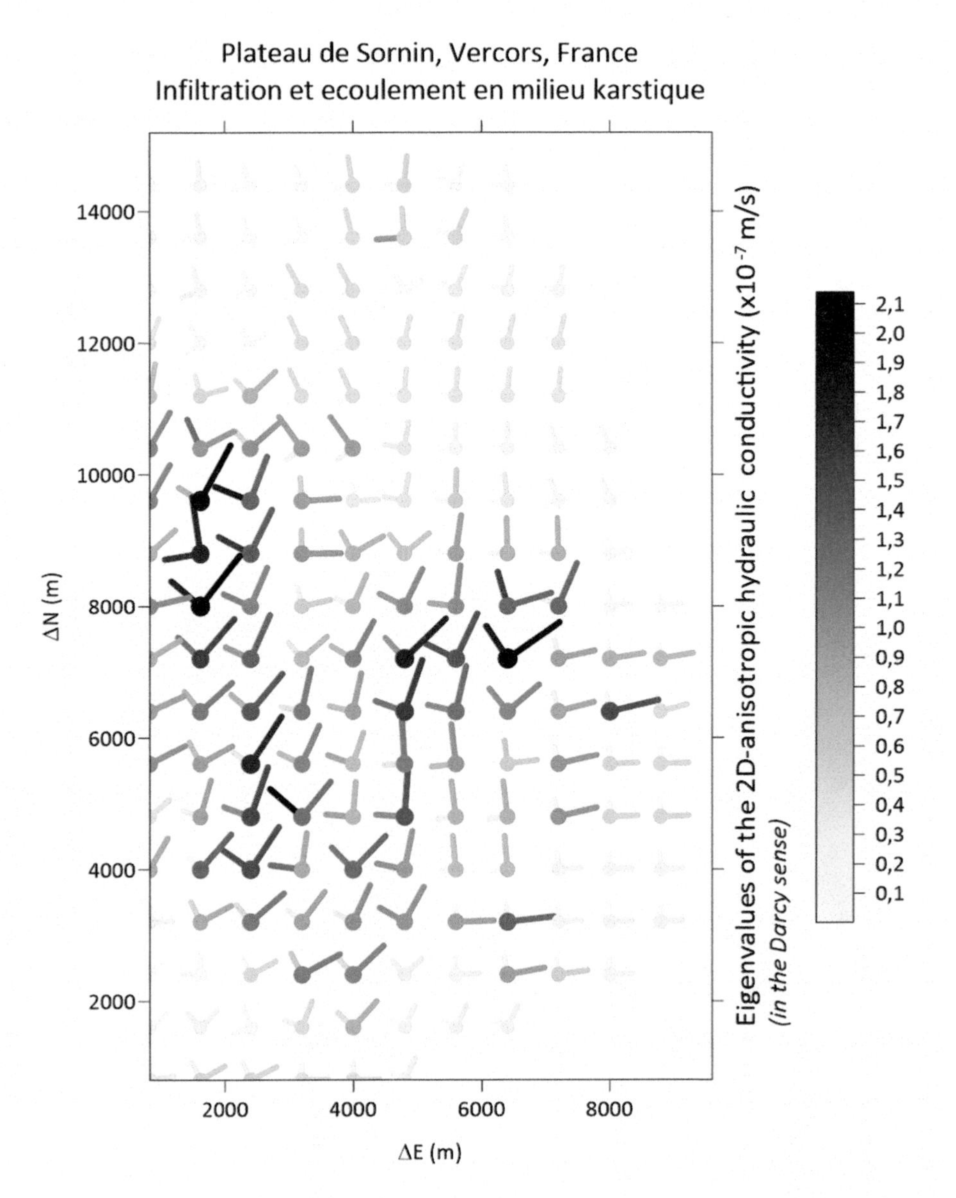

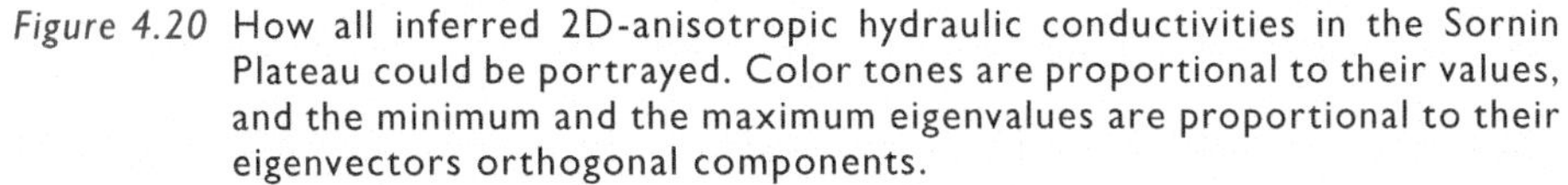

Figure 4.20 How all inferred 2D-anisotropic hydraulic conductivities in the Sornin Plateau could be portrayed. Color tones are proportional to their values, and the minimum and the maximum eigenvalues are proportional to their eigenvectors orthogonal components.

For 3D-hydraulic conductivity tensors, its composition is a bit more complicated because its description usually needs nine numbers plotted in a 3D space. In this case, it is only possible to map its three 2D projections into the three coordinate planes: EN, NZ, and ZE. However, to make sense to the observer, these planes must be simultaneously appreciated.

4.5.3 Interpolation of eigenvalues and eigenvectors

Sometimes unknown eigenvalues and eigenvectors of the hydraulic conductivity in a particular area may be assessed by interpolating (but never extrapolating) already known values in neighbor areas. However, only interpolate between similar structural patterns because and do not extrapolate. Therefore, interpolation only makes sense when similar patterns exist, as exemplified in Figure 4.21.

In these cases, the simplest way to define the elements of an interpolate second-order cartesian orthogonal tensor is performing everyday algebraic operations over the corresponding tensor elements of all neighboring tensors. As tensor elements are Euclidean vectors, real scalars' operations and multiplication preserve their mathematical character. However, it is possible to interpolate the eigenvectors' values only if matrices describe them. However, only interpolate the eigenvectors' direction cosines and never interpolate their corresponding angles to avoid incorrect results!

Figures 4.22 and 4.23 show two known anisotropic hydraulic conductivities.

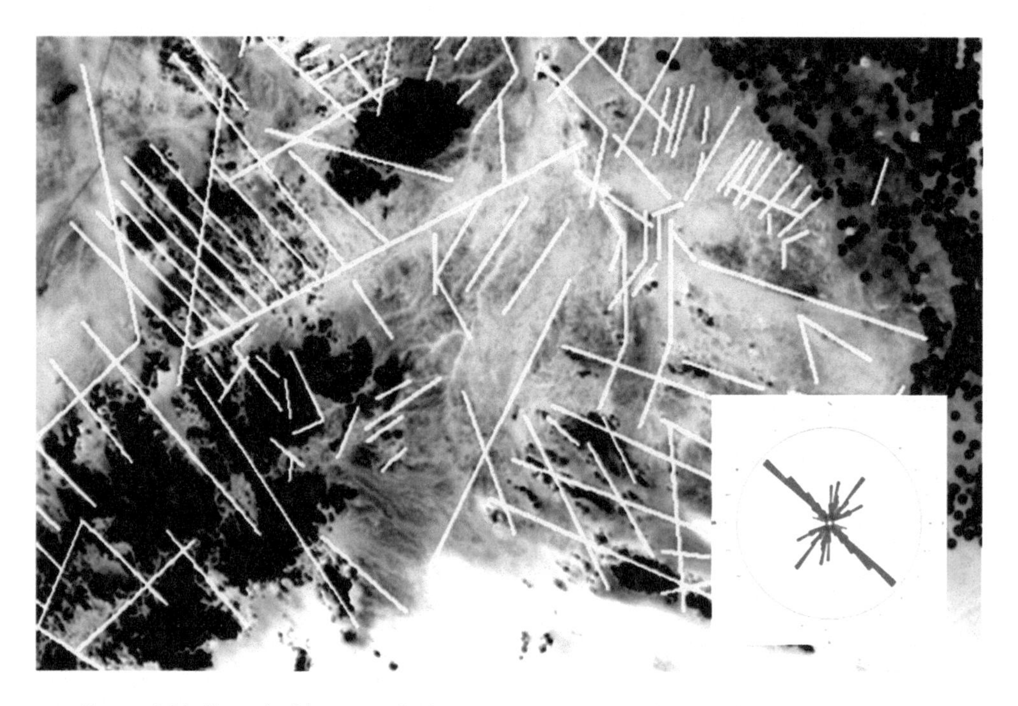

Figure 4.21 Detail of large-scale lineament trace map of the Wajid Group outcrop in the Wadi Al-Dawasir area showing lineament distribution and frequency (APUD, with modifications, from the excellent paper entitled. "Natural fracture system of the Cambro.Permian Wajid Group, Wadi Al-Dawasir, SW Saudi Arabia"; Mohammed Benaafia, Mustafa Hariria, Giovanni Bertottia, Abdulaziz Al-Shaibania, Osman Abdullatifa, Mohammad Makkawia; *Journal of Petroleum Science and Engineering*, vol. 175, April 2019, pp. 140–158).

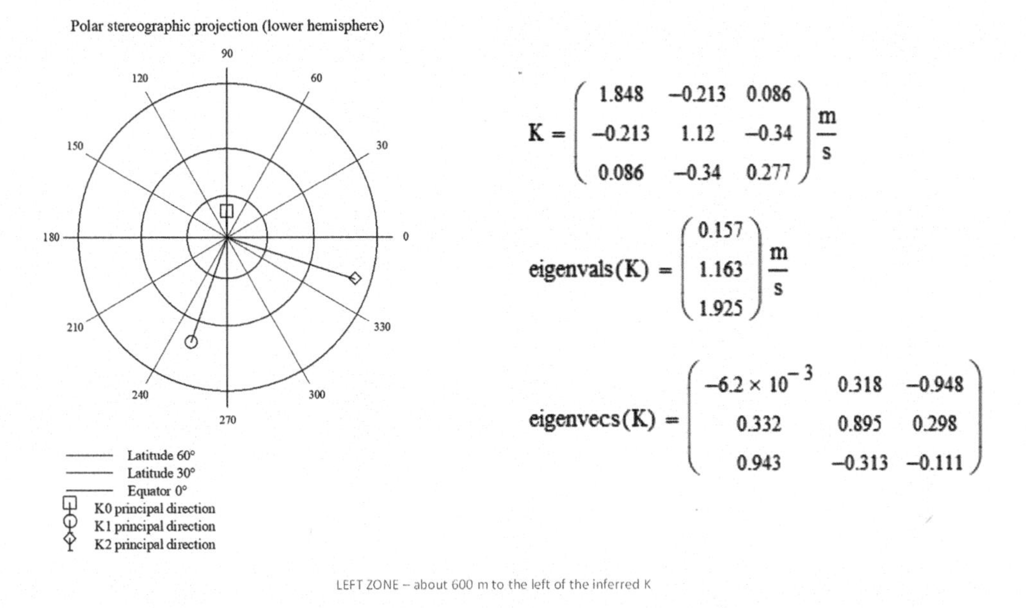

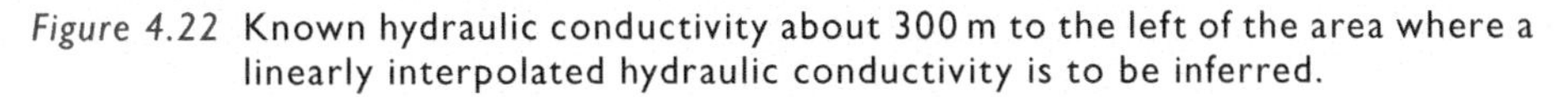

Figure 4.22 Known hydraulic conductivity about 300 m to the left of the area where a linearly interpolated hydraulic conductivity is to be inferred.

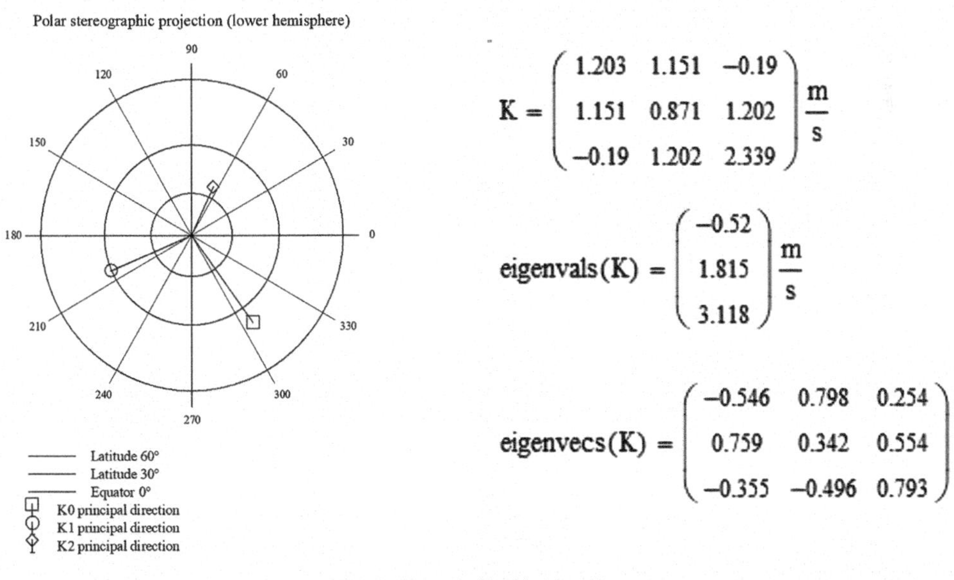

Figure 4.23 Known hydraulic conductivity about 600 m to the right of the area where a linearly interpolated hydraulic conductivity is to be inferred.

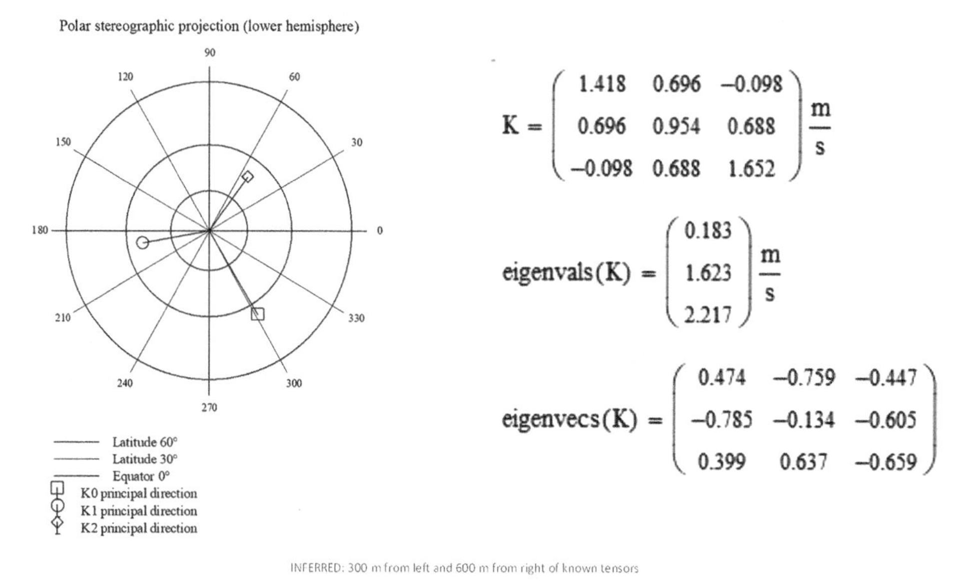

Figure 4.24 Inferred hydraulic conductivity second-order tensor in a particular area of a geologic unit portion based on the interpolation of the tensor eigenvalues and eigenvectors typifying other similar areas in the same geologic unit, but about 300 and 600 m, respectively, to the left and to the right.

Then, the inferred conductivity is:

$$K_{\text{inferred}} = \frac{2}{3} \cdot \begin{pmatrix} 1.203 & 1.151 & -0.19 \\ 1.151 & 0.871 & 1.202 \\ -0.19 & 1.202 & 2.339 \end{pmatrix} \text{m/s} + \frac{1}{3} \cdot \begin{pmatrix} 1.848 & -0.213 & 0.086 \\ -0.213 & 1.12 & -0.34 \\ 0.086 & -0.34 & 0.277 \end{pmatrix} \text{m/s}$$

Figure 4.24 shows the inferred anisotropic hydraulic conductivity.

4.6 Analyzing hydraulic head data

4.6.1 *Traditional analysis*

Hydraulic head measurements must be associated with specific times and localization of the observation points:

- The observation times must be chronological: yearly, monthly, daily, or hourly.
- The observation places must be referred to the map (E, N, Z) coordinates, duly related to a known reference frame system.

Therefore, all head measurements $H(\mathbf{r}, t)$ must be taken as discrete dependent variables, only defined at specific locations and some moments, despite the natural continuity of all implicit physical processes.

Different interpolation and extrapolation techniques extend these data to other moments and positions. Interpolated data within the observation period and contained in the monitored area are more consistent than extrapolated data.

The rise and fall of the water table elevation in a fractured rock aquifer's saturated zone mostly depend on the intensity and duration of the groundwater recharge and the value of the aquifer's effective specific porosity.

Bare outcrops of fractured rocks in cold and temperate climate limit water infiltration from direct precipitation runoff or dripping from soaked narrow soil horizons. However, groundwater recharge from long-lasting snow and ice melts is more significant. In a wet tropical climate, typified by deep chemical weathering, the soil-rock transition zone may be very thick and may acquire an open and significant secondary porosity. These horizons can then retain enough infiltrated rainwater to give rise to perched aquifers that gradually feed the fractured aquifer underneath (see Figure 4.25).

A rough estimate of the seasonal amount of the "rise and fall" of the water table in the saturated zone results from the division of the infiltrated rainfall height by the effective porosity. Note that the "effective" infiltrated rainfall height is only a small

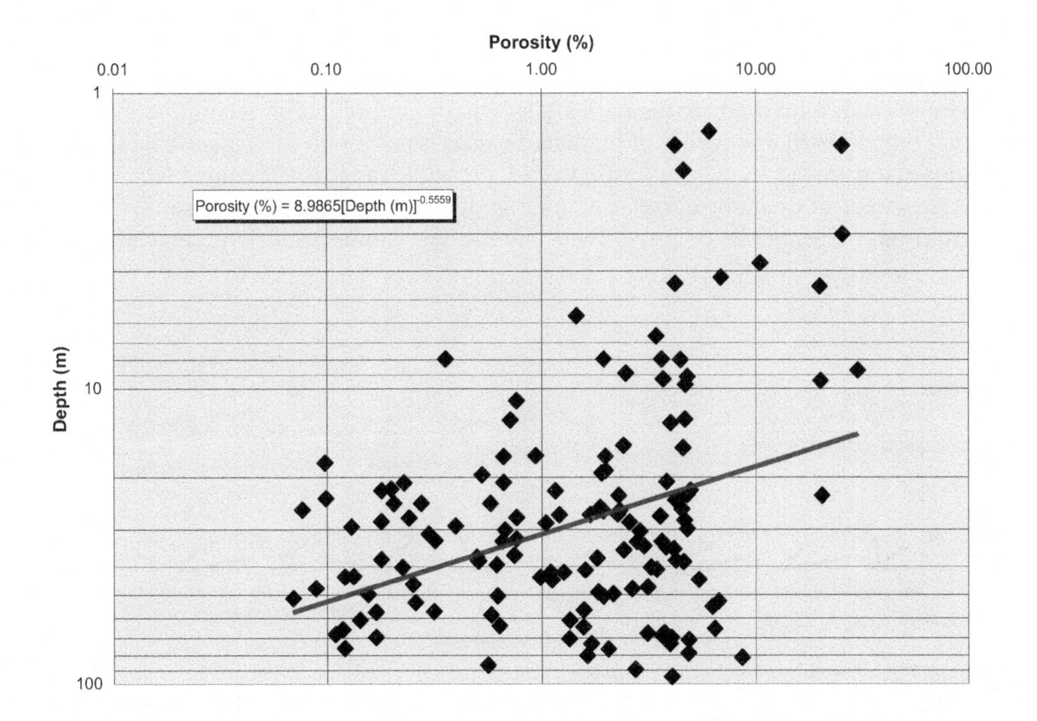

Figure 4.25 Plot of porosity against depth. Samples extracted from a very thick soil–rock transition zone in a deep weathered quartzite 120 km^2 plateau in an intertropical climate. A significant amount of infiltrated rainfall is temporarily stored each year at the soil–rock interface and slowly transferred to the fractured aquifer below. Annual recharge, statistically inferred from four years, varied from 500 to 1,000 L/s.

fraction of the total rainfall height: about 5%–25%. As fractured rocks' effective porosity is 10–50 times smaller than that of sedimentary rocks, water table level fluctuation in fractured rocks is correspondingly much higher than in sedimentary rocks (see Figure 4.26). Sometimes, depending on the regional climate's peculiarities, these perched aquifers may be fully depleted before the start of the next rainy season or may recharge more than one time a year.

Whenever possible, the time variation of the water table elevation should be monitored by self-powered and fully automatic data loggers. If this is not possible, recorded levels measured at different dates may be interpolated to allow synchronous comparisons. Almost every commercial spreadsheet program incorporates friendly piecewise-polynomials interpolation routines for time series analysis. Linear interpolation gives sharp corners, but cubic spline interpolation at regular time-intervals gives smooth curves continuous up to the second derivative. Moving average or exponential smoothing, integrated into commercial spreadsheets programs, may concomitantly be used to reduce data irregularities.

Fast Fourier Transform algorithms, included in some mathematical software, can filter and describe time series data in terms of "frequency domain," revealing all significant temporal cycles if they exist (see Figure 4.27).

Future or past values, even time series gaps, can be approximate by traditional interpolation or extrapolation algorithms. For almost periodic sequences, it may be possible to get plausible results using some type of forecasting technique, included in some mathematical software (Press et al., 1994; Sophocles, 1992; see Figure 4.28).

However, despite their mathematical cleverness, no prediction technique matches actual outcomes. If a sequence of predicted values is not required, it is sufficient to use traditional statistics to picture future values' probable range (see Figure 4.29).

Assessment of dynamic correlation among different time series may confirm their mutual interdependence or prove their cause-effect connection. Lag correlation or

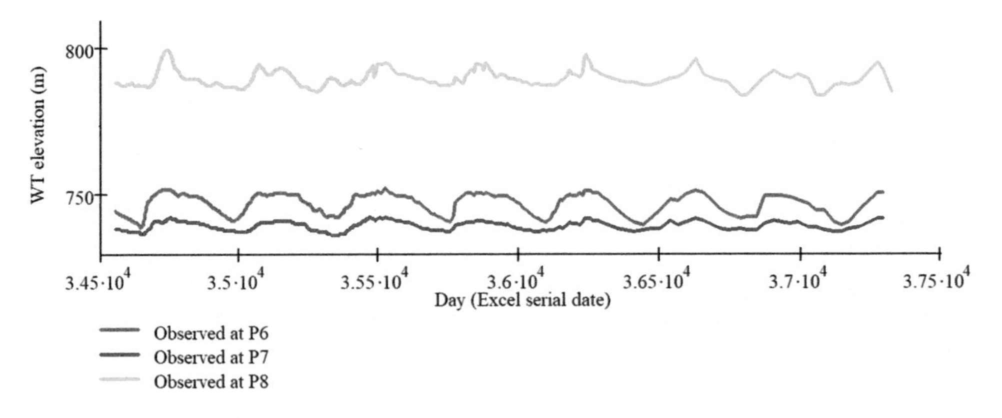

Figure 4.26 Water table "rise and fall" during seven years at three observation points in the saturated zone of a moderately weathered quartz-schist monocline ridge, defined by porosities lower than 1%. Yearly water table level oscillations, only due to seasonal rainfall variations, attained more than 10 m in two observation points.

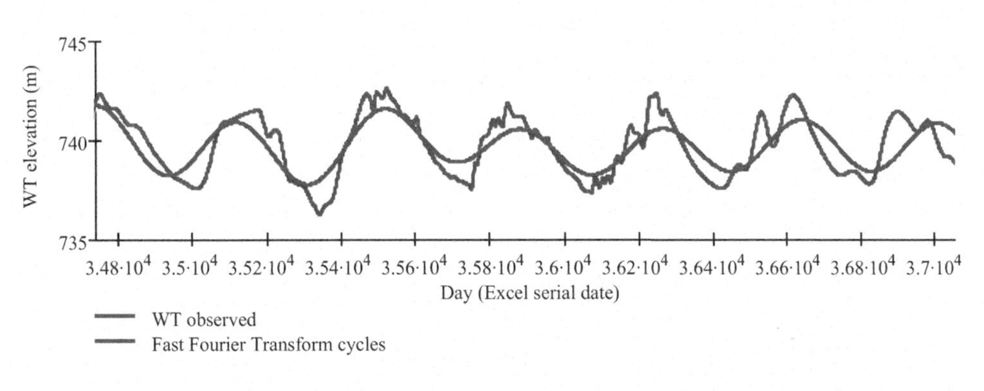

Figure 4.27 Fast Fourier Transform techniques applied to a time series of water table elevation in saturated schist forcefully removed all irregularities and secondary cycles and emphasized the seasonal cycles' main aspects.

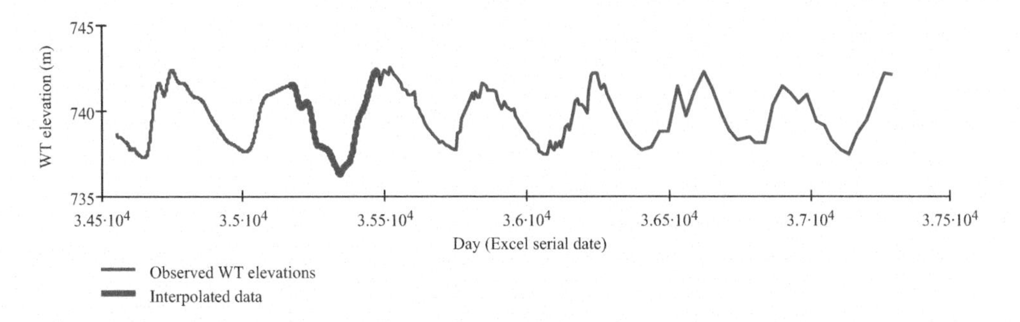

Figure 4.28 Time series of continuously monitored water table levels. The missing gap was filled using Burger's linear prediction method. Predicted values only "mimic" past values, correctly shifted ahead in time. Nevertheless, this type of adjusted time series may allow some statistical analysis.

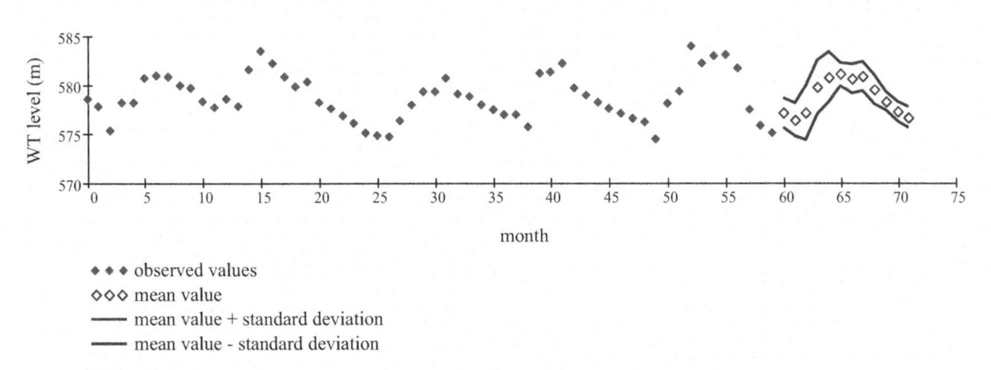

Figure 4.29 Time series of monthly water table elevations. It is possible to discern five complete repetitive cycles of 12 months each (a regular hydrologic year). Observed irregularities are due to several causes, including varying rainfall intensity. For each month, it is possible to compute five-year statistics, like averages and standard deviation. At the end of every month, the predicted levels may be systematically re-evaluated.

nonparametric correlation techniques, also available in some mathematical software, may give good results (see Figure 4.30).

The spatial interpolation (and extrapolation) of observed quantities monitored at regular or erratic sampling locations may be achieved by different types of built-in interpolation algorithms of commercial software specifically designed to produce contour maps from rectangular arrays of estimated values, shortly called "grids." Check, for example, the capabilities of SURFER—*from Golden Software, Inc.*

In a certain sense, almost all interpolation algorithms are weighted averages; that is, finite interpolation polynomials were the interpolating functions $f_i(\mathbf{r})$ that play the role of averaging weights. However, instead of being constructed as functions of the coordinates **r** of the estimation point, as for traditional piecewise approximants, these averaging weights are constructed as functions of the radial separation $\delta\mathbf{r}_{ij}$ (and sometimes also of the azimuth α_{ij}) between the estimation point i and a few points j around it, forming a *cluster* of influencing values (see Figure 4.31).

The main difference between weighted average algorithms resides in the way their weights are determined. As a direct consequence, different weighting criteria yield comparable but unequal estimations for the same data and grid geometry. Usually, for each type of interpolation algorithm, the influence of size, spatial distribution, and inherent variability of the input data on the resulting grid's character is adequately considered in the "help section" of commercial software.

Inverse Distance is one of the most straightforward and intuitive weighted average methods. Each observed value j must be pondered by the inverse of their separation from the estimation point i (frequently, distance squared). Then, the closer the observation point to the estimation point, the higher its contribution to the estimated value.

Natural Neighbor is an elegant and robust weighted average suggested for sparse and irregular observation points. Optimal averaging weights are proportional to the

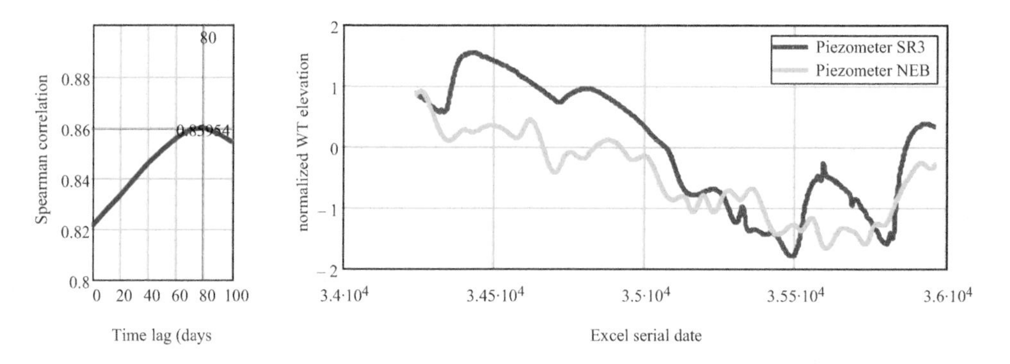

Figure 4.30 The left graph shows "Spearman's correlation coefficient" × "growing time series lag" used to compare the time evolution pattern of the two WT time-series shown in the right graph, one for the piezometer NEB, closed to deep pumping wells on fractured quartzite, and the other for piezometer SR3 far from the pumping wells, located at the top of a 350 m high plateau, more than 25 km apart. The right graph, comparing these time series "lagged" by 80 days, clearly show that both time series are effective high correlated.

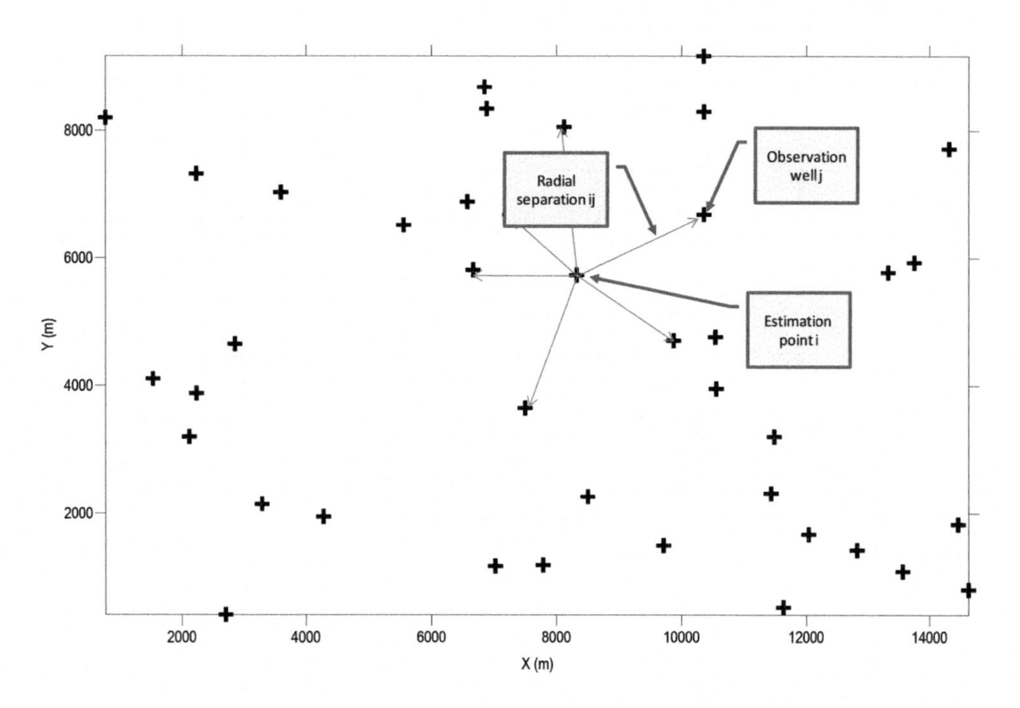

Figure 4.31 Example of a cluster of five observation wells j around an arbitrary estimation point i.

"influence area" of each observation point j clustered around the estimation point i (see, for example, the fundamentals of the Thiessen Method—inspiration source of the Natural Neighbor method—that are described in sections devoted to the subject of Descriptive Hydrology in standard textbooks).

Kriging is a more elaborated weighted average based on two underlying assumptions. The first one assumes that all observations can be decomposed into two parts. One part is a deterministic value, as a global average, or a linear trend value. The other part is a "residual" value, defined as the difference between the observed value and the deterministic value. The second assumption presumes that it is possible to statistically infer the character of a "measure" of the dissimilarity of two "residuals" as a function of their radial separation $\delta \mathbf{r}_{ij}$ up to a specific limit (and sometimes, as a function of the radial separation $\delta \mathbf{r}_{ij}$ and also of the azimuth α_{ij}). Based on an empirically constructed "dissimilarity measure" between closer observations, local averaging weights are optimized for every cluster of P points around an estimation point i (see Figure 4.31). The corresponding "objective function" to be minimized for each cluster is a statistical measure of the local dissimilarity between predicted and observed values (error variance).

Besides the set of observed values, additional information about the sample domain may be: ignored (simple kriging), treated as complementary random variables (cokriging), or incorporated as a paired spatial bias.

Table 4.2 Input data set

Order	*E (m)*	*N (m)*	*Z (m)*	*H (m)*
1	741,749.92	7,786,118.54	670.00	615.00
2	712,850.46	7,785,151.50	568.00	530.00
3	708,794.42	7,783,985.81	568.00	530.00
4	711,900.74	7,780,999.76	640.00	604.00
5	745,198.91	7,771,800.78	660.00	648.00
6	810,899.43	7,769,598.54	502.00	502.00
...	...	...	...	...
...	...	...	...	...
...	...	...	...	...
159	706,249.19	7,430,951.04	580.00	577.00
160	704,899.72	7,425,898.83	575.00	564.00
161	655,100.57	7,425,700.20	750.00	742.00
162	710,000.14	7,423,701.23	600.00	565.00
163	703,599.72	7,423,150.53	610.00	587.00
164	689,750.07	7,422,082.14	650.00	604.00

Finally, it is essential to note that kriging algorithms produce point estimations or generate averages of groups of point estimations around cell centers, consequently smoothing the map contours. Excellent introductory texts on Geostatistic and Kriging are available, for example, Kosakowski (2005) and Bohling (2005).

Example 4.2 below helps to grasp the fundamentals of almost all *griding* methods and illustrates the construction of a variant of an inverse distance squared algorithm.

Example 4.2

The stabilized values of the piezometric elevations H_{weel} at the roof of a half-confined groundwater body trapped in a sandstone formation were recorded during the perforation of 164 deep wells for oil prospecting (see Table 4.2). Based on this data record, it was possible to picture the piezometric surface's most probable configuration over a vast area of near 70,000 km^2.

An optimized inverse distance squared algorithm was chosen to interpolate (and extrapolate) the observed piezometric elevations H in a rectangular grid of 40×64 equidistant points at the center squares 5,000 m side. Based on that grid, the following procedure yields a contour map of H-equipotentials (see Figure 4.32).

Step A: Detrending

The coefficients of a deterministic 3D planar trend H_{trend} passing by the cloud of all m observed heads (E_{well}, N_{well}, H_{well}) were found by the traditional least squared method. The resulting best-fit planar trend was as follows:

$$H_{trend} = -2.124223 \times 10^{-4} E + 2.257132 \times 10^{-5} N + 540.19$$

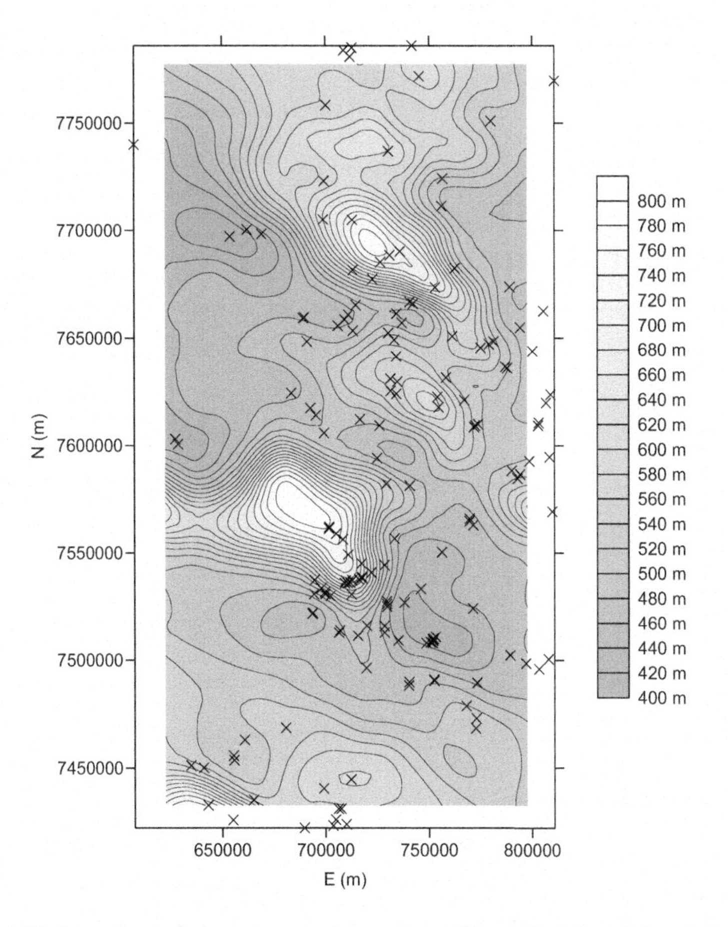

Figure 4.32 Location of the observation points (X symbols) and head H contours based on the grid estimates.

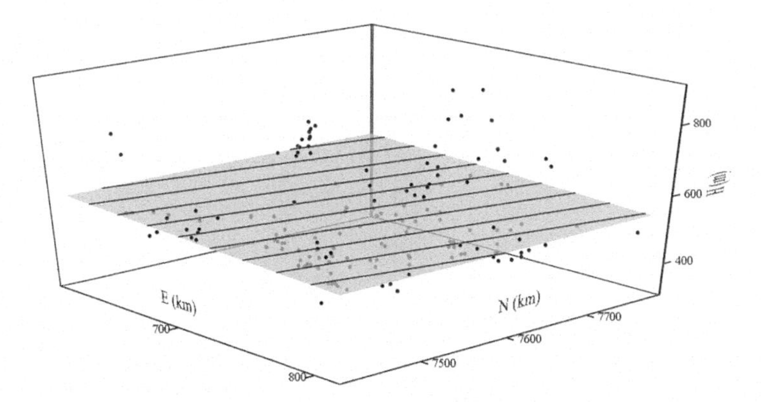

Figure 4.33 Plot of the observed heads and the planar trend.

The head residuals δH_{well} at each well, referred to the trend plane H_{trend}, were computed as the difference between the observed head H_{well} and the calculated trend H_{trend} at the well location (see Figure 4.33):

$$\delta H_{well} = \left(-2.124223 \times 10^{-4} E_{well} + 2.257132 \times 10^{-5} N_{well} + 540.19\right) - H_{well}$$

Step B: Proximity of observed wells

The proximity or closeness of each pair of monitoring wells, s and t, was measured by two quantities: their separation $\delta \mathbf{r}_{s,\,t}$ and corresponding azimuth $\alpha_{s,\,t}$:

$$\delta \mathbf{r}_{s,\,t} = \left[\left(E_{well_s} - E_{well_t}\right)^2 + \left(N_{well_s} - N_{well_t}\right)^2\right]^{\frac{1}{2}}$$

$$\alpha_{s,\,t} = a \tan\left(\frac{N_{well_s} - N_{well_t}}{E_{well_s} - E_{well_t}}\right)$$

Care must be exercised when computing $\alpha_{s,\,t}$ to avoid vague answers.

Step C: Parameters optimization

The residual head δH_i at each grid node i – defined by its coordinates E_i and N_i – was estimated by the inverse distance squares algorithm, taken only into account the observed residual heads δH_j at P nearby wells – defined by its coordinates E_j and N_j – around the estimation node i.

This algorithm previously optimizes four parameters:

- P: the number of neighboring wells around estimation nodes
- A: the value the squared distance
- B and C: the eccentricity and the orientation of the domain anisotropy

The resulting algorithm is shortly written as:

$$\delta H_j = \sum_j \left(w_{i,\,j} \cdot \delta H_k\right)$$

In the above equation, the subscript j refers to the cluster or nearby wells around the estimation point i. Define the weighting averages $w_{i,j}$ as a combination of the 3 expressions below.

Then, the weighting averages are defined by:

$$w_{i,j} = \frac{1}{E_1\,E_2}$$

Where the variables E_1 and E_2 result from the following expressions: the variable E_1:

$$E_1 = \frac{A\left(\delta r_{i,j}\right)^2}{\frac{\cos\left(C+\alpha_{i,j}\right)^2}{B^2} + B^2\sin\left(C+\alpha_{i,j}\right)^2} + 1$$

$$E_2 = \sum_{j=1}^{j} \frac{1}{\frac{A\left(\delta r_{i,j}\right)^2}{\frac{\cos\left(C+\alpha_{i,j}\right)^2}{B^2} + B^2\sin\left(C+\alpha_{i,j}\right)^2} + 1}$$

It is easy to apprehend that the averaging weights $w_{i,j}$ vary between zero and one. Also, their total sum is always unity. Its value rapidly falls with separation (see Figure 4.34).

A tractable approach helps to optimize each parameter P, A, B, and C, beginning with the more simple expression and finalizing with the more complex. By that choice, a loss of mathematical rigor is inevitable, but not too compromising.

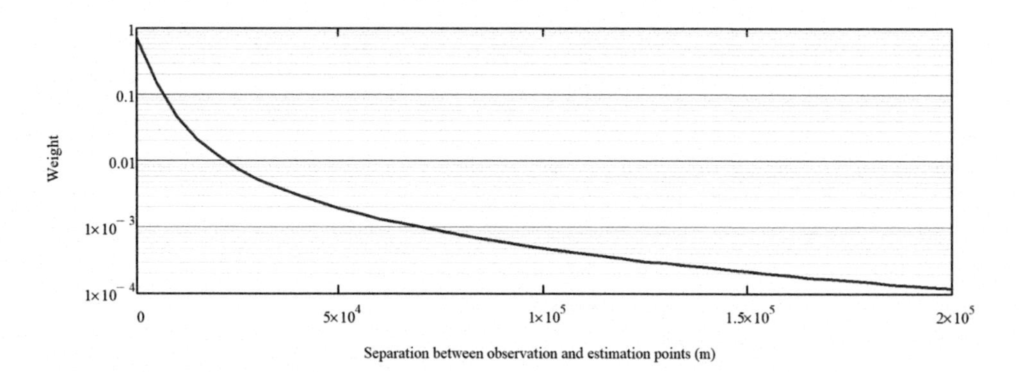

Figure 4.34 Weight value declines as radial separation increases. The graph was plotted after optimization of the four parameters P, A, B, and C.

The objective function is the sum of all squared interpolation errors between observed and estimated residuals at all observation points, that is, at all the wells. As the first parameter to be optimized was A, the sum of squared errors $\sum_A$ was:

$$\sum_A = \sum_{s=0}^{m} (\text{estimated } \delta H_s - \text{observed } \delta H_s)^2$$

The residual head at any well s considers all m observed values at other locations, except at the target well. Then, to estimate the residual heads at points t based on the observed heads at all points s, except at the target well, the following expression was used:

$$\delta H_t = \frac{\sum_{s=0}^{m} \left[(s \neq t) \left[\frac{1}{1 + A\left(\delta \mathbf{r}_{s,t}\right)^2} \delta H_s \right] \right]}{\sum_{s=0}^{m} \left[\frac{1}{1 + A\left(\delta \mathbf{r}_{s,t}\right)^2} \right]}$$

Now, the original expression of $\sum_A$ was modified to enhance the weight of the neighborhood of the estimation points:

$$\sum_A = \sum_{s=0}^{m} \left[\left(\frac{\text{mean } \delta \mathbf{r}}{\text{mean } \delta s} \right)^2 \cdot (\text{estimated } \delta H_s - \text{observed } \delta H_s) \right]^2$$

In the above expression:

- mean $\delta \mathbf{r}$ = average of the distances between all pairs of different wells
- mean δr_s = average of the distances between the well s and all remaining wells

Finally, the minimization of the squared error sum $\sum_A$ defined the unknown A value, as graphically shown in Figure 4.35.

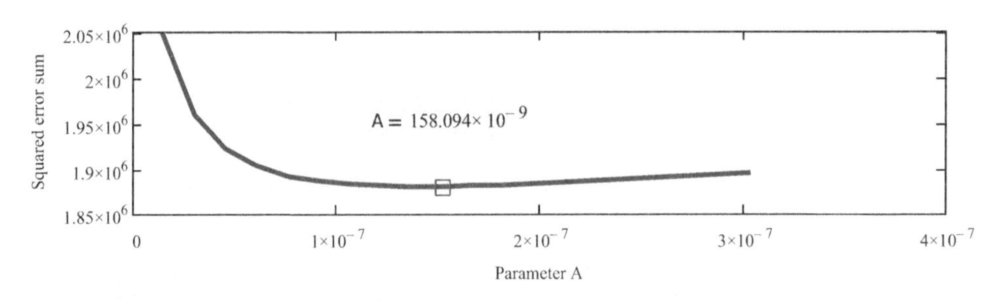

Figure 4.35 Variation of the squared errors sum $\sum_A$ with parameter A (minimization by the conjugate gradient method).

The next parameters to be optimized were B and C. These parameters control the inherent anisotropy of the observed values concurrently due to the geometric pattern of the sampling points and to the geologic heterogeneity of the monitored domain. However, to estimate residual heads at points t based on the observed heads at all points s, except at the target well, a more complex expression was adopted, including not only the separation $\delta \mathbf{r}_{s,\,t}$ but also the corresponding azimuth $\alpha_{s,\,t}$:

$$\delta H_t = \frac{\sum_{s=0}^{m}\left[(s \neq t)\left[\frac{1}{1 + A\left(\delta r_{s,\,t}\right)^2 \dfrac{1}{\sin\left(\alpha_{i,\,j} + C\right)^2 B^2 + \dfrac{1}{B^2}\cos\left(\alpha_{i,\,j} + C\right)^2}}\delta H_s\right]\right]}{\sum_{s=0}^{m}\left[\frac{1}{1 + A\left(\delta r_{s,\,t}\right)^2 \dfrac{1}{\sin\left(\alpha_{i,\,j} + C\right)^2 B^2 + \dfrac{1}{B^2}\cos\left(\alpha_{i,\,j} + C\right)^2}}\right]}$$

To enhance the influence of the observations in the far-field and eventually detect an existing anisotropic pattern, the sum of the squared error $\sum_{BC}$ was also modified as follows:

$$\sum_{BC} = \sum_{s=0}^{m}\left[\left(\frac{\text{mean } \delta s}{\text{mean } \delta r}\right)^2 \cdot \left(\text{estimated } \delta H_s - \text{observed } \delta H_s\right)\right]^2$$

Figure 4.36 shows the graphical expression of B and C and the values that minimize the squared error sum $\sum_{BC}$ simultaneously.

Figure 4.37 shows the polar graph of the optimized anisotropy ellipse. However, this plot may give a false impression of the effects of the detected anisotropy. As the anisotropy factor multiplies the distance squared in the denominator of the algorithm's expression, their effects on the grid values are inverted. In other words, the grid-based contours "shrink in" and "stretch out" in the opposite directions of the major and minor axis of the best fit ellipsis, respectively.

Besides the sample pattern's influence, a qualified hydrogeologist may infer additional geological influences on the detected anisotropy.

The last parameter to be optimized was the quantity P of neighboring wells around estimation points. Thus, applying the same objective function (by already now computed with the optimized parameters A, B, and C) to each cluster of points around the estimation points, but successively increasing P from the nearest observation well to those more distant closer, it was possible to see that eight (8) was the optimized number for P (see Figure 4.38).

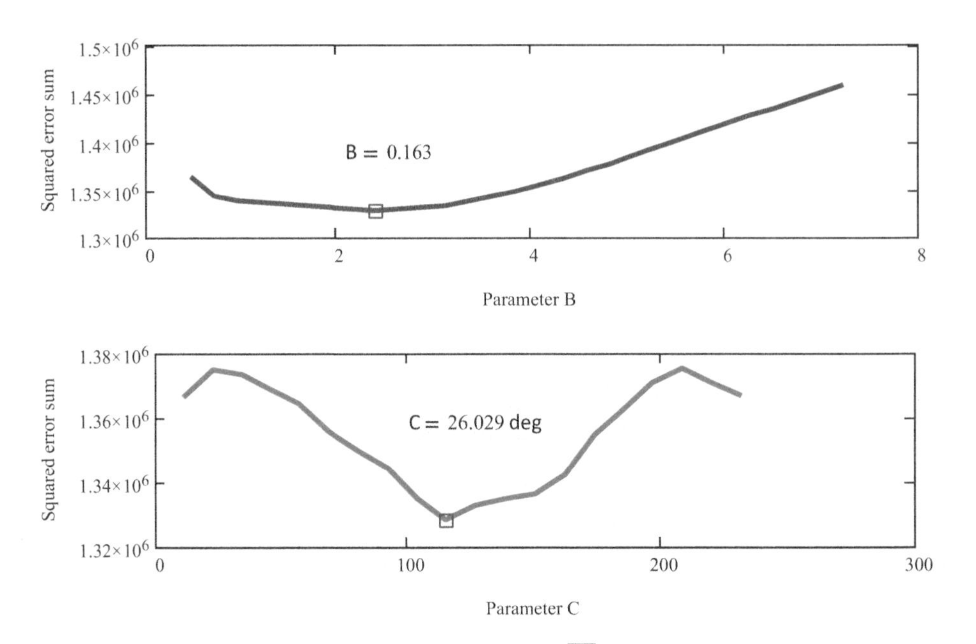

Figure 4.36 Variation of the squared errors sum $\sum_{BC}$ with the optimized parameters B and C (minimization by the conjugate gradient method).

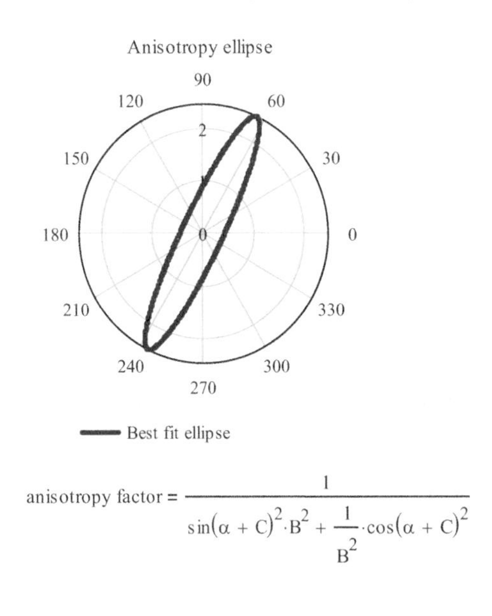

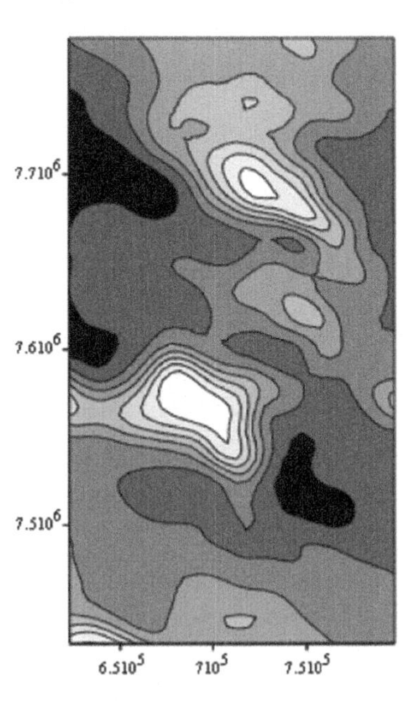

Figure 4.37 Best fit ellipse that reflects the inherent anisotropy of the observed values. It is mathematically not possible to separate what is due to the geometric pattern of the sampling points from what is due to the geologic heterogeneity of the monitored domain.

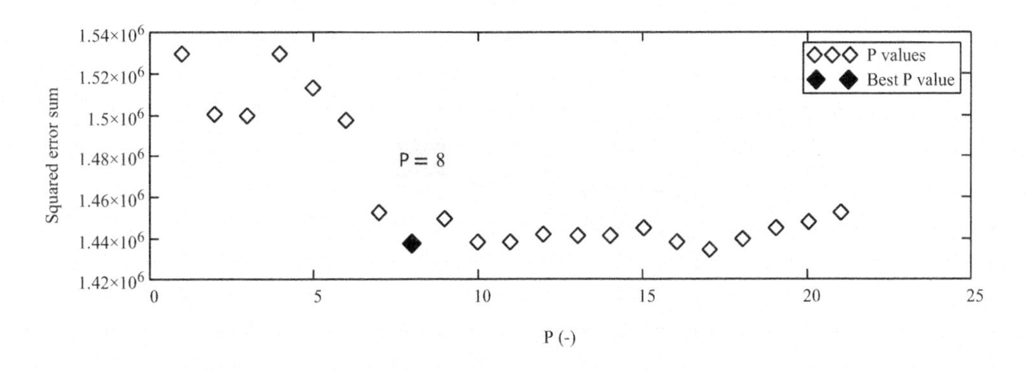

Figure 4.38 Variation of the squared errors sum $\sum_{P}$ with the quantity P of neighboring wells around estimation points. Eight (8) determines the first minimum.

Finally, the final piezometric levels result from the deterministic residual trend values' addition to the optimized grid values.

If necessary, a smoothing filter may be applied to the grid nodes (i, j) to reduce the "bull's-eyes" inevitably associated with inverse distance algorithms.

4.6.2 *Integrated hydraulic head analysis*

Figure 4.39 shows the relief[9] and the significant subvertical fracture field's lineaments around an underground mine[10] on fractured and karstified dolomites (this figure only shows the main features).

The next paragraphs and figures exemplify exploring hydraulic data, particularly the variation of the significant fractures' interconnection degrees.

Figure 4.40 outlines the main talwegs (deep black) of a buried epikarst, under a thick sedimentary cover, inferred by exhaustive terrestrial gravimetry.[11] This epikarst is the upper part of a karst system that stores the incoming water before it percolates to underlying aquifers. This figure also shows (but less accurate) the thalwegs (light gray) of the depressed groundwater table conditioned by the underground pumping of the mine (2018/02/14) that was estimated by a numerical algorithm.[12] The NW and NE fractures (see Figure 4.39) control the WT thalwegs.

The lower groundwater thalwegs, defined after processing the WT contour lines by a numerical algorithm, are nearly parallel or perpendicular to the upper epikarst thalwegs, defined by the previous terrestrial gravimetry survey (see Figure 4.40). Both thalwegs conform remarkably well to the mineralized zone dip and strike, as shown in Figure 4.41. It is essential to emphasize that these two unequal analysis approaches – *electrical resistivity investigation and numerical treatment of the WT contour lines* – revealed similar morphological buried features. That last remark validates the gravimetric survey and the reliable quality of the monitored water table data. The brecciated dolomite, just over 50 m thick, plunges to NW. The steep dipping shear zones, including the hanging wall and the mineralized zone concentrated at the footwall, favors preferential percolation.

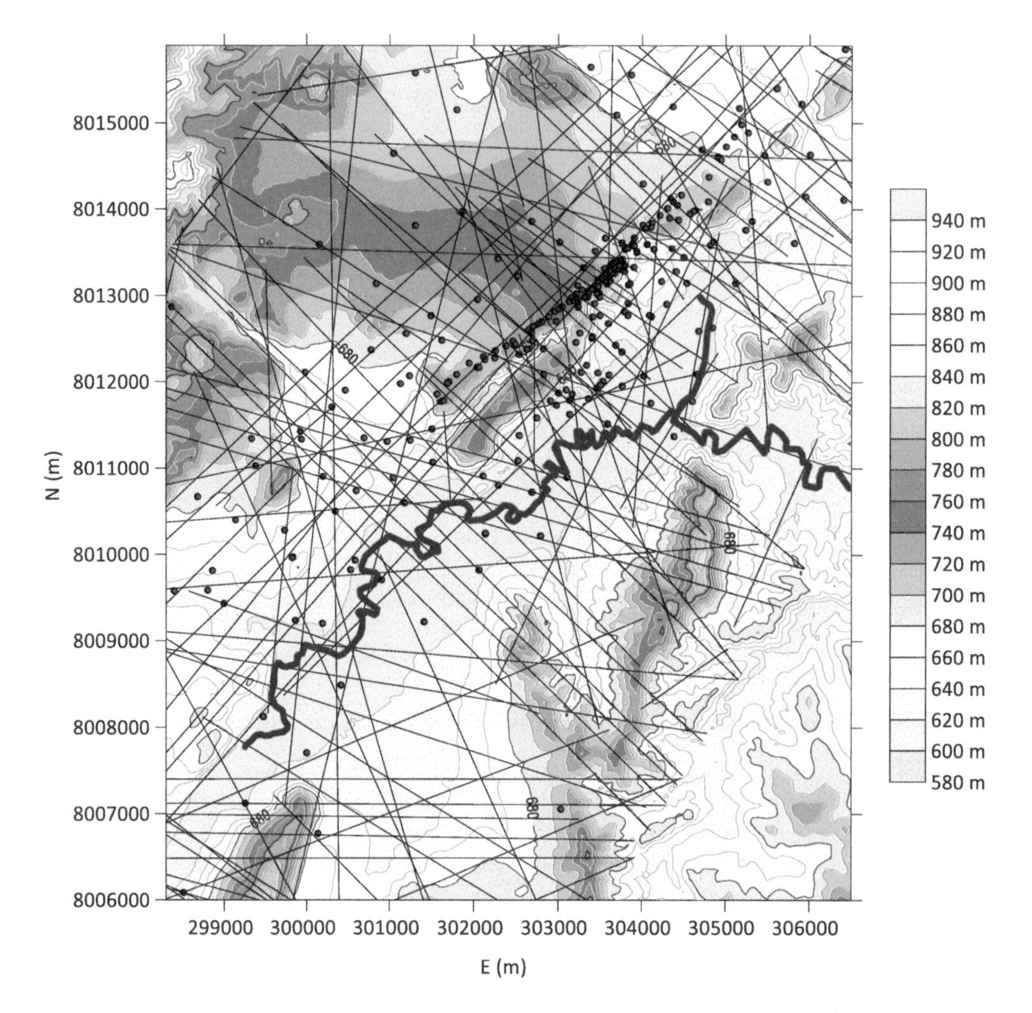

Figure 4.39 Relief of the area of influence of the underground mine of Vazante. This figure includes the WT contour lines, the fracture field, and the control piezometers' location in 2018.

Water table contour grids may be filtered by different algorithms (as low-pass, high-pass, contrast amplification) to enhance some of their peculiarities. A simple filter applied to a node calculates a weighted average of its hydraulic head and the hydraulic heads at its neighboring nodes. If there is enough and well-distributed field data, a simple normalized Laplacian operator may reveal the strike of subvertical fractures, as shown in Figure 4.42, applied to the Vazante Mine during the first years of the dewatering of the underground mine.

Poor hydraulic interconnection between apparently persistent discontinuities may also hamper groundwater flow. Indeed, a model exclusively based on the discontinuities' geometry without considering the extent of their hydraulic connectivity may

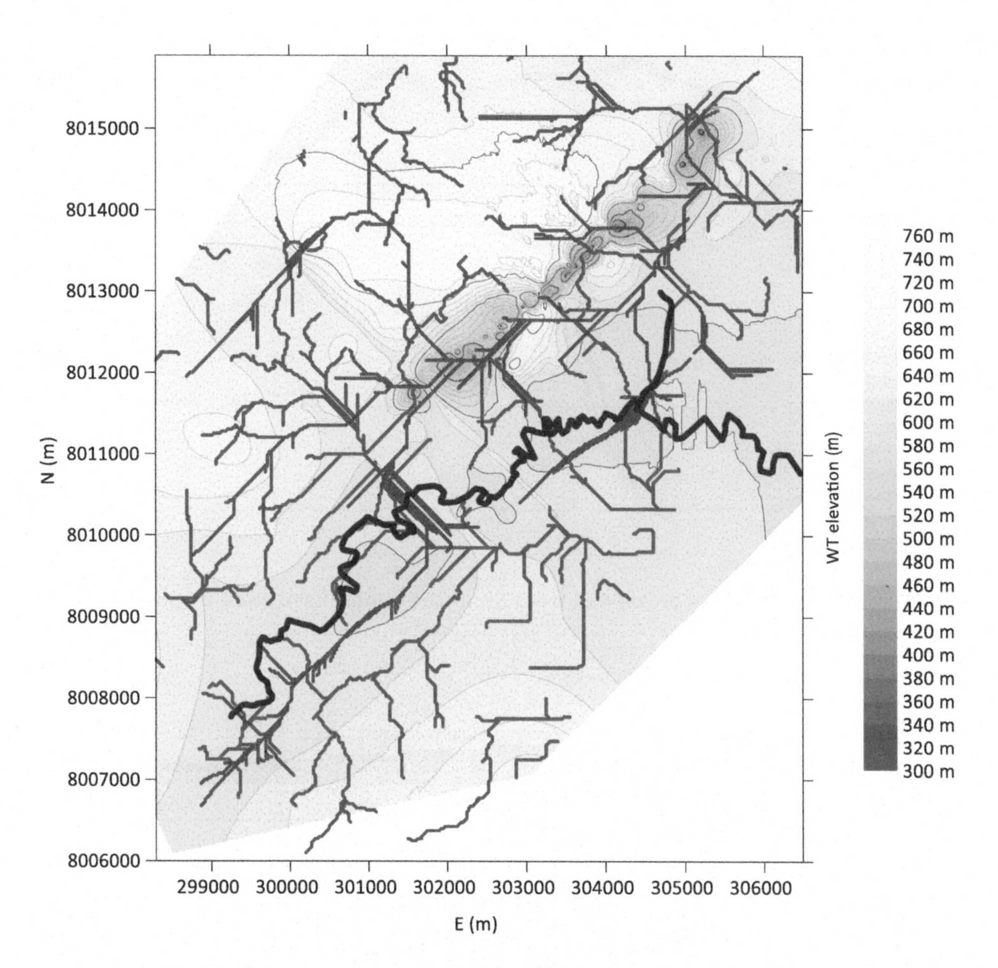

Figure 4.40 Groundwater contour lines of the lowered by Vazante underground mine's pumping equipment. This figure also shows the upper thalwegs (blue lines) of the epikarst and, but less accurate, the groundwater's lower thalwegs (red lines).

overestimate the hydraulic behavior of the rock mass. Intercrossed major discontinuities are frequently interconnected and are not blocked by much smaller ones from hundreds to kilometers long.

Preferential corridors of groundwater flow generally induce similar and quasi-simultaneous water table measurements of piezometers located on their flow paths. Therefore, the dynamic correlation of the variations of WT time series facilitates the comparative judgment of any pair of piezometers' comportment. As comparisons of time-ordered data are essential to get realistic results, one must use nonparametric correlation measures as, for instance, Spearman or Kendal's τ rank correlation coefficients. High dynamic correlation values may imply high hydraulic connectivity. Then, apply these tests, one by one, for all pairs of piezometers under analysis.

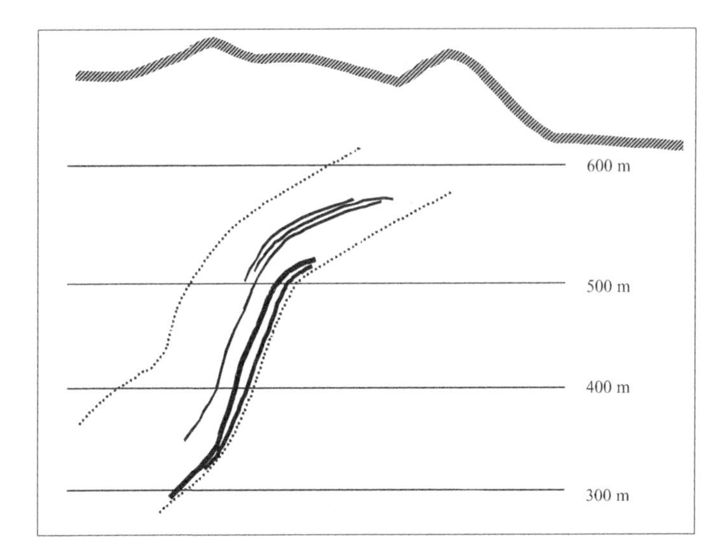

Figure 4.41 Cross-section viewed from SE of the mineralized zone dipping NW.

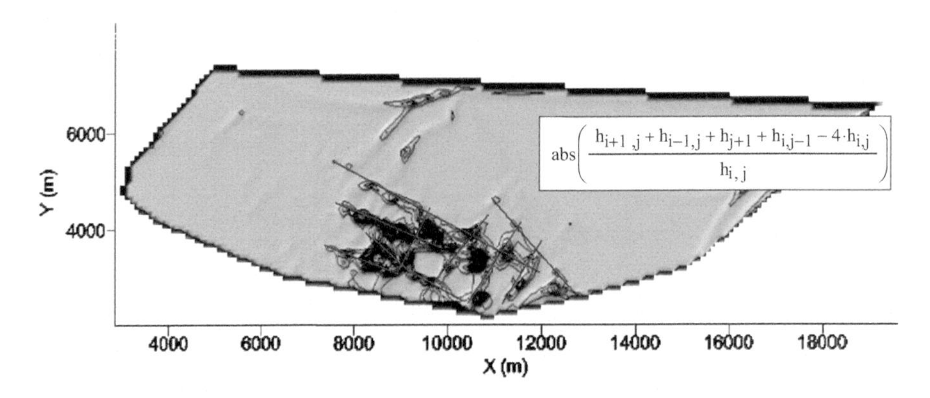

Figure 4.42 A normalized Laplacian operator applied to the hydraulic head contour map (100 × 100 m grid) of the central part in 2009 enhance the anisotropic trends of hydraulic interconnected subvertical fractures (some of them marked out by straight lines).

The Spearman rank correlation coefficient test is a nonparametric hypothesis testing. It is a variant of Pearson's rank correlation coefficient, and instead of applying the Pearson's formula to the original data, it applies to the ranks of the data. Some commercial mathematical software[13] includes a Spearman subroutine to allow rapid comparisons of pairs of WT time series. The reader is encouraged to know more about these specialized software packages.

Figure 4.42 maps the results obtained from testing the dynamic correlation between all pairs of piezometers of the Vazante Mine. In this map, straight lines join piezometers that intercept fractures characterized by a high (rank correlation $r > 0.99$, thicker traces), medium (rank correlation 0.99 > r > 0.98, dash traces), and low hydraulic connectivity (rank correlation 0.98 > r > 0.97, thin dash traces). It is important to remember that these lines do not picture groundwater flow trajectories but only tie by straight line interconnecting piezometers. The prevalence of the NE orientation is remarkable, especially along with the mineralized range. There also exist some connections along NW. The concentration of interconnected piezometers, chiefly if highly correlated, anticipates the necessity of preventing high water inrush when crossing these zones during mine-galleries development.

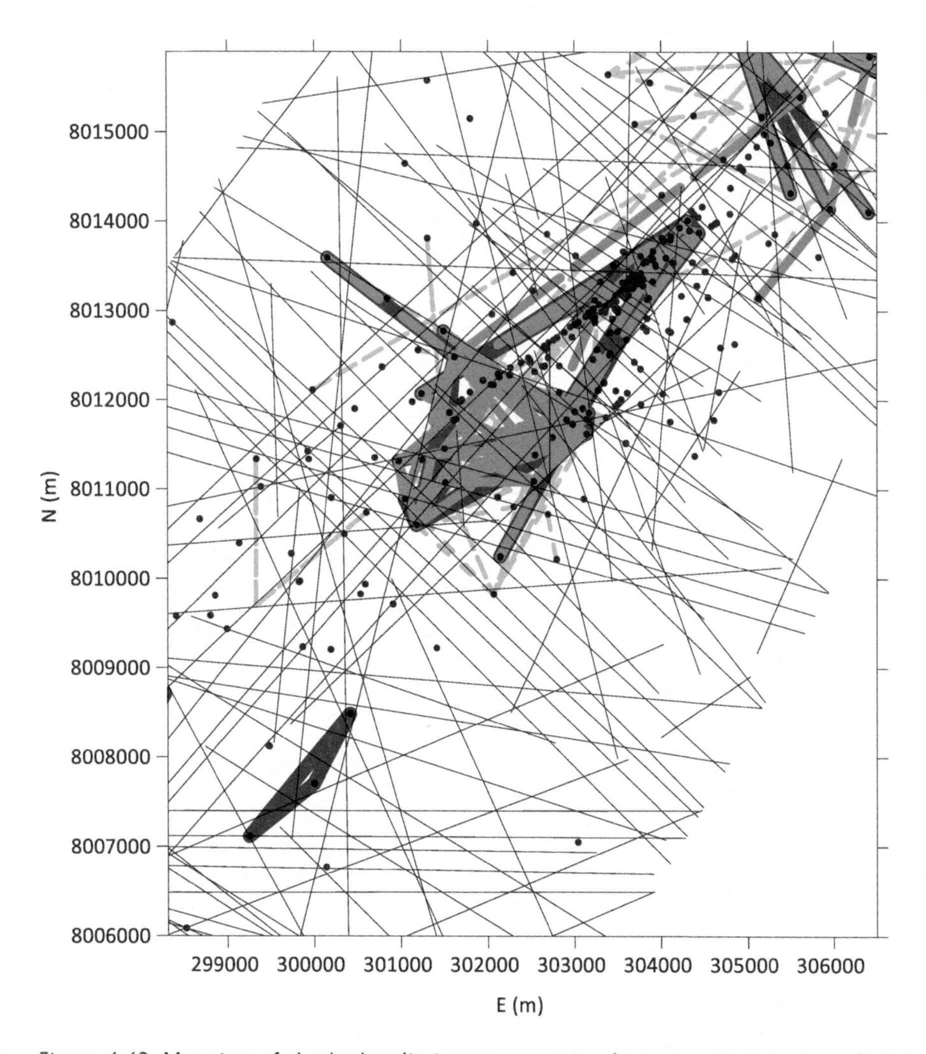

Figure 4.43 Mapping of the hydraulic interconnection between piezometers. It is important to remember that these lines do not picture groundwater flow paths but only unite interconnected piezometers. The concentration of interconnected piezometers, chiefly if highly correlated, anticipates the necessity of preventing high water inrush when crossing these zones during mine-galleries development.

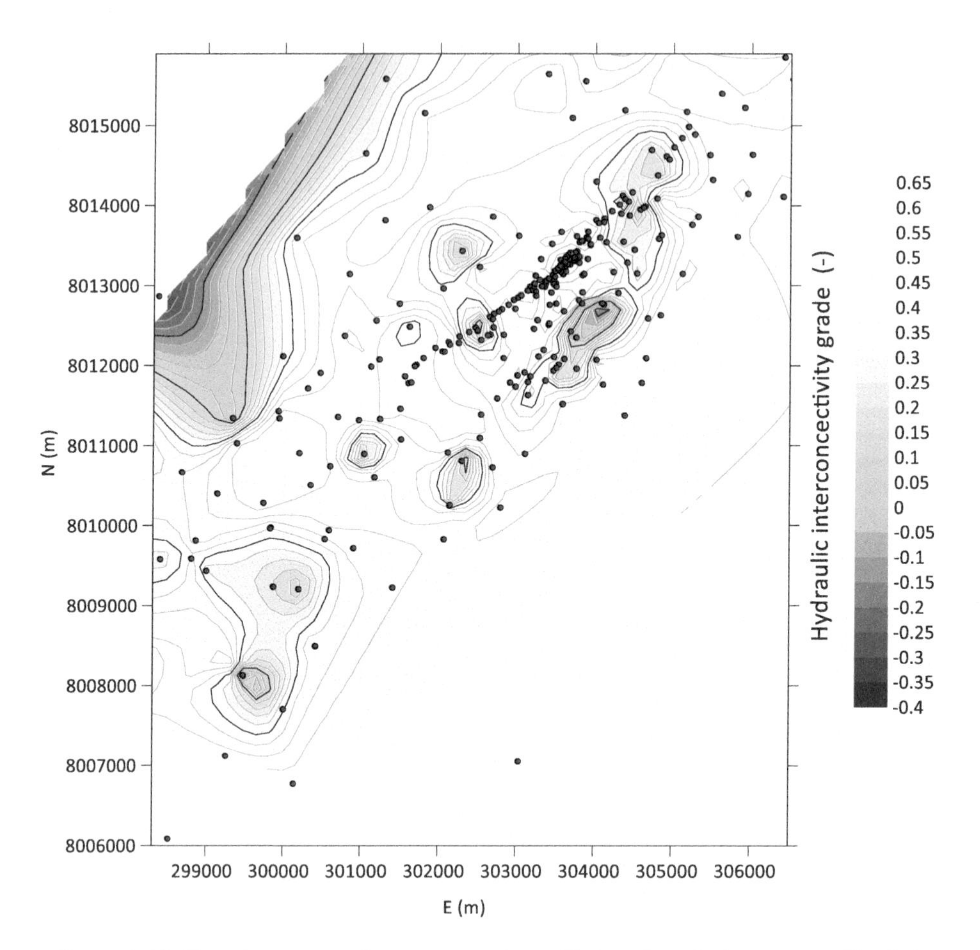

Figure 4.44 Map the average Spearman rank coefficient for all piezometer. The lighter the color, the higher the average hydraulic interconnection, anticipating the necessity of preventing high water inflows when crossing them during the mine galleries development.

Figure 4.43 maps the average Spearman rank coefficient for all piezometer, a procedure that gathers useful information for galleries development. Effectively, the lighter the color in the map, the higher the positive Spearman coefficient. Then, the higher the average hydraulic interconnection, anticipating the necessity of preventing high water inflows when crossing them during the mine-galleries development (Figure 4.44).

Notes

1 Roughly appreciated by its relative areal extent κ as the sum of clogged segments $\sum \delta l_c$ divided by the sum of the lengths of all traces $\sum \delta l$, that is, $\kappa = \sum \delta l_c / \sum \delta l$.

2 "Hydrogeology of the Mercosur aquifer system in the Paraná and Chaco-Paraná Basins, South America, and comparison with the Navajo-Nugget aquifer system, USA"; Arújo L. M., França, A. B., Potter, P. E.; *Hydrogeology Journal*, vol. 7, no. 3, pp. 317–336, Springer, June 1999.

3 Hantush, M. S., Jacob, C. E., "Non-steady radial flow in an infinite leaky aquifer", *Transactions of the American Geophysical Union*, vol. 36, no. 1, pp. 95–100, 1955. To avoid hard calculations the reader may use AQTESOLV software to solve its problems.
4 Cooper, H. H., Jacob, C. E., "A generalized graphical method for evaluating formation constants and summarizing well field history", *Transactions of the American Geophysical Union*, vol. 27, pp. 526–534, 1946. To avoid hard calculations the reader may use AQTESOLV software to solve its problems.
5 As conceptually defined in Chapter 2, Section 2.3.5.2.
6 Zwikker, C., *Physical Properties of Solid Materials*, Pergamon Press., London, 1954.
7 Al-Ambar N., *Infiltration et Ecoulement en Milieu Karstique*, Etude Statistique des Données Climatologiques et Hydrogéologiques, Hydrogéologie du Bassin Versant du Furon, Vercors, France; Université Scientifique et Médicale de Grenoble, 1979.
8 Arnaud, C. L., Lucas, K. L., *Contribution à l'Etude du Plateau de Sornin-Vercors*, Isère, Grenoble, Institut de Géologie, 1972.
9 Digital model relief with 14,875 coordinates SRTM.
10 Vazante Mine, Minas Gerais State, Brazil.
11 Geoph 8, Johannesburg, South Africa.
12 Surfer – Golden Software (https://www.goldensoftware.com/products/surfer).
13 MATHCAD (https://www.mathcad.com) or MATLAB (https://www.mathworks.com/).

References

Bohling, G., 2005, "Kriging," Kansas Geological Survey, C&PE 940, October.

Franciss, F. O., 1994, First Underground Refrigerated LPG Caverns in Brazil, Eurock 94, Rock Mechanics in Petroleum Engineering, Delft, Netherlands.

Kosakowski, G., 2005, *Introduction to Applied Geostatistics*, Paul Scherrer Institut, Villigen.

Press, W. H. et al., 1994, Linear Prediction and Linear Predictive Coding, Chapter 13, in *Numerical Recipes*, Cambridge University Press, 2nd Ed.

Sophocles, J. O., 1992, *Yule-Walker Algorithm and Burg's Method in Optimum Signal Processing*, Macmillan, New York (Mathcad 15 includes a complete algorithm concerning Burg's Method.).

Chapter 5

Finite differences

5.1 Preliminaries

Groundwater flow systems concerning fractured rock masses are generally inhomogeneous, irregularly anisotropic, and particularized by mixed boundary conditions. The numerical simulation of their hydraulic heads' spatial and temporal variation by a single polynomial approximant may bear unacceptable errors. However, fragmenting these complex systems into simple adjoining subsystems, each described by its polynomial approximant, may reduce the solution errors to acceptable levels. These restricted "piecewise" approximants, properly allocated for each subsystem, must fit the boundary conditions connected to their adjacent subsystems. Finite difference and finite element methods make use of that approach.

In the last decades, the decrease in size and cost, concurrently with the increase of speed and memory of commercial computers as well as the development of friendly programming languages, induced most universities, research centers, and independent consultant engineers, geologists, geophysicists, hydrogeologists to explore the possibilities of cost-effective numerical techniques more frequently. Practical problems that require solving partial differential equations applied to relatively complex domains may only be acceptably using realistic numerical techniques. Among them, finite difference algorithms are attractive as they are intuitive and straightforward and suited to spreadsheets or other mathematical programming types. There is a vast literature on this subject, including technical papers and textbooks. The reader is encouraged to gain more knowledge on that subject (Bear, 1979; Wang and Anderson, 1982; Lam, 1994; Press et al., 2007). However, according to the author's plan, this book only discusses the essential concepts of consistent finite difference algorithms, including 2D and 3D complete algorithms applied for hard rocks.

The present example concerns a prediction exercise about how a localized discharge inflow of 8.000 m^3/h of pressurized water at the end of a development gallery will spread throughout the fracture net of the rock mass after a sudden closure. This forced injection aims to recover the original groundwater condition, planned to be done after the end of the underground mine exploitation, and after backfilling all open stopes with suitable material.

Figure 5.1 shows a flowrate map obtained by the finite difference, second-order tensor model approach applied to fractured hard rocks. It exemplifies the application of the methodology detailed in the next items. This model was calibrated considering the local hard rock's anisotropic permeability and the fracture systems' hydraulic properties as

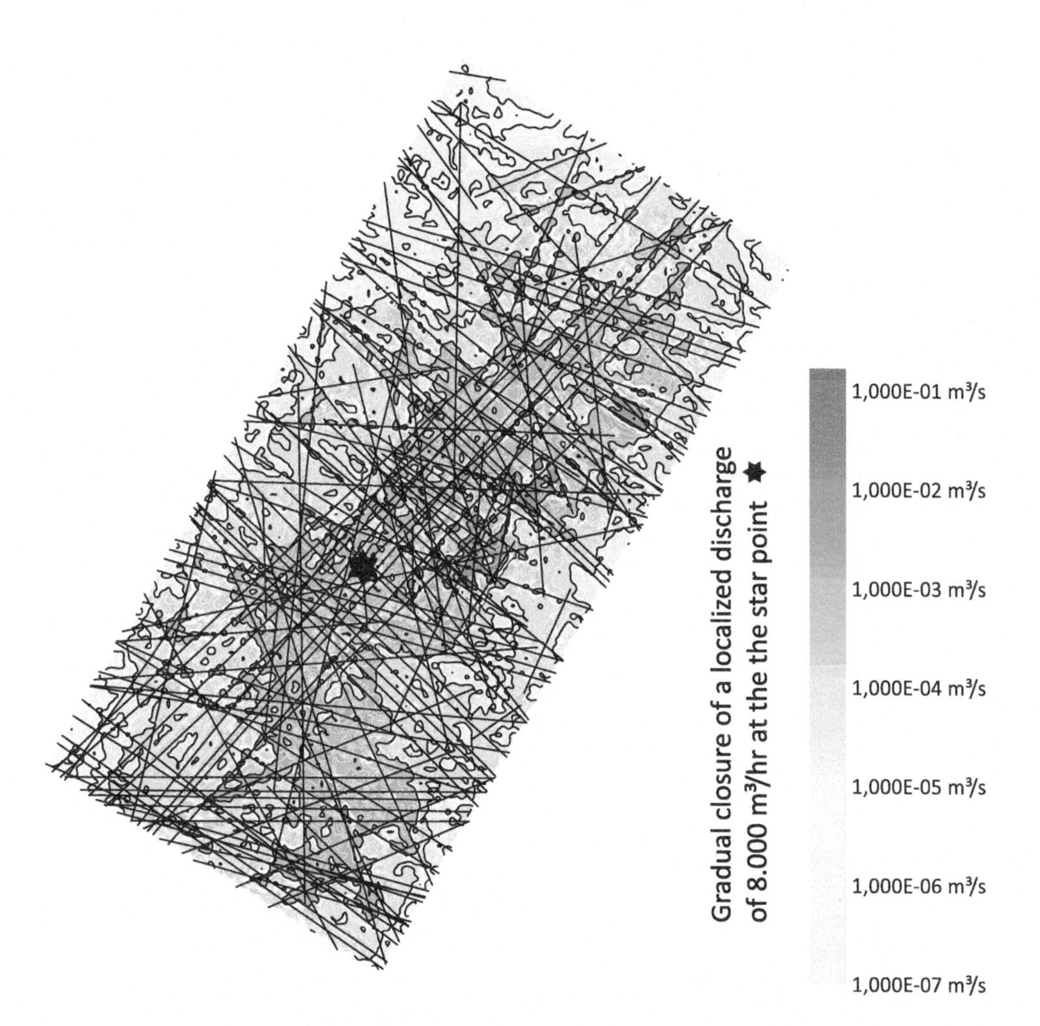

Figure 5.1 This example concerns a prediction exercise about the probable flowrate field following the almost instantaneous closure of a localized discharge of 8.000 m^3/h of pressurized water at the point signaled by a star on the map, aiming to recover the original groundwater condition.

functions of their litho-structural features. The boundary equations included data from all piezometers and all flowrates induced by groundwater pumping (February 2018).

5.2 Modeling hydrogeological systems

5.2.1 Concepts

A limited rock mass may be viewed as an isolated geological system formed by an assemblage of interrelated lithologic and structural features, from minute to large scales, making up an integrated whole, containing around 90%–99% of solid matter and a small proportion of fluid matter in the remaining space. If its fluid phase is in focus, it is labeled a hydrogeological system, generally too complicated to be fully appreciated. However, simplified models may describe only some of its main aspects and behavior but from different points of view, such as a "groundwater management model," in this case, formulated to help decisions about cost-effective operation granting environmental safety.

A numerical simulation devised to predict groundwater flow requires two paired models. The first one, called "conceptual" and necessarily conceived before the second, explains the various modes of occurrence and properties of the fluid phase employing verbal, iconic, and quantified descriptors based on the interpretation of many kinds of field and laboratory investigations, as geological mapping, borehole core samples, geochemical indicators, geophysical parameters, hydrological data. The second one, the "mathematical" model, results from applying hydraulics principles to the conceptual model. It describes the physical and chemical processes involving the percolation of the interconnected and unrestrained fluid phase, including its carrying capacity, in purely mathematical terms and simulates the groundwater hydromechanical behavior for different possible scenarios. Its retro and predictive ability must always be measured against some monitored space-time values, and if detected errors exceed predefined levels, both models must be improved cyclically or replaced by better descriptions. However, depending on its target value, this model may turn to be sophisticated.

Construction of a conceptual model is an *inductive* process involving inferences and correlations from on a finite set of field observations and investigations results. It is not a deterministic outcome from a precise algebraic formula, 100% correct, for any accurate mathematical model. Its degree of realism depends on the hydrogeological system's inherent complexity and the financial constraints of time and energy devoted to building it. It may be more or less acceptable according to the needed output details.

Contrasting to the conceptual model, the mathematical model is a formal *deductive* process, presumably well-formulated, whose reliability is entirely reliant on the conceptual model's premises. Consequently, the conceptual model is by far the most crucial part of a numerical simulation process.

Conceptual models vary on how the main structural features that convey groundwater are treated, their properties are estimated, and their hydraulic performances are described. Conductive fractures and conduits within a subsystem may be modeled as a network of separated but interconnected discontinuities or fused in a single anisotropic entity. Their hydraulic properties may derive from the integration of small-scale statistical data or the interpretation of large-scale tests' results. Corresponding hydraulic behavior may be portrayed for each discontinuity or groups of fractures or by a combination of both sorts of descriptions.[1]

5.2.2 Guidelines for conceptual models

Generally, hard rock masses are inhomogeneous at many scales, and their conceptual models, depending on the type of the numerical approach, may be challenging to construct. However, conceptual models for pseudo-continuous solutions compliant to permeabilities described by second-order tensors are relatively simple to implement for most practical applications. There are no fixed rules to build them but only a few guidelines. Indeed, considering all comments and observations presented in Chapter 1, one may proceed as follows:

- If a dependable topographic map is not available at a scale judged appropriate for the intended simulation, including watercourses, ponds, springs, seeps, and other hydrogeological features of interest, the modeler may compile cartographic and surveys data to generate the model *base map*, preferably as a suitable digital elevation model.
- If not available or not adequate, the modeler may produce a *hydrogeological map* compliant to the compiled *base map* relying on existing geological and hydraulic information properly supplemented by fieldwork, geophysics, geochemistry, borehole sampling, hydraulic field tests.
- As reliable field data are usually scarce, the model boundaries conditions are preferably described for "natural" Dirichlet and Neumann conditions, for instance, lakes, rivers, practically impervious geological contacts, or impervious contacts.
- The modeler must agree with the end-users regarding the network parameters and the cell dimensions to accommodate the details of the numerical simulation and, if needed, subdivide the flow domain into distinct units quasi-homogeneous domains, having almost similar hydraulic properties based on its hydrogeological features and properties, as well as, more or less detailed, according to what is essential.
- For the intact rock matrix within each cell or group of cells, the modeler must evaluate its absolute or relative average 3D permeability tensor based on laboratory tests, published data, or experience.
- For the fractured rock mass within each cell or group of cells, the modeler must evaluate its absolute or relative average 3D permeability tensor, taking into account the intact rock matrix properties and the rock discontinuities. As before, that evaluation can rely on field tests,[2] published data, or experience. All major discontinuities whose trace length exceeds half to a quarter of the base map's smaller horizontal dimension will be considered in the next step.
- Based on geological and structural features, field tests, published data, and experience, the modeler must evaluate the absolute or relative average 3D permeability tensor of the significant discontinuities not included in the previous step. It is essential to keep in mind that their eigenvalues and eigenvectors must implicitly translate their hydraulic transmissivity and interconnectivity at the simulation scale.
- The modeler must consider the influence of significant discontinuities and apply the *summation rule* (see Chapter 1, Section 1.5.4.2) to explicitly integrate the 3D permeability tensor characterizing each cell or group of cells.
- It is necessary to calibrate the 3D permeability tensors of the grid cells by optimizing the numerical simulation based on known time series of hydraulic head values at observation points in the flow domain and its boundaries, as well as, on known time series of flowrates at sources, sinks, and installed wells. That calibration becomes easier if based on relative rather than absolute previously inferred values.
- To improve his model, the modeler must recalibrate the simulation based on time series of the flow domain's gross mass balance, but only if adequately known.
- It is essential to use the same premises and choice criteria of all sensible parameters for the whole model to assure a relative consistency between all the adopted premises, values, and calibration methods. After a few continuous adjustments to factual data, the calibrated simulation may replicate the probable groundwater flow with some dependability.

5.3 Finite difference basics

5.3.1 Difference equations

Finite differences may replace the derivatives occurring in partial differential equations, thus converting these partial differential equations into simpler *difference equations*. As a result, continuity and boundary condition differential equations may be substituted by equivalent difference equations.

On the other hand, any 3D problem domain can be discretized, partitioned into rectangular cuboids defined by the size of their edges and their geometrical centers (see Figure 5.2). After this, continuity difference equations may be applied at all cuboids' centers within the problem domain. Additional supplementary difference equations consist of specified heads and flowrates at all cuboids at the boundaries of the problem domain (Dirichlet and Neumann conditions). These equations, taken simultaneously, generate algebraic equations whose numerical treatment gives the approximate solution of the partial differential equation at all problem domain centers. A similar argument applies to 2D or 1D problems.

5.3.2 Finite differences

A continuous function may now be only valued at regularly spaced points in space-time. Still, with that restriction, its first-order and second-order partial derivatives can be estimated by appropriate finite differences at these points (r, t).

For example, consider a segment of the sloping water table in a horizontal aquifer divided into four cells, 100 m wide (see Figure 5.3). The water table elevation, taken as equivalent to the hydraulic head h by the Dupuit's assumption, is only described at abscissas 150, 250, 350, and 450 m. However, based on simple geometry, it is possible to

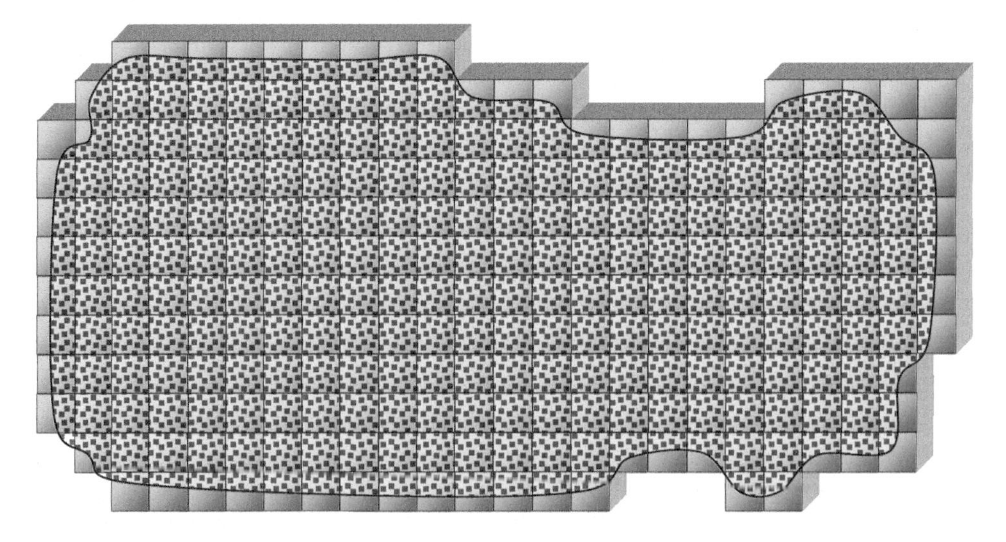

Figure 5.2 Example of rough discretization of a horizontal aquifer.

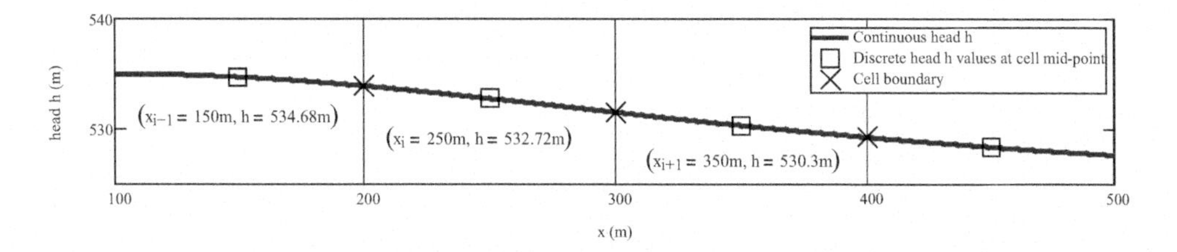

Figure 5.3 Trace of the sloping water table of a horizontal aquifer on an x-cross section divided into four cells, 100 m wide. Three generic contiguous points are defined by (x_{i-1} = 150 m, h = 534.68 m), (x_i = 250 m, h = 532.72 m), and (x_{i+1} = 350 m, h = 530.30 m).

estimate the first and the second-order partial x-derivatives of h at the cell boundaries and cell midpoints, respectively.

Denoting by δx the 100 m equal intervals, the first-order partial x-derivative at the cell boundary at $x_{i-1/2} = 200$ m, just in the middle of $x_{i-1} = 150$ m and $x_i = 250$ m, can be estimated by:

$$\left(\frac{\partial}{\partial x}h\right)_{i+\frac{1}{2}} = \frac{h_{i+1} - h_i}{\partial x} + O\left(\partial x^2\right) \tag{5.1}$$

The additional term $O(\delta x^2)$ in the above equation stands for the approximation error. By a similar rule, the first-order partial x-derivative at $x_{i+1/2} = 300$ m, just in the middle of $x_i = 250$ m and $x_{i+1} = 350$ m, can be estimate by:

$$\left(\frac{\partial}{\partial x}h\right)_{i-\frac{1}{2}} = \frac{h_i - h_{i-1}}{\partial x} + O\left(\partial x^2\right) \tag{5.2}$$

For the above expressions (Equations 5.1 and 5.2), called central approximation formulas, the error is proportional to δx^2. Consequently, dividing the interval δx by two, that is, making new δx = δx/2, would decrease the truncation error to a quarter of the previous one.

The x – ∂h/∂x plot of the numerical first-order partial derivatives at the cell boundaries 200, 300, and 400 m is shown in Figure 5.4.

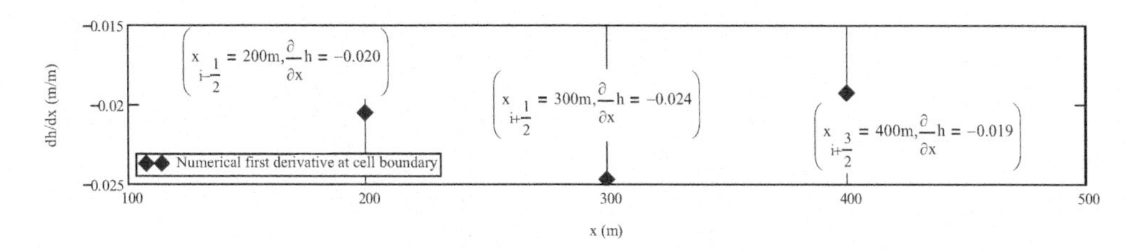

Figure 5.4 x – ∂h/∂x plot of the numerical of the first-order partial derivatives at cell boundaries $x_{i-1/2} = 200$ m, $x_{i+1/2} = 300$ m, and $x_{i+3/2} = 400$ m.

The second-order partial x-derivative at the cell boundary 300 m, in the middle of $x_i = 250$ m and $x_{i+1} = 350$ m, may also be estimated by a central approximation:

$$\left(\frac{\partial^2}{\partial x^2}h\right)_i = \left(\frac{\partial}{\partial x}\frac{\partial}{\partial x}h\right)_i = \frac{\frac{h_{i+1}-h_i}{\partial x} - \frac{h_i - h_{i-1}}{\partial x}}{\partial x} + O\left(\partial x^2\right) \tag{5.3}$$

As before, the term $O(\delta x^2)$ corresponds to the order of magnitude of the approximation error. The $x - \partial^2 h/\partial x^2$ plot of the numerical second-order partial derivatives at cell mid-points 250 and 350 m is shown in Figure 5.5.

In general, an n-order finite difference equation applied at a point x_i may be formally deduced from a backward or a central or a forward Taylor series expansion about this point by ignoring the terms containing higher-order derivatives. The truncation error is proportional to the order of magnitude of the first term of the series discarded part. For a first-order finite difference equation, the truncation error is $O(\delta x)$ for a backward or a forward expansion and $O(\delta x^2)$ for a central expansion.

The first-order and second-order partial derivative approximations estimated by appropriate finite differences, as discussed above, can be applied to any continuous parameter over the problem domain. The numerical algorithm must be constructed, keeping in mind that the hydraulic conductivity is implicitly considered a continuous parameter in the general continuity equation. As a necessary consequence, the hydraulic conductivity must be assumed as a continuous property to assure dependable algorithms, even if highly variable. As an example of that possibility, see Figure 5.6 one of the obtained results of a numerical sensitivity analysis devised to simulate the head decay during the downward percolation of water through a layer of laminated sediments.

5.3.3 Difference equations for steady-state systems

Finite difference equations that are algebraically equivalent to the continuity differential equations for steady-state systems can also be directly derived. As an example, consider a rectangular subsystem centered at point i, defined by edges δx, δy, and δz (see Figure 5.7).

For constant density ρ, the Poisson equation applied to a 1D steady-state system all along the x-axis can be written as:

$$\frac{\partial}{\partial x}\left[k\left(\frac{\partial}{\partial x}h\right)\right] = Q \tag{5.4}$$

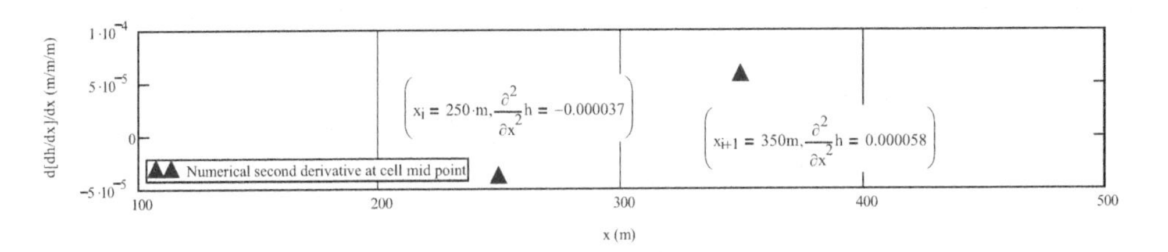

Figure 5.5 $x - \partial^2 h/\partial x^2$ plot of the numerical second-order partial derivatives at cell mid-point $x_i = 250$ m and $x_{i+1} = 350$ m.

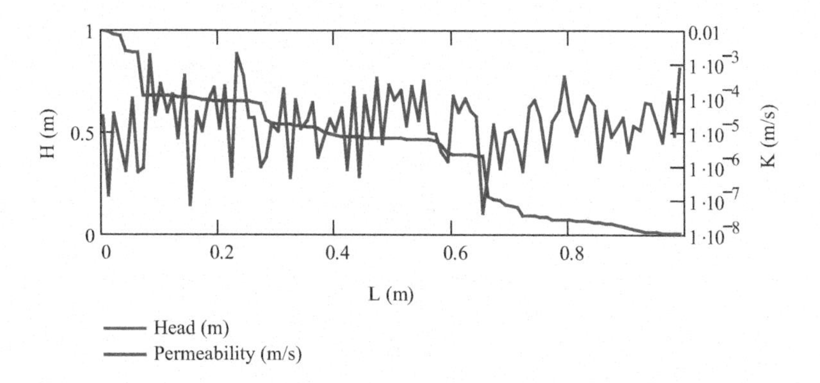

Figure 5.6 One of the numerical results of a sensitivity analysis about the head decay associated with the vertical percolation of water in an encapsulated integral core sample, measuring 1 m length, extracted from a thick layer of laminated sediments. Heads are read on the left Y-axis. Jagged vertical permeabilities are read on the right Y-axis (log scaled). The permeability of each 1 cm thick lamina varied randomly between 10^{-8} and 10^{-2} m/s.

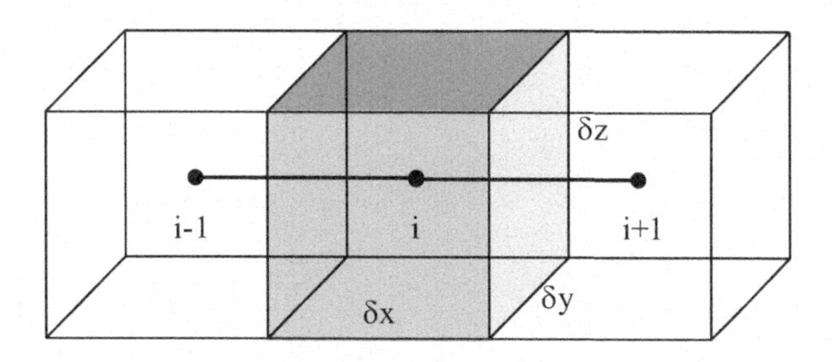

Figure 5.7 Subsystem with edges δx, δy, and δz centered at point x_i. Water enters left area δyδz at point $x_{i-1/2}$ and leaves right area δyδz at point $x_{i+1/2}$ under hydraulic gradients $(h_i - h_{i-1})/\delta x$ and $(h_{i+1} - h_i)/\delta x$, respectively.

Considering the hydraulic conductivity k as a continuous variable, the left member of the above equation can be expanded by the product rule, and the above expression takes the form:

$$\frac{\partial}{\partial x}k\left(\frac{\partial}{\partial x}h\right)+k\left(\frac{\partial}{\partial x}\frac{\partial}{\partial x}h\right)=Q \tag{5.5}$$

Consider now three contiguous generic points: x_{i-1}, x_i, and x_{i+1}. The finite difference approximation for the first term of the left member must be applied to the central point x_i to reduce the truncation error and guarantee a stable algorithm, as done by taken the average of the approximations $x_{i-1/2}$ and $x_{i+1/2}$.

$$\left[\frac{\partial}{\partial x}k\left(\frac{\partial}{\partial x}h\right)\right]_i=\frac{\left[\frac{\partial}{\partial x}k\left(\frac{\partial}{\partial x}h\right)\right]_{i-\frac{1}{2}}+\left[\frac{\partial}{\partial x}k\left(\frac{\partial}{\partial x}h\right)\right]_{i+\frac{1}{2}}}{2}+O\left(\partial x^2\right) \tag{5.6}$$

Then, the equivalent finite difference equation for the above expression is:

$$\left[\frac{\partial}{\partial x}k\left(\frac{\partial}{\partial x}h\right)\right]_i = \frac{\left(\frac{k_{i+1}-k_i}{\delta x}\right)\left(\frac{h_{i+1}-h_i}{\delta x}\right)+\left(\frac{k_i-k_{i-1}}{\delta x}\right)\left(\frac{h_i-h_{i-1}}{\delta x}\right)}{2}+O\left(\partial x^2\right) \quad (5.7)$$

By the same arguments, the finite difference equation for the second product of the left member is:

$$k\left(\frac{\partial}{\partial x}\frac{\partial}{\partial x}h\right)=k_i\frac{\left(\frac{h_{i+1}-h_i}{\delta x}\right)-\left(\frac{h_i-h_{i-1}}{\delta x}\right)}{\delta x}+O\left(\delta x^2\right) \quad (5.8)$$

Simplifying and collecting terms, the finite difference equivalent for the considered continuity equation (Equation 5.4) is:

$$\frac{\left(\frac{k_{i+1}+k_i}{2}\right)\left(\frac{h_{i+1}-h_i}{\delta x}\right)-\left(\frac{k_i+k_{i-1}}{2}\right)\left(\frac{h_i-h_{i-1}}{\delta x}\right)}{\delta x}=Q \quad (5.9)$$

Multiplying both members of the above expression by the subsystem volume, and considered that $\delta\delta y\delta z=\delta S$ and $\delta x\delta y\delta z=\delta V$, yields:

$$\left[\frac{\left(\frac{k_{i+1}+k_i}{2}\right)\left(\frac{h_{i+1}-h_i}{\delta x}\right)-\left(\frac{k_i+k_{i-1}}{2}\right)\left(\frac{h_i-h_{i-1}}{\delta x}\right)}{\delta x}\right]\delta S=Q\,\delta V \quad (5.10)$$

The above expression has a clear physical meaning. The left member measures the difference between the entering and leaving discharges along the x-axis through the two opposite faces of area δS. The right member measures the amount of water volume $Q \cdot \delta V$ added or drained by an occasional source (or sink) of strength Q per unit volume.

It is crucial to keep in mind that this algorithm considers the hydraulic conductivity as a continuous variable but only valued at the cell centers. In this case, the hydraulic conductivity at midpoints $x_{i-1/2}$ and $x_{i+1/2}$ must be estimated by the *arithmetic mean*. However, if one assumes that the hydraulic conductivity takes constant values inside each cell, forming a series circuit along the x-axis, it is recommended to approximate the hydraulic conductivity at midpoints $x_{i-1/2}$ and $x_{i+1/2}$ by a *harmonic mean* (Bear, 1972). A simple exercise to enhance these two types of means' numerical effects is shown in Figure 5.8. Considering the premises that support the pseudo-continuous concept for random fractures, implicitly assuming averages between cell mid-points, the author prefers the arithmetic mean based on his modeling experience.

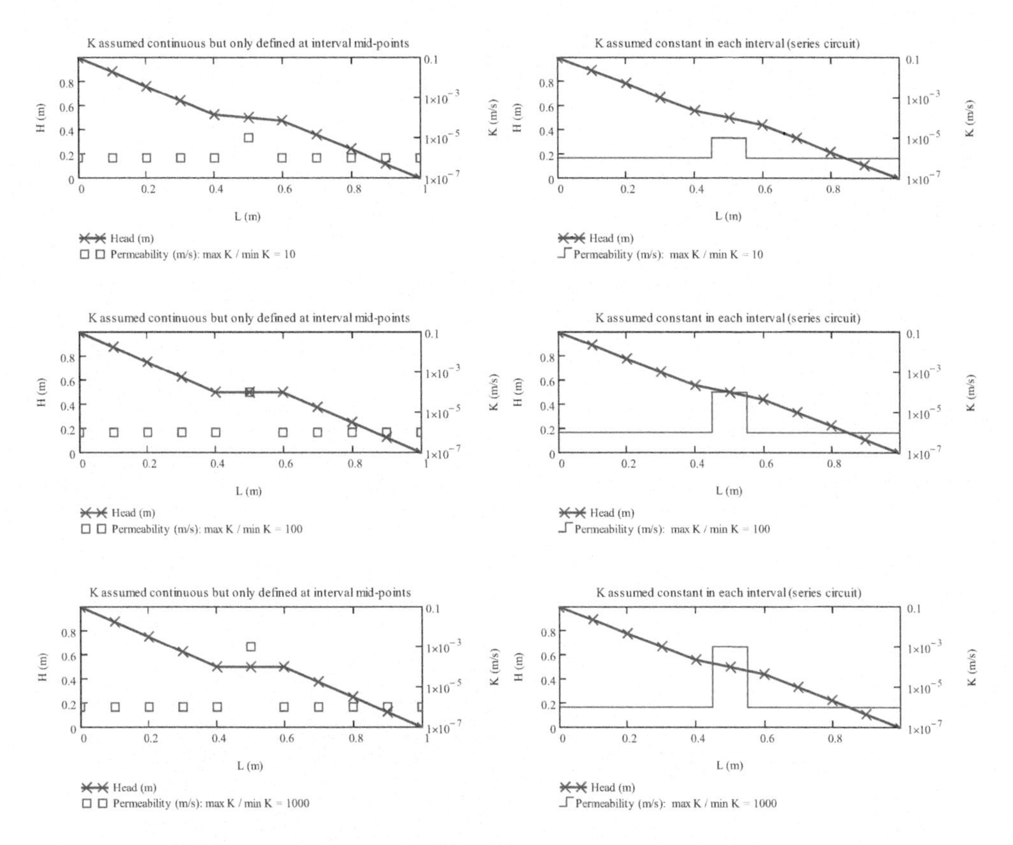

Figure 5.8 The five pairs of L × H plots above give an idea about the influence of the type of mean, arithmetic, or harmonic, used in the finite difference algorithm. Compare the head decay's numerical simulation results for a steady upward percolation of water from the beginning to the end of a homogenous soil column with height L = 1 m under a constant head of ΔH = 1 m. The soil permeability is 10^{-6} m/s in all cases except for a 10 cm layer in the column center that is N times more pervious than the rest of the soil. The left and right plots correspond, respectively, to the arithmetic and harmonic means for N equal to 10, 100, and 1,000, as indicated in the plots. For N = 10, there is no appreciable difference between the left and right plots. For N = 100 or 1,000, the left plots show negligible decays for the central cell, as expected, while the right plots remain practically unchanged.

5.3.4 Difference equations for unsteady-state systems

For unsteady-state systems, a 1D full continuity equation system all along the x-axis can be written as:

$$\frac{\partial}{\partial x}\left[k\left(\frac{\partial}{\partial x}h\right)\right]=S_V\left(\frac{\partial}{\partial t}h\right)+Q \qquad (5.11)$$

The term S_V is the specific volumetric storage.

As done with the space-variable x, the time variable t is also discretized into many steps of equal time intervals δt counted from a starting time t_0. Then, the product $n \cdot \delta t$ measures the time interval $(t - t_0)$.

The first-order partial t-derivative $\partial h/\partial t$ occurring in the right member of the above governing equation may be approximated at point x_i and at time step $n + 1/2$, that is just in the middle of the time steps n and $n + 1$, then:

$$\left[S_V\left(\frac{\partial}{\partial t}h\right)\right]^{n+\frac{1}{2}} + Q = S_V\left[\frac{(h_i)^{n+1} - (h_i)^n}{\delta t}\right] + Q + O\left(\delta t^2\right) \tag{5.12}$$

In the above equation, the pseudo-exponents n and $n + 1$ stand for the values of the head h_i at the moments $(n) \cdot \delta t$ and $(n + 1) \cdot \delta t$ (see Figure 5.9).

The finite expression for space-continuity at x_i, on the left member of the equation, is undefined at the t-step $n + 1/2$. However, this value may be taken as the arithmetic

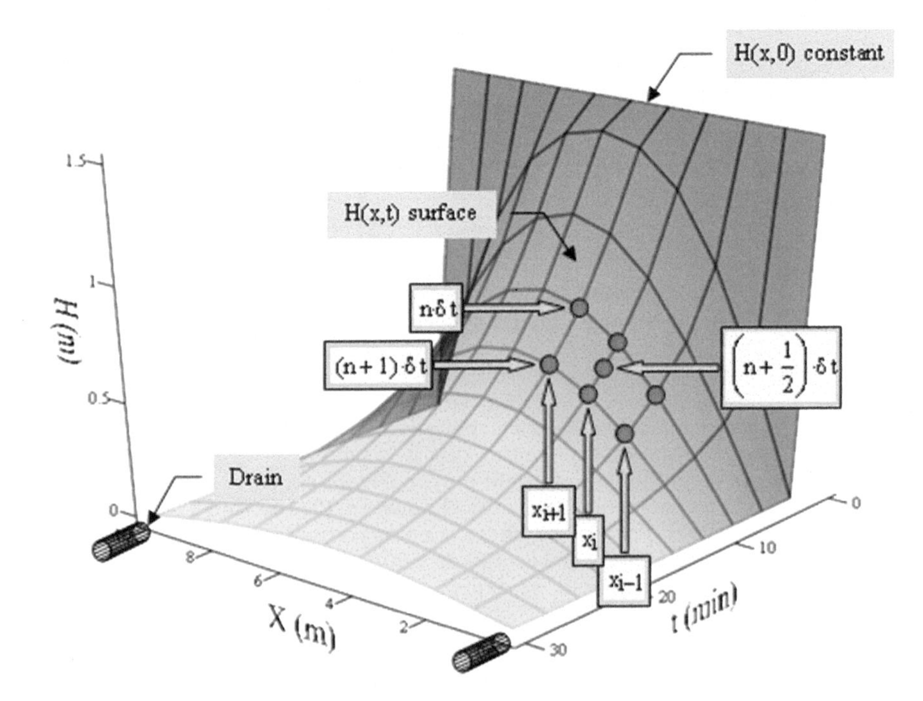

Figure 5.9 Transient surface H(x, t) estimated by a Crank–Nicolson scheme. The first-order partial t-derivative $\partial h/\partial t$ is approximated at point x_i and at time step $n + 1/2$, that is, just in the middle of the t-steps n and $n + 1$. The finite expression for the space continuity is taken as the arithmetic average of the finite expressions at point x_i at t-steps n and $n + 1$.

average of the corresponding finite expressions at time steps n and n + 1. Then, for each t-step n + 1/2, the equivalent finite difference equation for continuity is:

$$\frac{1}{2}\left[\begin{array}{l}\left(\frac{k_{i+1}+k_i}{2}\right)\left[\frac{(h_{i+1})^n-(h_i)^n}{\delta x}\right]-\left(\frac{k_i+k_{i-1}}{2}\right)\left[\frac{(h_i)^n-(h_{i-1})^n}{\delta x}\right]\ldots \\ +\left(\frac{k_{i+1}+k_i}{2}\right)\left[\frac{(h_{i+1})^{n+1}-(h_i)^{n+1}}{\delta x}\right]-\left(\frac{k_i+k_{i-1}}{2}\right)\left[\frac{(h_i)^{n+1}-(h_{i-1})^{n+1}}{\delta x}\right]\end{array}\right]\delta S\ldots=0$$
$$-\left[S_V\left[\frac{(h_i)^{n+1}-(h_i)^n}{\delta t}\right]+Q\right]\delta V \tag{5.13}$$

As before, the pseudo-exponents stand for the t-steps n and n+1. This algorithm is unconditionally stable and second-order accurate in time and space. It generates for each t-step a set of simultaneous equations that must be solved to find the head values h corresponding to that t-step. That smart solution was conjointly proposed by John Crank (1916–2006), a British mathematical physicist dedicated to the numerical solution of partial differential equations, and Phyllis Nicolson (1917–1968), also a British mathematician associated with the development of the Crank–Nicolson scheme in the mid-20th century.

5.3.5 Difference equations for boundary conditions

If the finite difference equations are applied to the interior nodes of a steady-state groundwater system for which all the heads at the exterior nodes, that is, at the boundary nodes, are already specified (Dirichlet boundary conditions), then the number of unknown heads and algebraic equations are the same. However, the number of unknowns exceeds the number of algebraic equations if the discharge rates at one or more exterior nodes are specified as an alternative for heads (Neumann boundary conditions). In this case, supplementary equations are needed to solve the problem.

For example, consider the Neumann prescription applied to a boundary point x_{i-1}. In this case, the hydraulic exit gradient $\partial h/\partial x_{i-1}$ at that point can be approximated by a second-order accurate backward difference:

$$\left(\frac{\partial}{\delta x}h\right)_{i-1}=\frac{3h_{i-1}-4h_i+h_{i+1}}{2\cdot\delta x}+O\left(\delta x^2\right) \tag{5.14}$$

Recall that a quadratic interpolation polynomial collocated at three successive and collinear point x_{i-1}, x_i, and x_{I+1} separated by equal intervals δx yields this same algorithm.

In Equation 5.15, the heads h_{i-1}, h_i, and h_{l+1} refer to the successive three collinear points x_{i-1}, x_i, and x_{l+1}. These points are located along an interior perpendicular to the boundary at x_{i-1}. The corresponding Neumann condition is:

$$k\frac{3h_{i-1}-4h_l+h_{i+1}}{2\delta x}\delta_x\delta_z=-q_{i-1} \tag{5.15}$$

Then, it is possible to express h_{i-1} in terms of q_{i-1}, h_i, and h_{i+1}:

$$h_{i-1}=\frac{1}{3}\left(4h_i-h_{i+1}-2\frac{\delta x}{\delta x\ \delta z}\frac{q_{i-1}}{k}\right) \tag{5.16}$$

Replacing h_{i-1} occurring in all algebraic equations for the left-hand expression of Equation 5.16 equates the number of equations and unknowns. After solving this set for the interior heads, the Neumann condition at remaining border heads h_{i-1} is directly calculated.

For unsteady simulations, both conditions may be time-dependent values, if this is the case.

There are other ways to apply boundary conditions, including approximated solutions for refined and non-regular boundaries (see, for example, Lam, 1994).

5.3.6 Simultaneous difference equations

5.3.6.1 Preliminaries

For extensive groundwater systems, described by inhomogeneous and anisotropic subsystems, full of different kinds of wells, drains, natural springs, and sinkholes, particularized by many types of boundary and initial conditions, the resulting set of simultaneous difference equations may involve several thousands of unknowns. According to their hydraulic properties, boundary, and initial conditions, solving these equations yield the head values at the discrete cuboid centers within the problem domain or its borders.

Only two iterative techniques, known as Jacobi and Gauss–Seidel methods, will be considered next. Carl Gustav Jacob Jacobi (1804–1851), Johann Carl Friedrich Gauss (1777–1855), and Philipp Ludwig von Seidel (1821–1896) were outstanding and inspired 19th-century German mathematicians. These methods are easily programmable with the help of commercial mathematical and graphical software. See some specialized sites: www.office.microsoft.com/excel, www.ibm.com/software/lotus, www.ptc.com, www.mathworks.com, www.wolfram.com.

5.3.6.2 Gauss–Seidel iterative routine

A simple exercise, Start of Example 5.1, provides an easy way to grasp the essence of the Gauss–Seidel iteration routine.

Example 5.1

In this 2D simple exercise, groundwater flows within a homogeneous, isotropic, horizontal, and confined sandstone aquifer trapped by an upper and a bottom layer of impervious shale. Two almost parallel and vertical diabase dikes but practically impervious isolate 1,000 m broad sandstone strata. A piece of that band, measuring 1,000 × 1,000 m in planar view, may be taken as a simple groundwater system to be numerically simulated (see Figure 5.10).

Now, consider a 2D square lattice defined by 20 × 20 squares of 50 m size (see Figure 5.11).

Denoting by $h_{i,j}$ the head value at the center (i, j) of each square and by δx its side dimension, the finite difference equation is simply:

$$h_{i,j} = \frac{1}{4}\left(h_{i-1,j} + h_{i+1,j} + h_{i,j-1} + h_{i,j+1} + \frac{q_{i,j}}{T}\delta x^2 \right)$$

That equation applied at the 400 interior nodes of the 2D square lattice generates a set of simultaneous equations characterized by a band diagonal matrix for the head coefficients. In this case, the Gauss–Seidel method may be used with success.

The heads along the left and right sides of the grid, respectively, defined by $x_1 = 0$ m and $x_{21} = 1{,}000$ m, are prescribed by the Dirichlet condition (see Figure 5.12). The upper side and lower side of the grid, defined by $y_1 = 0$ m and $y_{20} = 1{,}000$ m, are no-flow borders except at point ($x_{18} = 850$ m, $y_1 = 0$ m), where 50 m^3/h of water leaks through a "filtered" fissure in the frontal dike. The sandstone constant transmissivity is 3.5×10^{-4} m^2/s.

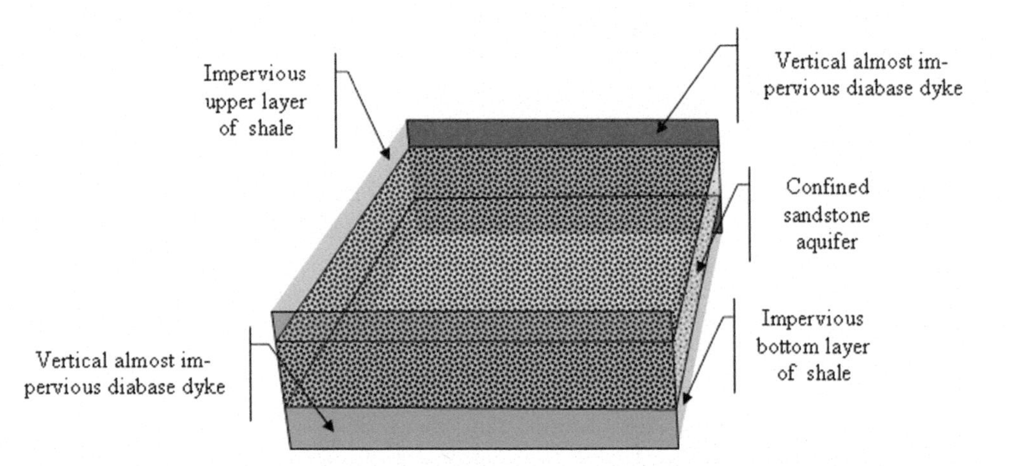

Figure 5.10 Confined sandstone aquifer, "sandwiched" between an upper and a lower shale layers. Two parallel and vertical impervious diabase dikes, one in the front and the other in the rear of the perspective, isolate the aquifer laterally.

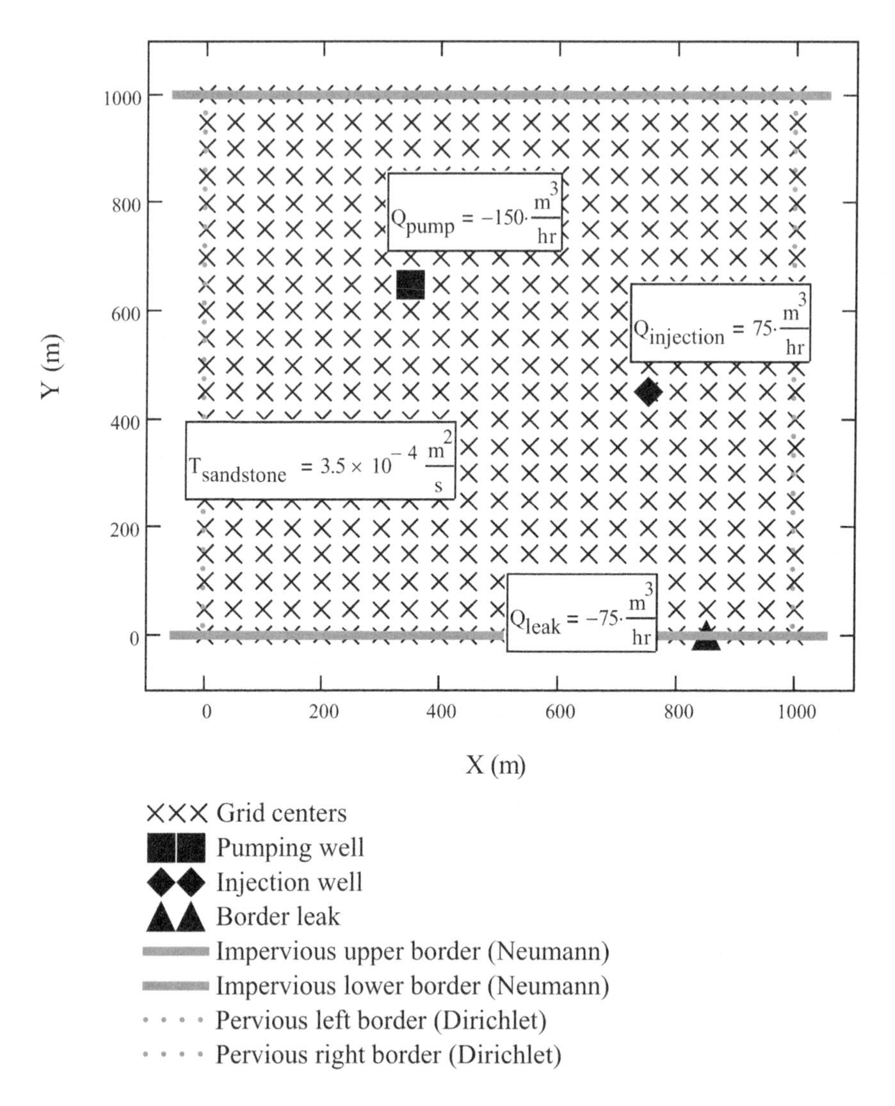

Figure 5.11 Top view: a 2D square lattice formed by 400 squares of 50 m size. To each grid center (i, j), defined by coordinates (x_i, y_j), corresponds a generic head $h_{i,j}$. Four adjacent heads are described by four triplets: $(x_{i-1}, y_j, h_{i-1,j})$, $(x_{i+1}, y_j, h_{i+1,j})$, $(x_i, y_{j-1}, h_{i,j-1})$, and $(x_i, y_{j+1}, h_{i,j+1})$. Subscripts i and j run from 1 to 21.

Moreover, 150 and 75 m^3/h of water are pumped and injected into at two wells, respectively, located at the points ($x_8 = 350$ m, $y_{14} = 650$ m) and ($x_{16} = 750$ m, $y_{10} = 450$ m). A uniform withdrawal and accretion simulate the pumping and injection wells on the top of both cells, numerically equal to the pumping and the injection rate divided by the top area δx^2. For the breach in the frontal dike, the constant leakage corresponds to the total leak divided by the lateral area $\delta x \delta z$.

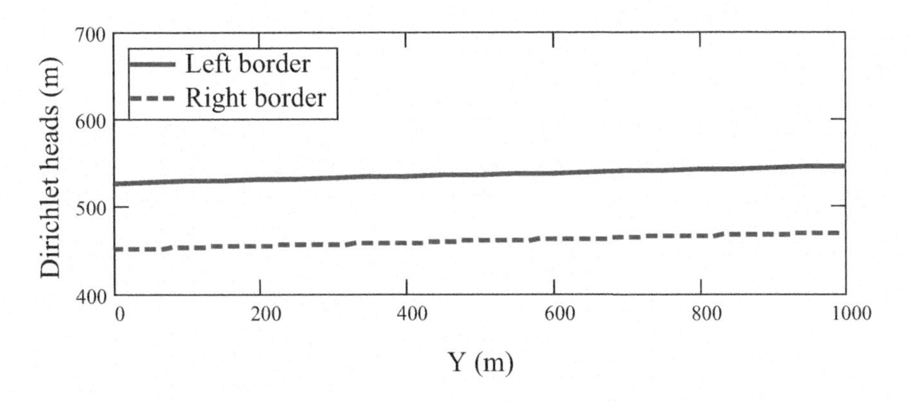

Figure 5.12 Specified head along the left and right borders (Dirichlet boundary conditions).

To start a Gauss–Seidel iteration, an initial inexact solution array for all unknown heads $h_{i,j}$ must be initially guessed. There are many ways to do that. For example, it is possible to fit a surface h_0 (x, y) to the specified boundary heads. Another approach is to find a simplified solution employing an analytical method. Then, for all interior points (i, j), where $q_i = 0$, except for the pump point (8, 14), injection point (16, 10), and near border points along the rows (i, 2) and (i, 20), the general finite difference equation gives the head value $h_{i,j}$:

$$h_{i,j} = \frac{1}{4}\left(h_{i-1,j} + h_{i+1,j} + h_{i,j-1} + h_{i,j+1}\right)$$

The heads $h_{8,14}$ at the pump point (8, 14) and $h_{16,10}$ at injection point (16, 10) are given by:

$$h_{8,14} = \frac{1}{4}\left[h_{7,14} + h_{9,14} + h_{8,13} + h_{8,15} + \frac{\left(Q_{pump}\right)_{8,14}}{T}\right]$$

$$h_{16,10} = \frac{1}{4}\left[h_{15,10} + h_{17,10} + h_{16,9} + h_{16,11} + \frac{\left(Q_{injection}\right)_{16,10}}{T}\right]$$

In these expressions, Q_{pump} and $Q_{injection}$ are negative and positive quantities, respectively.

For the near border points along the rows (i, 2) and (i, 20), the general expressions for $h_{i,j}$ include unknown heads along the boundaries (i, 1) and (i, 21). At these border points, the corresponding exit gradients may be approximated by three-point formulas:

$$\left(\frac{\partial}{\partial y}h\right)_{i,1} = \frac{-h_{i,3} + 4h_{i,2} - 3h_{i,1}}{2\delta y} + O\left(\delta y^2\right)$$

$$\left(\frac{\partial}{\partial y}h\right)_{i,21} = \frac{3h_{i,21} - 4h_{i,20} + h_{i,19}}{2\delta y} + O\left(\delta y^2\right)$$

Now, for the impervious boundaries, all exit gradients normal to the no-flow borders are zero. Then, equating these approximations for $\partial h/\partial y$ to zero at points (i, 1) and (i, 21) it is possible to express $h_{i,1}$ and $h_{i,21}$ in terms of the interior heads, except for the leak at point (18, 1):

$$h_{i,1} = \frac{1}{3}\left(4h_{i,2} - h_{i,3}\right)$$

$$h_{i,21} = \frac{1}{3}\left(4h_{i,20} - h_{i,19}\right)$$

Substituting $h_{i,1}$ and $h_{i,21}$ for the left side of these expressions into the general finite difference equation at points along the rows (i, 2) and (i, 20), then removing these unknowns, except near the leak point (18, 1), we get the following equations:

$$h_{i,2} = \frac{1}{4}\left[h_{i-1,2} + h_{i+1,2} + \frac{1}{3}\left(4h_{i,2} - h_{i,3}\right) + h_{i,3}\right]$$

$$h_{i,20} = \frac{1}{4}\left[h_{i-1,20} + h_{i+1,20} + h_{i,19} + \frac{1}{3}\left(4h_{i,20} - h_{i,19}\right)\right]$$

Simplifying and collecting we get:

$$h_{i,2} = \frac{3}{8}h_{i-1,2} + \frac{3}{8}h_{i+1,2} + \frac{1}{4}h_{i,3}$$

$$h_{i,20} = \frac{3}{8}h_{i-1,20} + \frac{3}{8}h_{i+1,20} + \frac{1}{4}h_{i,19}$$

At the leak point (18, 1), Neumann condition states that:

$$\frac{-h_{18,3} + 4h_{18,2} - 3h_{18,1}}{2\delta y}\delta x\ \delta z = \frac{\left(Q_{leak}\right)_{18,1}}{k}$$

Considering that $\delta x = \delta y$ and that $k \cdot \delta z = T$, this equation simplifies to:

$$-h_{18,3} + 4h_{18,2} - 3h_{18,1} = -2\frac{\left(Q_{leak}\right)_{18,1}}{T}$$

In that expression, Q_{leak} is also a negative quantity, from which:

$$h_{18,1} = \frac{1}{3}\left[4h_{18,2} - h_{18,3} + \frac{2(Q_{leak})_{18,1}}{T}\right]$$

Substituting $h_{18,1}$ for the left side of that expression into the general finite difference equation at point (18, 2) removes that unknown:

$$h_{18,2} = \frac{1}{4}\left[h_{17,2} + h_{19,2} + \frac{1}{3}\left[4h_{18,2} - h_{18,3} + \frac{2(Q_{leak})_{18,1}}{T}\right] + h_{18,3}\right]$$

Simplifying and collecting:

$$h_{18,2} = \frac{3}{8}h_{17,2} + \frac{3}{8}h_{19,2} + \frac{1}{4}h_{18,3} + \frac{1}{4}\frac{(Q_{leak})_{18,1}}{T}$$

Having expressed all interior heads in terms of its interior neighbors, the iteration procedure starts from the initial solution estimate $[h_{i,j}]^0$, running from i=1 to 21 and from j=2 to 20 (because columns (1, j) and (21, j) are Dirichlet prescribed) as follows:

- At all interior points (i, j), except at the pump point (8, 14), injection point (16, 10), and near border points along rows (i, 2) and (i, 20):

$$\left(h_{i,j}\right)^{k+1} = \frac{1}{4}\left[\left(h_{i-1,j}\right)^{k+1} + \left(h_{i+1,j}\right)^{k} + \left(h_{i,j-1}\right)^{k+1} + \left(h_{i,j+1}\right)^{k}\right]$$

- At pump point (8, 14) and injection point (16, 10):

$$\left(h_{8,14}\right)^{k+1} = \frac{1}{4}\left[\left(h_{7,14}\right)^{k+1} + \left(h_{9,14}\right)^{k} + \left(h_{8,13}\right)^{k+1} + \left(h_{8,15}\right)^{k} + \frac{\left(Q_{pump}\right)_{8,14}}{T}\right]$$

$$\left(h_{16,10}\right)^{k+1} = \frac{1}{4}\left[\left[\left(h_{15,10}\right)^{k+1} + \left(h_{17,10}\right)^{k} + \left(h_{16,9}\right)^{k+1} + \left(h_{16,11}\right)^{k}\right] + \frac{\left(Q_{injection}\right)_{16,10}}{T}\right]$$

- At all near border points (i, 2) and (i, 20), except near the leak point (18, 1):

$$\left(h_{i,2}\right)^{k+1} = \frac{3}{8}\left(h_{i-1,2}\right)^{k+1} + \frac{3}{8}\left(h_{i+1,2}\right)^{k} + \frac{1}{4}\left(h_{i,3}\right)^{k}$$

$$\left(h_{i,20}\right)^{k+1} = \frac{3}{8}\left(h_{i-1,20}\right)^{k+1} + \frac{3}{8}\left(h_{i+1,20}\right)^{k} + \frac{1}{4}\left(h_{i,19}\right)^{k+1}$$

- Near the leak point (18, 1):

$$\left(h_{18,2}\right)^{k+1} = \frac{3}{8}\left(h_{17,2}\right)^{k+1} + \frac{3}{8}\left(h_{19,2}\right)^{k} + \frac{1}{4}\left(h_{18,3}\right)^{k} + \frac{1}{4}\frac{\left(Q_{leak}\right)_{18,1}}{T}$$

In the above iteration formulas, the pseudo-exponents k or k + 1 stand, respectively, for the approximations of the head values $h_{i,j}$ after k or k + 1 iteration. Moreover, it takes for granted that these systematic iterations are orderly performed so that iterations k + 1 for $h_{i-1,j}$ and $h_{i,j-1}$ are always calculated before the iteration k + 1 for $h_{i,j}$, thus avoiding the need for extra storage.

Iteration stops when the highest relative change after iteration k + 1, calculated by $\left[\left(\left(h_{i,j}\right)^{k+1} - \left(h_{i,j}\right)^{k}\right)/\left(h_{i,j}\right)^{k}\right]$, taking into account all iterated values $h_{i,j}$ or only at some specific and critical points is smaller than a prescribed small limit ε:

$$\max\left[\frac{\left(h_{i,j}\right)^{k+1} - \left(h_{i,j}\right)^{k}}{\left(h_{i,j}\right)^{k}}\right] < \varepsilon$$

At the end of the iteration procedure, all head values along with the border points (i, 1) and (i, 21), except for the leak point, are calculated by the Neumann condition:

$$h_{i,1} = \frac{1}{3}\left(4h_{i,2} - h_{i,3}\right)$$

$$h_{i,21} = \frac{1}{3}\left(4h_{i,20} - h_{i,19}\right)$$

At the leak point (18, 1):

$$h_{18,1} = \frac{1}{3}\left[4h_{18,2} - h_{18,3} + \frac{2\left(Q_{leak}\right)_{18,1}}{T}\right]$$

The resulting contour heads are shown in Figure 5.13.

The Gauss–Seidel iteration procedure is an organized improvement of a 19th-century old method credited to Jacobi, where, unlike Gauss–Seidel, old values $(h_{i-1,j})^k$ and $(h_{i,j-1})^k$ are never replaced for recently iterated values within the same iteration loop, as formally pointed out in the general Jacobi iterative formula bellow:

$$\left(h_{i,j}\right)^{k+1} = \frac{1}{4}\left(h_{i-1,j}\right)^{k} + \left(h_{i+1,j}\right)^{k} + \left(h_{i,j-1}\right)^{k} + \left(h_{i,j+1}\right)^{k} + \frac{q_{i,j}}{T}\delta x^2$$

For most practical applications, both methods converge. However, according to the author's personal experience, there are cases for which Jacobi converges, and Gauss–Seidel does not (Varga, 1962).

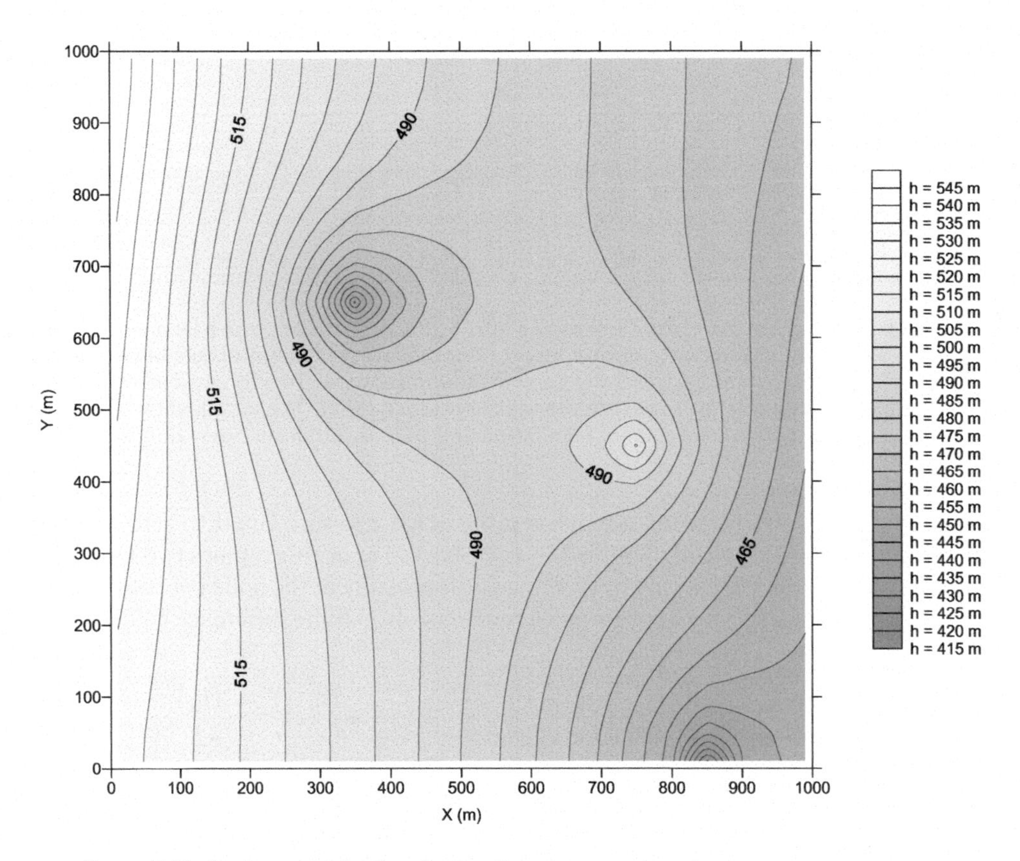

Figure 5.13 Equipotentials distribution for the exercise above.

5.3.6.3 Crank–Nicolson iterative routine

The next exercise, Example 5.2, as done for the Gauss–Seidel method, provides an easy way to apply the Crank-Nicolson iterative routines for each t-step unsteady simulations.

Example 5.2

In this elementary exercise, consider the temporary uplift pressure relief under a 10 m wide airport concrete pavement, as shown in Figure 5.14.

Any point of the contact joint between pavement and subbase may be located by its abscissa x, varying from 0 to 10 m. To apply a finite difference algorithm, that joint may be discretized into 20 equal cells, each one measuring $\delta x_i = 0.5$ m, where the subscript i runs from 1 to 20. A maximum uplift pressure $(h_i)^0 = 1.5$ m at all cell nodes occurs after an exceptional inundation and before full drain operation at time t_0; this defines the problem's initial conditions. Immediately after the start of the full drainage, the

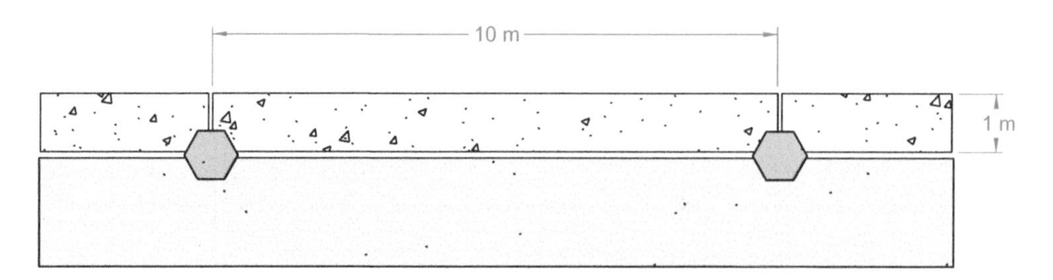

Figure 5.14 Schematic cross section of a 10 m wide airport concrete pavement. A pervious and continuous contact joint demarcates the interface between the pavement base and the cement-bound subbase. Two longitudinal drains at its extremities assure the uplift pressure relief. The initial water pressure after an exceptional inundation and before full drain operation is 1.5 m.

two longitudinal drains permanently keep the uplift pressure equal to zero at both extremities of the pavement, identified by cells δx_1 and δx_{20}; this typifies two Dirichlet boundary conditions. Denoting by $(h_i)^n$ the uplift pressure at the node i of each cell δx_i at the t-step $t_n = t_0 + n \cdot \delta t$, the appropriate Crank–Nicolson finite difference equation is:

$$(h_i)^n - (h_i)^{n-1} = FN\left[\frac{(h_{i-1})^{n-1} - 2(h_i)^{n-1} + (h_{i+1})^{n-1}}{2} + \frac{(h_{i-1})^n - 2(h_i)^n + (h_{i+1})^n}{2}\right]$$

The non-dimensional factor FN is called the Fourier number and is defined by:

$$FN = \frac{T\,\delta t}{S_C \delta x^2}$$

Now, the pseudo-exponents n and n+1 denote the t-step order instead of the iteration cycle index as in the Gauss–Seidel method. The variables T, S_c, δx, and δt stand, respectively, for contact transmissivity, contact storativity, grid cell length, and time interval.

Applying the finite difference equation at the center of each contact cell generates a set of 20 simultaneous linear equations that have to be solved for each t-step after t_0 either by an iterative routine, easily programmable, or by a more computer-efficient non-iterative method.

Keeping always the heads at both edge cells equal to zero, that is $(h_1)^1 = (h_{20})^1 = 0$, the iterative routine formula determines the heads $(h_i)^1$ for the first t-step $t_1 = t_0 + \delta t$:

$$(h_i)^1 = \frac{1}{2}\frac{FN\left[(h_{i-1})^0 - 2(h_i)^0 + (h_{i+1})^0 + (h_{i-1})^1 + (h_{i+1})^1\right] + 2(h_i)^0}{1 + FN}$$

Note that for the Crank–Nicolson numerical scheme, a low F_n number is not required for stability; however, it is required for numerical accuracy. The first iteration stops when the highest relative change after the last iteration cycle is smaller than a prescribed small limit ε.

For the consecutive t-steps $t_n = t_{n-1} + n \cdot \delta t$ defined by n = 2, 3, 4 …, the extreme values must always be equal to zero, that is $(h_1)^n = (h_{20}) = 0$. Then, the iteration formulas for the subsequent iterative routines are:

$$(h_i)^2 = \frac{1}{2}\frac{FN\left[(h_{i-1})^1 - 2(h_i)^1 + (h_{i+1})^1 + (h_{i-1})^2 + (h_{i+1})^2\right] + 2(h_i)^1}{1+FN}$$

$$(h_i)^3 = \frac{1}{2}\frac{FN\left[(h_{i-1})^2 - 2(h_i)^2 + (h_{i+1})^2 + (h_{i-1})^3 + (h_{i+1})^3\right] + 2(h_i)^2}{1+FN}$$

$$(h_i)^{n+1} = \frac{1}{2}\frac{FN\left[(h_{i-1})^n - 2(h_i)^n + (h_{i+1})^n + (h_{i-1})^{n+1} + (h_{i+1})^{n+1}\right] + 2(h_i)^n}{1+FN}$$

$$(h_i)^n = \frac{1}{2}\frac{FN\left[(h_{i-1})^{n-1} - 2(h_i)^{n-1} + (h_{i+1})^{n-1} + (h_{i-1})^n + (h_{i+1})^n\right] + 2(h_i)^{n-1}}{1+FN}$$

If correctly applied, the Gauss–Seidel method may be used in each iterative routine to find $(h_i)^n$ at the n t-step.

Considering $T = 1/3 \times 10^{-6}\,m^2/s$, $S_c = 2 \times 10^{-5}$, $\delta x = 0.5\,m$, and a time interval $\delta t = 1\,s$, the corresponding Fourier number is FN = 0.667. After performing 190 iterative routines, the resulting x-h curves for t-steps of 0, 20… 160 s are shown in Figure 5.15. The resulting x-t-h surface corresponding to the uplift pressure decay is shown in Figure 5.16.

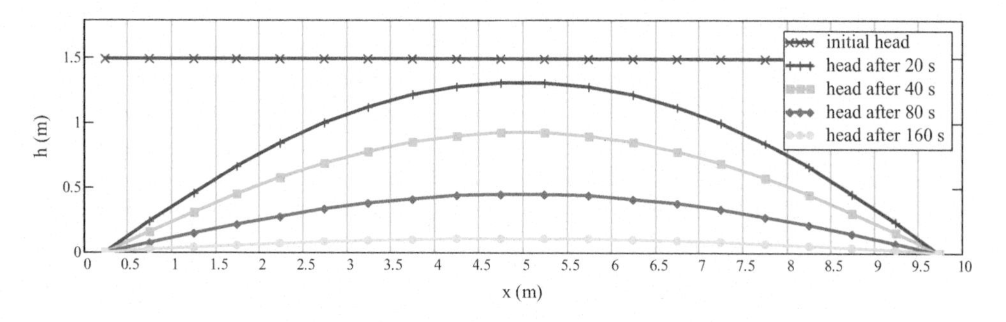

Figure 5.15 Plot of x-h curves for t-steps of 0 s, 20 s … 160 s. Note that for homogeneous parameters, the problem's solution is symmetric; this means that it could be solved considering only half-pavement and prescribing zero-flow at x_5 (Neumann boundary condition).

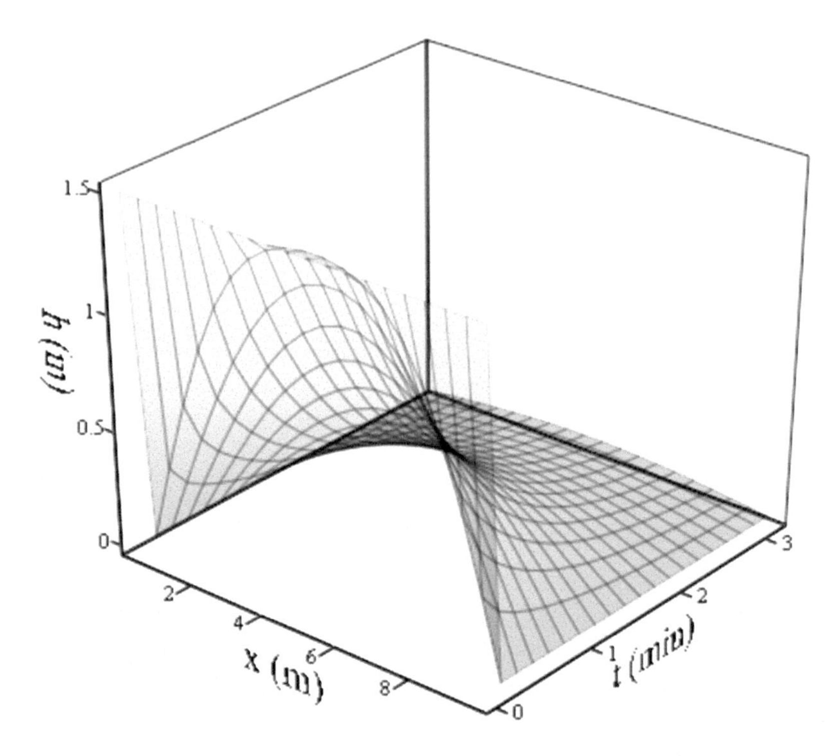

Figure 5.16 Plot of the x-t-h surface for t-steps continuously running from 0 to 190 s.

5.4 Finite differences algorithms for hard rocks

5.4.1 Preliminaries

Finite differences algorithms for general anisotropic models are simple to deduce from basic concepts and, for practical purposes, may reasonably simulate the hydraulic behavior of anisotropic rocks, including its discontinuities, even if geometrically complex.

This item presents some 3D finite differences algorithms to solve direct groundwater flow problems through randomly fractured rock masses. Related 2D and 1D algorithms are easily derived from these 3D solutions (see Chapter 2).

Whenever possible, it is convenient to reduce an intricate fractured rock mass system into an assemblage of quasi-homogeneous subsystems having quasi-uniform characteristics and dominant anisotropy. Each quasi-homogeneous subsystem may be further subdivided into smaller subsystems but preserving common properties depending on the desired details. To refine, even more, it is always possible to individualize properties for each elementary subsystem.

Before applying finite difference algorithms, it is essential to keep in mind that, as any system is a small part of a larger one, its boundary conditions must replicate the flow regime connecting both to preserve mass input–output balance. Wrong assumptions may produce a reliable pattern of the equipotentials but leading to incorrect predictions of hydraulic gradients and flow rates. Besides acceptable boundary and initial conditions hypotheses, practical solutions of direct problems require a sound knowledge of the spatial distribution of the hydraulic anisotropic transmissivity [T] or conductivity [k].

5.4.2 Steady-state solutions

5.4.2.1 Dupuit's approximation

5.4.2.1.1 UNCONFINED FLOW

Assume the following premises:

- Pervious media anisotropic and inhomogeneous referred to an orthogonal frame XYZ (axis z: zenithal).
- Subsystems or cells are defined by upright prismatic columns but having gently inclined bases (*see* Figure 5.17).
- Hydraulic gradients estimated by finite differences

NB: Estimates by finite differences are equivalent to approximate local hydraulic heads by a second-order interpolation polynomial collocated at the center of mass of the middle subsystem and the adjacent subsystems.

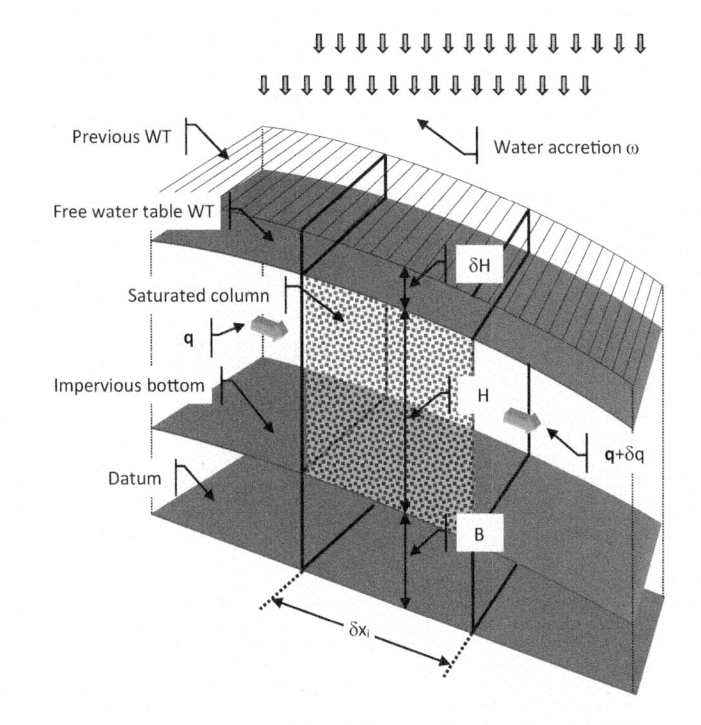

Figure 5.17 Subsystem defined by a vertical parallelepiped of which the base is an inclined parallelogram.

Consider the following list of symbols:

- i, j: indices of the coordinate (x_i, y_j) of the middle cell (vertical z-axis pass through this point)
- $i-1$: index one step to the West of i.
- $i+1$: index one step to the East of i.
- $j-1$: index one step to the South of j.
- $j+1$: index one step to the North of j.
- [k] (L/T): anisotropic hydraulic conductivity, defined by the second-order tensor:

$$k = \begin{pmatrix} k_{xx} & k_{xy} \\ k_{yx} & k_{yy} \end{pmatrix} \tag{5.17}$$

- **J** (–): local hydraulic gradient, defined by the vector:

$$\mathbf{J} = \begin{pmatrix} J_x \\ J_y \end{pmatrix} \tag{5.18}$$

- **q** (L/T): local specific discharge, defined by the vector:

$$\mathbf{q} = \begin{pmatrix} q_x \\ q_y \end{pmatrix} = \begin{pmatrix} q_{xx} + q_{xy} \\ q_{yx} + q_{yy} \end{pmatrix} \tag{5.19}$$

- q_x (L/T): component of the specific discharge **q** along the x-direction resulting from the two parallel sub-components' addition: $q_{xx}+q_{xy}$.
- q_y (L/T): component of the specific discharge **q** along the y-direction resulting from the addition of the two parallel sub-components: $q_{yx}+q_{yy}$.
- k_{xx} (L/T): component of [k], responds to the discharge vector qxx parallel to the x-axis induced by the hydraulic gradient component $\mathbf{J}_x$.
- k_{xy} (L/T): component of [k], responds to the discharge vector q_{xy} parallel to the x-axis induced by the hydraulic gradient component $\mathbf{J}_y$.
- k_{yy} (L/T): component of [k], responds to the discharge vector qyy parallel to the y-axis induced by the hydraulic gradient component $\mathbf{J}_y$.
- k_{yx} (L/T): component of [k], responds to the discharge vector qyx parallel to the y-axis induced by the hydraulic gradient component $\mathbf{J}_y$. It is essential to remind that as $k_{xy}=k_{yx}$ implies $q_{xy}=q_{yx}$.
- B (L): base elevation that may have a variable but a relatively gentle inclination, referred to an arbitrary horizontal datum.
- H (L): hydraulic head measured from the base elevation B; if measured from the arbitrary horizontal datum: $H_{datum}=B+H$.
- δx (L): the base dimension of the prismatic column along the x-direction.
- δy (L): the base dimension of the prismatic column along the y-direction.
- Q_x (L^3/T): discharge across the vertical x-cross section: $Q_x=q_x \cdot H \cdot \delta y$.
- Q_y (L^3/T): discharge across the vertical y-cross section: $Q=q_y \cdot H \cdot \delta x$.

- $Q_{x,y}$ (L^3/T): sum of discharges Q_x and Q_y: $Q_{x,y}=Q_x+Q_y$.
- ω (L/T): vertical unit-area recharge.

It is possible to deduce the finite difference algorithm by applying the continuity principle for constant water density.

Consider the coordinates (x_i, y_j) of the prismatic cell's central point conveniently defined by the indices (i, j). Steady flow Q_i and Q_j continuously traverse the orthogonal cross sections $H \cdot \delta y_i$ and $H \cdot \delta x_j$ of that cell. Total $Q_{i,j}$ flow is $Q_{i,j}=Q_i+Q_j$ and according to the Darcy's Law may be calculated by:

$$Q_{i,j} = -\begin{pmatrix} \delta y & 0 \\ 0 & \delta x \end{pmatrix} \begin{bmatrix} (k_{xx})_{i,j} & (k_{xy})_{i,j} \\ (k_{yx})_{i,j} & (k_{yy})_{i,j} \end{bmatrix} \cdot H_{i,j} \begin{bmatrix} \left[\frac{\partial}{\partial x}(B+H)\right]_{i,j} \\ \left[\frac{\partial}{\partial y}(B+H)\right]_{i,j} \end{bmatrix} \begin{pmatrix} 1 \\ 1 \end{pmatrix} \tag{5.20}$$

Then, Equation 5.21 can be expanded as:

$$Q_{i,j} = -H_{i,j} \left[\begin{bmatrix} (k_{xx})_{i,j}\cdot\delta y & (k_{xy})_{i,j}\cdot\delta y \\ (k_{yx})_{i,j}\cdot\delta x & (k_{yy})_{i,j}\cdot\delta x \end{bmatrix} \cdot \begin{bmatrix} \left(\frac{\partial}{\partial x}B\right)_{i,j} \\ \left(\frac{\partial}{\partial y}B\right)_{i,j} \end{bmatrix} + \begin{bmatrix} (k_{xx})_{i,j}\cdot\delta y & (k_{xy})_{i,j}\cdot\delta y \\ (k_{yx})_{i,j}\cdot\delta y & (k_{yy})_{i,j}\cdot\delta x \end{bmatrix} \cdot \begin{bmatrix} \left(\frac{\partial}{\partial x}H\right)_{i,j} \\ \left(\frac{\partial}{\partial y}H\right)_{i,j} \end{bmatrix} \right] \cdot \begin{pmatrix} 1 \\ 1 \end{pmatrix} \tag{5.21}$$

Recalling the derivation rules and substituting H^2 for V, the nonlinear terms can be "linearized." Then, Equation 5.21 takes the form:

$$Q_{i,j} = -\left[\begin{bmatrix} (k_{xx})_{i,j}\cdot\delta y & (k_{xy})_{i,j}\cdot\delta y \\ (k_{yx})_{i,j}\cdot\delta x & (k_{yy})_{i,j}\cdot\delta x \end{bmatrix} \cdot \begin{bmatrix} (V_{i,j})^{\frac{1}{2}}\left(\frac{\partial}{\partial x}B\right)_{i,j} \\ (V_{i,j})^{\frac{1}{2}}\left(\frac{\partial}{\partial y}B\right)_{i,j} \end{bmatrix} + \begin{bmatrix} (k_{xx})_{i,j}\cdot\delta y & (k_{xy})_{i,j}\cdot\delta y \\ (k_{yx})_{i,j}\cdot\delta x & (k_{yy})_{i,j}\cdot\delta x \end{bmatrix} \cdot \begin{bmatrix} \frac{1}{2}\left(\frac{\partial}{\partial x}V\right)_{i,j} \\ \frac{1}{2}\left(\frac{\partial}{\partial y}V\right)_{i,j} \end{bmatrix} \right] \cdot \begin{pmatrix} 1 \\ 1 \end{pmatrix} \tag{5.22}$$

The immediate neighborhood of the central prismatic cell at (i, j) is the four prismatic cells located at (i – 1, j) and (i + 1, j) plus (i, j – 1) and (i, j + 1). The four adjacent faces are defined at the middle intervals (i + 1/2, j) and (i – 1/2, j) plus (i, j + 1/2) and (i, j – 1/2). To approximate the discharges $Q_{i+1/2,j}$, $Q_{i-1/2,j}$, $Q_{i,j+1/2}$, and $Q_{i,j-1/2}$ that traverse these faces by difference equations, one must apply the rules previously discussed in this chapter.

Now, continuity requires for each cell at (i, j) that:

$$\left(Q_{i+\frac{1}{2},j} \quad -Q_{i-\frac{1}{2},j} \right) + \left(Q_{i,j+\frac{1}{2}} \quad -Q_{i,j-\frac{1}{2},j} \right) - \omega_{i,j} \cdot \delta x \cdot \delta y = 0 \tag{5.23}$$

Approximating these discharges by their corresponding finite difference equations in Equation 5.23, collecting and solving for $V_{i,j}$, one obtains an appropriate algorithm for a Gauss–Seidel iteration routine:

$$\left(V_{i,j} \right)^{k+1} = \left[\left(A_{DU} \right)_{i,j} \right]^{-1} \left[\left(B_{DU} \right)_{i,j} \right]^{k} \tag{5.24}$$

In the above equation, k stands for the iteration cycle.

For each cell (i, j), the term $[A_{DU}]_{i,j}$ remains constant and is calculated by:

$$\left(A_{DU} \right)_{i,j} = \left[\left(C_{RU} \right)_{i,j} \cdot \begin{pmatrix} \frac{1}{2\delta x} \\ \frac{1}{2\delta y} \end{pmatrix} + \left(C_{LD} \right)_{i,j} \begin{pmatrix} \frac{1}{2\delta x} \\ \frac{1}{2\delta y} \end{pmatrix} \right] \cdot \begin{pmatrix} 1 \\ 1 \end{pmatrix} \tag{5.25}$$

The coefficients $[C_{RU}]_{i,j}$ and $(C_{LD})_{i,j}$ may result from arithmetic means involving the adjacent cells:

$$\left(C_{RU} \right)_{i,j} = \begin{bmatrix} \frac{\left(k_{xx} \right)_{i+1,j} + \left(k_{xx} \right)_{i,j}}{2} \delta y & \frac{\left(k_{xy} \right)_{i,j+1} + \left(k_{xy} \right)_{i,j}}{2} \delta y \\ \frac{\left(k_{yx} \right)_{i+1,j} + \left(k_{xy} \right)_{i,j}}{2} \delta x & \frac{\left(k_{yy} \right)_{i,j+1} + \left(k_{yy} \right)_{i,j}}{2} \delta x \end{bmatrix} \tag{5.26}$$

$$\left(C_{LD} \right)_{i,j} = \begin{bmatrix} \frac{\left(k_{xx} \right)_{i,j} + \left(k_{xx} \right)_{i-1,j}}{2} \delta y & \frac{\left(k_{xy} \right)_{i,j} + \left(k_{xy} \right)_{i,j-1}}{2} \delta y \\ \frac{\left(k_{yx} \right)_{i,j} + \left(k_{yx} \right)_{i-1,j}}{2} \delta x & \frac{\left(k_{yy} \right)_{i,j} + \left(k_{yy} \right)_{i,j-1}}{2} \delta x \end{bmatrix} \tag{5.27}$$

Other types of means may be used. However, as previously pointed out, arithmetic means better fit pseudo-continuous solutions for fractured rock masses.

For each cell (i, j), the term $\{[B_{DU}]_{i,j}\}^k$ changes after each k iteration until the solution stabilizes. It is calculated by:

$$\left[\left(B_{DU}\right)_{i,j}\right]^k = \left[\begin{matrix} \left(C_{RU}\right)_{i,j}\begin{bmatrix} \frac{1}{2}\frac{\left(V_{i+1,j}\right)^k}{\delta x} \\ \frac{1}{2}\frac{\left(V_{i,j+1}\right)^k}{\delta y} \end{bmatrix}\cdots \\ +\left(C_{LD}\right)_{i,j}\begin{bmatrix} \frac{1}{2}\frac{\left(V_{i-1,j}\right)^k}{\delta x} \\ \frac{1}{2}\frac{\left(V_{i,j-1}\right)^k}{\delta y} \end{bmatrix} \end{matrix}\right]\begin{pmatrix} 1 \\ 1 \end{pmatrix}\cdots$$
$$+\left[\left[\left(v_{i,j}\right)^{\frac{1}{2}}\right]^k \left(C_C\right)_{i,j}\begin{bmatrix} \frac{\left(B_{i+1,j}\right)^k - \left(B_{i-1,j}\right)^k}{2\delta x} \\ \frac{\left(B_{i,j+1}\right)^k - \left(B_{i,j-1}\right)^k}{2\delta y} \end{bmatrix}\begin{pmatrix} 1 \\ 1 \end{pmatrix} + \omega_{i,j}\delta x\delta y\right] \tag{5.28}$$

Note the term $[V^{1/2}]_{i,j}$ in the second expression of the right-hand side. Boundary conditions must also be written in terms of V instead of H.

For each cell (i, j) the term $[C_C]_{i,j}$ of Equation 5.28 remains constant and is calculated by:

$$\left(C_C\right)_{i,j} = \begin{bmatrix} \left(k_{xx}\right)_{i,j}\delta y & \left(k_{xy}\right)_{i,j}\delta y \\ \left(k_{yx}\right)_{i,j}\delta x & \left(k_{yy}\right)_{i,j}\delta x \end{bmatrix} \tag{5.29}$$

After solving the problem, V's solution is transformed back to H; that is, $H = V^{1/2}$.

Example 5.3, a simple theoretical exercise, helps to understand the use of the algorithm.

Example 5.3

Imagine a pervious sandstone layer, 250 m thick, gentle dipping over an almost impervious shale layer in an inter-tropical zone. Subvertical fractures split the upper layer in many large mosaics (see Figure 5.18). The pervious upper layer is subjected to copious and persistent rains that attain more than 2,000 mm/year and last for several months during the seasonal rainfall. Almost 10%20% of the total rainfall infiltrates deeply into this layer during the wet months.

Due to the intense chemical attack in this hot climate, the weathered cap's soluble elements are progressively removed by groundwater flowing to natural exit springs at the upper layer base. If these subvertical fractures are not very pervious, compared

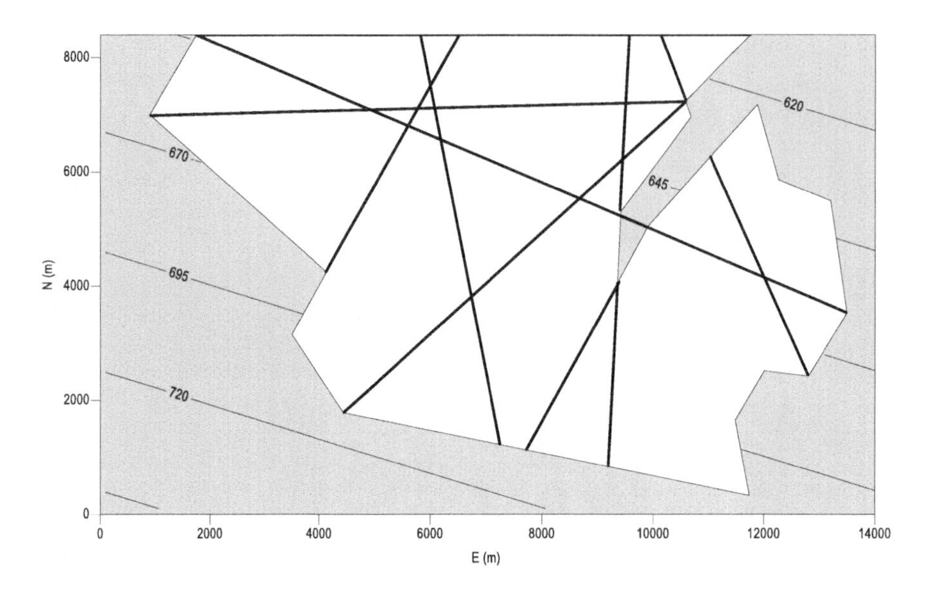

Figure 5.18 Gentle dipping pervious dominant sandstone layer 250 m thick over an almost impervious shale layer. Subvertical fractures split the upper layer into many large mosaics.

with the permeability of the remaining rock mass, these springs are dispersed around the sandstone-shale contact periphery, mostly in the reentrant relief forms. However, due to the long-term weathering, these fractures become progressively eroded, more pervious, and "capture" an ever-growing part of the groundwater flow. By progressive backward erosion along the fracture traces, springs gradually concentrate at the lowest points of the subvertical fractures. This geodynamic process has positive feedback. The steady-state behavior of the groundwater flow was simulated for four different scenarios to analyze the effect of different fracture transmissivities within the rock mass.

The input data for this problem are:

- *Boundary conditions*: A pervious subvertical fracture delimits the North boundary of the cap layer. Suppose that three piezometers located in that fracture gave a seasonal series of WT measurements during the wet season. Then, it was possible to approximate an average Dirichlet condition at that border (see Figure 5.19).
- Dirichlet conditions for the remaining boundaries were taken at each point (i, j) as having the variable ground elevation (top of the shale layer).
- Average unit-area vertical recharge: An exceptional continuous recharge was simulated: w = 14.964 mm/day
- Eigenvalues of the hydraulic conductivity of the top layer

$$k_R = \begin{pmatrix} 3\times10^{-6} \\ 1\times10^{-6} \end{pmatrix} \text{m/s}$$

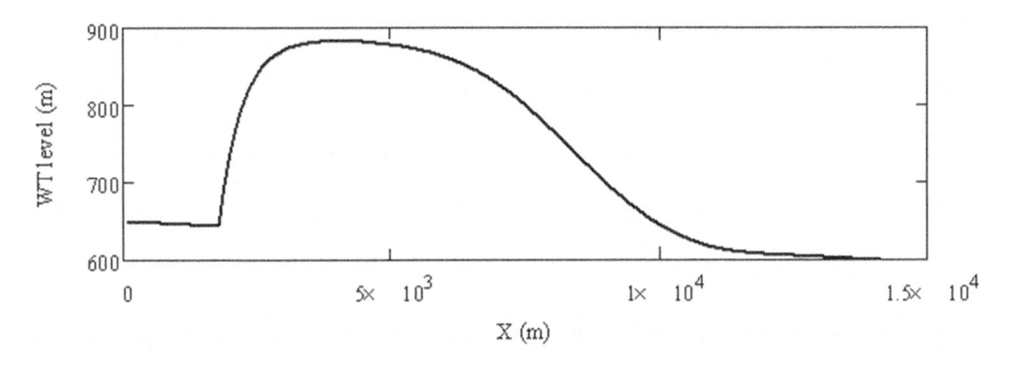

Figure 5.19 Approximate Dirichlet condition at the very pervious subvertical fractures at the North border of Figure 5.18.

- Eigenvectors of the hydraulic conductivity of the top layer:
  The direction of max k_R: along 106° clockwise from the y-axis.
  The direction of min k_R: perpendicular to max k_R.
- Hydraulic conductivity of any cell (i, j) referred to the xyz frame:

$$(K_R)_{i,j} = \begin{pmatrix} 2.848\times10^{-6} & -5.299\times10^{-7} \\ -5.299\times10^{-7} & 1.152\times10^{-6} \end{pmatrix} \text{m/s}$$

- Fractures:
  Numerical simulations were made for four scenarios discussed more ahead. Data concerning the fractures for Case 1 are condensed in Table 5.1.
- Model discretization:
  The model was discretized into 140 × 80 = 11,200 prismatic cells. The subvertical fractures were also discretized to fit these cells best. It is important to retain that it is always possible to discretized curved traces for Dupuit's approach or fracture surfaces for a true 3D model. However, in these scenarios, the transmissivity tensor varies locally with the trace xy-slope or the surface elements' geometry, and the modeler has more work to do.
  The equivalent hydraulic conductivity of each cell of the sandstone layer was estimated, agreeing to the methodology presented in Chapter 2. Therefore, the diagonal and off-diagonal components of the resulting equivalent hydraulic conductivity for each cell (i, j) were estimated by the formulas below. In these formulas, the indices (i, j) are purposely suppressed to alleviate their graphical appearance:
- Diagonal terms:

$$K_{xx} = (k_R)_{xx} + \left[\sum_{n=0}^{9}\left[(k_F)_{xx}\right]_n\right] \tag{5.30}$$

Table 5.1 Case I fractures data

Fracture number	*Fracture start and end coordinates*	*X (m)*	*Y (m)*	*The slope of the fracture trace in the XY plane (° positive counter-clockwise)*	*Fracture length (m)*	*Isotropic fracture transmissivity very roughly proportional to its length (m^2/s)*
0	Start	1,737	8,400			
	End	11,760	8,400	0.00	10,100	2.73E − 03
1	Start	6,518	8,397			
	End	4,125	4,245	60.26	4,837	1.31E − 03
2	Start	905.3	6,985			
	End	10,580	7,226	1.18	9,702	2.63E − 03
3	Start	4,450	1,786			
	End	10,570	7,211	41.52	8,147	2.21E − 03
4	Start	12,790	2,444			
	End	11,040	6,257	−65.23	4,295	1.16E − 03
5	Start	5,821	8,400			
	End	7,250	1,234	−79.00	7,335	1.99E − 03
6	Start	10,580	7,255			
	End	10,130	8,400	−65.56	1,208	3.27E − 04
7	Start	13,480	3,537			
	End	1,735	8,404	−22.55	1,2780	3.46E − 03
8	Start	9,388	4,070			
	End	7,724	1,134	60.46	3,448	9.33E − 04
9	Start	9,359	4,056			
	End	9,195	842.37	86.53	3,306	8.95E − 04
10	Start	9,400	5,324			
	End	9,566	8,400	86.31	3,106	8.41E − 04

$$K_{yy} = (k_R)_{yy} + \left[\sum_{n=0}^{9} \left[(k_F)_{xx} \right]_n \right] \tag{5.31}$$

Off-diagonal terms:

$$K_{xy} = (k_R)_{xy} + \left[\sum_{n=0}^{9} \left[(k_F)_{xx} \right]_n \right] \tag{5.32}$$

$$k_{yy} = k_{xy} \tag{5.33}$$

In these formulas, index R stands for rock mass. It was admitted that the hydraulic conductivity of the rock mass implicitly includes the effect of the minor discontinuities. Index F stands for fracture and sub-index n for the fracture number for a total of 10 fractures (see Table 5.1). The extremities of the parallel segments of the same fracture were joined by extra cells as to improve continuity (see Figure 5.20).

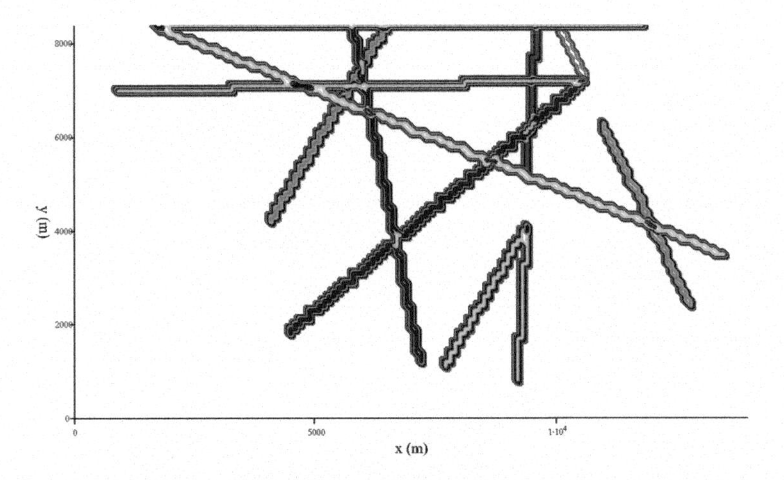

Figure 5.20 Final outline of the cells containing fractures. Shades are proportional to the geometric mean of the eigenvalues of each cell (numerical values not shown).

Simulation results:In this simple theoretical exercise, simulations were run for four cases considering different fracture transmissivities, as presented in Table 5.2. Note for cases 1 to 3 the extremely high contrast between the fracture transmissivities and the rock mass hydraulic conductivity.

Figures 5.21 to 5.24 show the simulation results.

For cases 2 and 3, it was necessary to modify the parameters to guarantee a stable algorithm, as discussed next.

Now, it is time to consider the stability of the condensed algorithm of Equations 5.24 and 5.25. In effect, as the contrast between the hydraulic transmissivity of the discontinuities and the hydraulic conductivity of the rock mass, implicitly including minor discontinuities, grows up, the finite algorithm may become unstable depending on the cell dimensions, the number of fractures but, most of all, conditioned by the anisotropy of the integrated fractured rock hydraulic conductivity, including the anisotropic rock mass permeability k_R and the anisotropic fractured rock transmissivities k_F. Without

Table 5.2 Simulated cases classified by the maximum and minimum ratio between fracture transmissivity (m^2/s) and the geometric mean of the sandstone eigenvalues (m/s)

Case	*Relation [maximum fracture transmissivity (m^2/s)]/ [geometric mean of k_R eigenvalues (m/s)]*	*Relation [minimum fracture transmissivity (m^2/s)]/ [geometric mean of k_R eigenvalues (m/s)]*
0	1	1
1	2.015×10^3 m	190.526 m
2	1.832×10^4 m	1.732×10^3 m
3	1.832×10^5 m	1.732×10^4 m

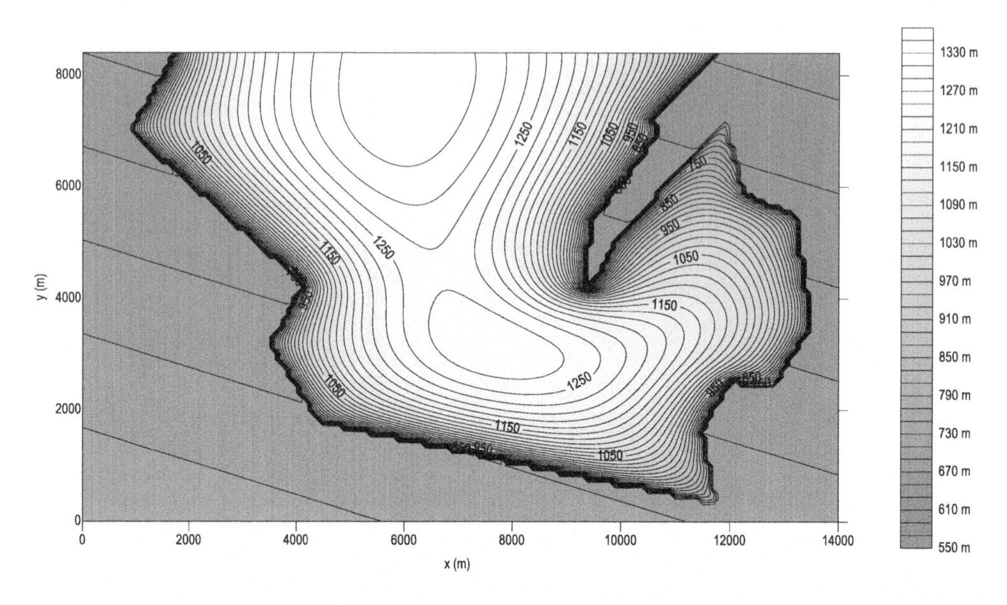

Figure 5.21 Case 0 – Groundwater flow pattern without fractures to compare the effect of growing transmissivity values. Compare with all the next figures.

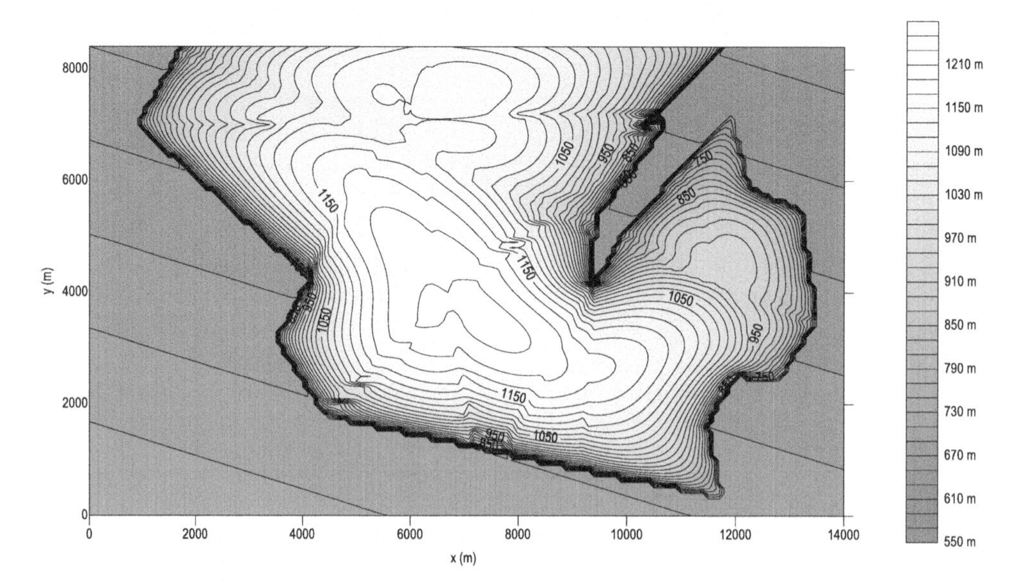

Figure 5.22 Case 1 – Groundwater flow pattern for the data presented in Table 5.1. The maximum and minimum ratio between fracture transmissivity (m^2/s) and the sandstone eigenvalues' geometric mean eigenvalues (m/s) were, respectively, 2.015×10^3 m and 190.526 m.

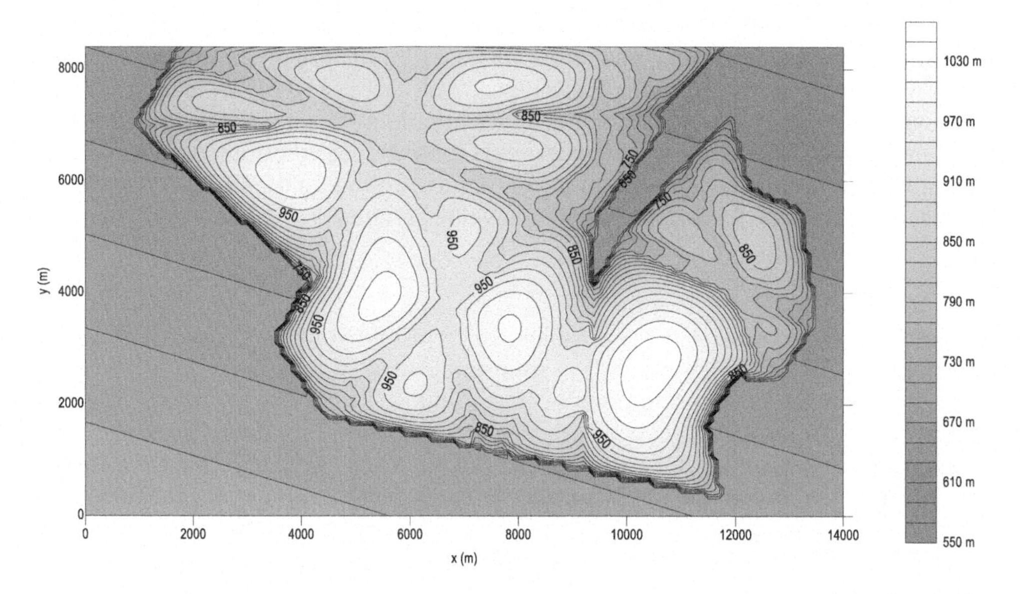

Figure 5.23 Case 2 – Groundwater flow pattern considering the same data presented in Table 5.1 but now for ratios, respectively, equal to 1.832×10^4m and 1.732×10^3m, that is 1,000 more pervious than the previous result.

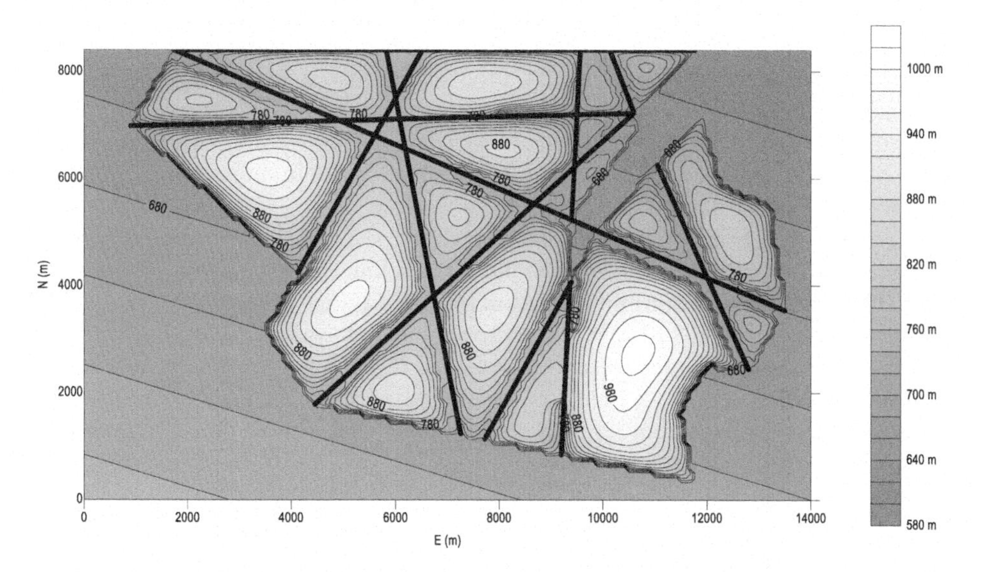

Figure 5.24 Case 3 – Groundwater flow pattern considering the same data presented in Table 5.1 but now for ratios, respectively, equal to 1.832×10^5m and 1.732×10^4m, again 1,000 more pervious than the previous result. In that figure, the fracture xy-traces are in their correct geographic position to show the effect of the bias introduced by locating each fracture in the middle of the prismatic cells.

any doubt, instabilities are caused by the off-diagonal terms $[k_R]_{xy}$ and $[k_F]_{xy}$ and arise at the neighborhood of cells signaled by heavily contrasted hydraulic conductivities. In the previous example, they only arose for cases 2 and 3; see Figures 5.23 and 5.24. However, simulations corresponding to Figure 5.21, case 0, and Figure 5.22, case 1, ran without any stability problem. Fortunately, without any substantial harm, restricted to some smoothing effect of the equipotentials contours near the critical cells, these instabilities are easily cured. In fact, after computing K_{xy} by equation 5.22, it is sufficient to test and substitute for K_{xy} the maximum value between $[k_R]_{xy}$ and K_{xy}:

$$\text{best } k_{xy} = \max\left[\left(k_R\right)_{xy}, K_{xy}\right] \tag{5.34}$$

It is important to recall that the off-diagonal terms of hydraulic conductivity tensor of each cell, that is $[k_R]_{xy}$ and $[k_F]_{xy}$, are usually small if compared with the corresponding diagonal terms. Additionally, $[k_R]_{xy}$ and $[k_F]_{xy}$ may be positive or negative quantities and independent of each other. For that reason, Equation 5.34 must be applied to relative values, that is, taking into account their positive or negative values, but never their absolute values. Sometimes both terms, that is $[k_R]_{xy}$ for the rock mass and $[k_F]_{xy}$ for individual fractures, may be positive or negative, and sometimes one of them is positive and the other negative. In fact, for a group of fractures within the same cell, the resulting summation $\sum [k_F]_{xy}$ for all fractures of that group may be very low, sometimes lower than $[k_R]_{xy}$, although the absolute values of the individual terms $[k_F]_{xy}$ of each fracture may occasionally be very high. Therefore, $[k_R]_{xy}$ and the sum $K_{xy} = \left\{[k_R]_{xy} + \Sigma[k_F]_{xy}\right\}$ may be quite of the same order but set apart by too much different eigenvectors. Finally, it must be said that the anisotropic character of the resulting tensor, mostly dependent on the numerical values of the diagonal tensor elements, is relatively preserved by the recommended correction. Then, by observing the stability criterion condensed in Equation 5.34, it is possible to analyze an intricate group of vertical fractures. As an example, consider the 20 intercrossed random fractures show in Figure 5.25.

To enhance the fracture transmissivities' effect, the intrinsic permeability of the rock mass k_R was considered isotropic and set to a low value: 10^{-30} m/s. On the other hand, the maximum and minimum fracture transmissivities were set remarkably high: $8.749 \times 10^{10}\,m^2/s$ and $1.174 \times 10^7\,m^2/s$, respectively. As these values differ, more than 10^{40} from k_R, one may be tempted to set it to zero; but this is not a good idea, as it leads to numerical overflow. Figure 5.26 shows the simulated head distribution contour plot and the impervious base's border. Figure 5.27 shows the contour plot of the total discharge $Q = Q_E + Q_N$ that traverse each cell.

5.4.2.1.2 CONFINED FLOW

Assume the same premises as for unconfined flow. Also consider the same list of symbols with the modifications below:

- H_c (L): the thickness of the confined layer measured from the base elevation B.
- Q_x (L^3/T): discharge across the vertical x-cross section: $Q_x = q_x \cdot H_c \cdot \delta y$.

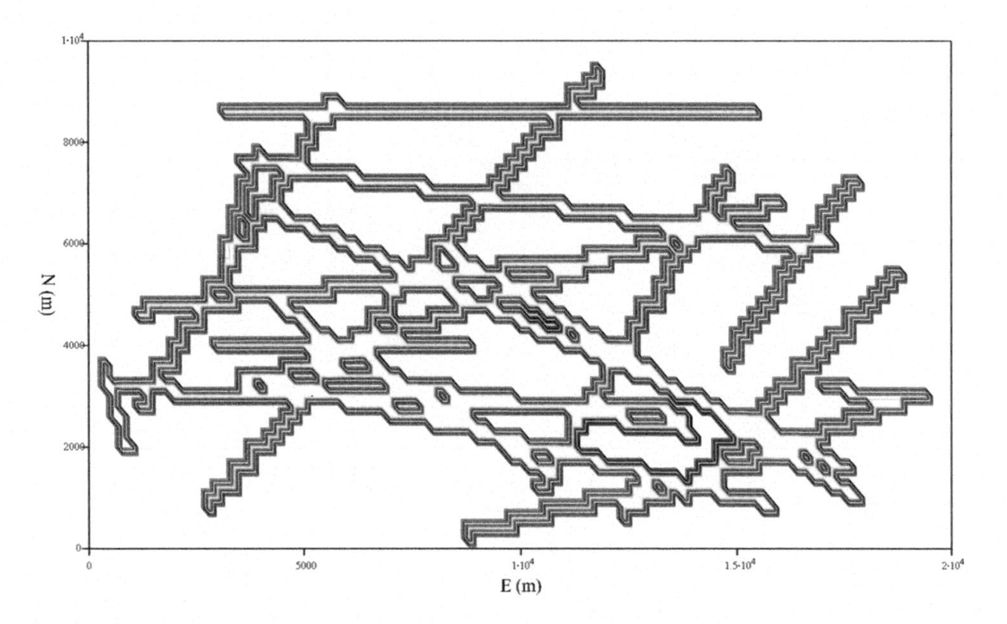

Figure 5.25 Final outline of 20 cells containing fractures. Shades are proportional to the geometric mean of the eigenvalues of each cell (numerical values not shown). To get better results the extremities of the segments for the same fracture were joined by extra cells. Cell dimensions: 200 × 200 m.

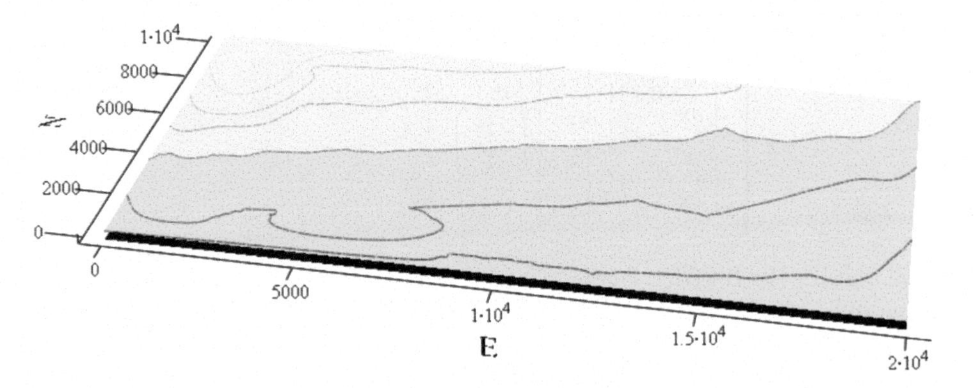

Figure 5.26 Contour plot of the simulated head distribution. Note the border of the inclined impervious base. The average water column above base was 245 m.

- Q_y (L^3/T): discharge across the vertical y-cross section: $Q_y = q_y \cdot H_c \cdot \delta x$.
- Ω (L/T): unit-area vertical leakage or infiltration at the top and/or at the base of the confined layer; both parameters are proportional to the head difference between the confined layer and the base of the overlying layer and/or the top of the underlying layer.

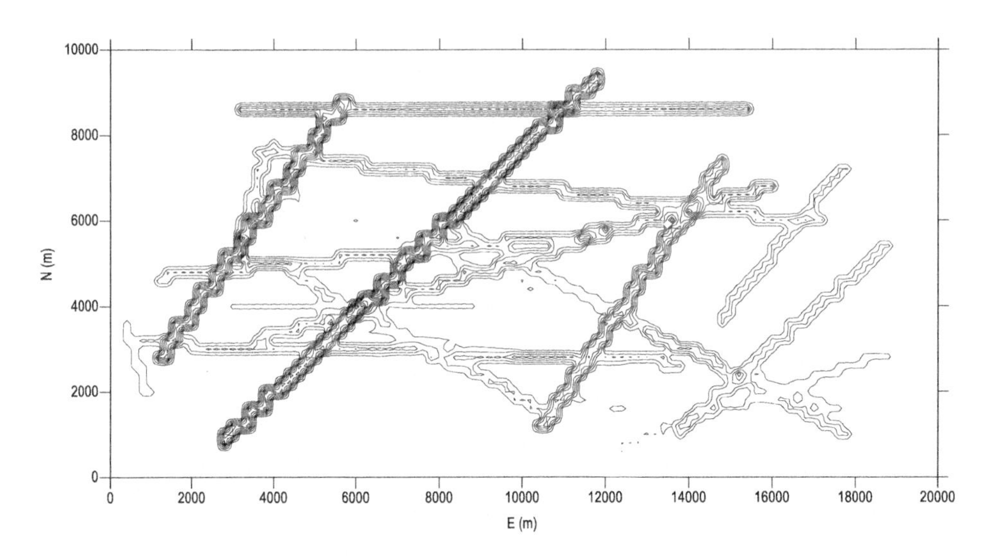

Figure 5.27 Contour plot of the total discharge $Q = Q_E + Q_N$ that traverse each cell containing fractures. Numerical values not plotted to alleviate the drawing.

For constant water density, the total flow $Q_{i,j}$ across a prismatic cell estimated by the Darcy's Law is:

$$Q_{i,j} = -\begin{pmatrix} \delta y & 0 \\ 0 & \delta x \end{pmatrix} \begin{bmatrix} (k_{xx})_{i,j} & (k_{xy})_{i,j} \\ (k_{yx})_{i,j} & (k_{yy})_{i,j} \end{bmatrix} \cdot (H_c)_{i,j} \begin{bmatrix} \left[\dfrac{\partial}{\partial x}(B+H)\right]_{i,j} \\ \left[\dfrac{\partial}{\partial y}(B+H)\right]_{i,j} \end{bmatrix} \cdot \begin{pmatrix} 1 \\ 1 \end{pmatrix} \tag{5.35}$$

Recall that $B + H = B + (H_C + P/\rho g)$.

Continuity requires for each cell:

$$\left(Q_{i+\frac{1}{2},j} \quad -Q_{i-\frac{1}{2},j} \right) + \left(Q_{i,j+\frac{1}{2}} \quad -Q_{i,j-\frac{1}{2}} \right) \tag{5.36}$$

Approximating discharges by the corresponding finite difference equations in Equation 5.36, collecting and solving for $H_{i,j}$, one obtains an appropriate algorithm for a Gauss–Seidel iteration routine:

$$(H_{i,j})^{k+1} = \left[(A_{DC})_{i,j}\right]^{-1} \left[(B_{DC})_{i,j}\right]^{k} \tag{5.37}$$

In the above equation, k stands for the iteration cycle.

For each cell (i, j), the constant term $[A_{DC}]_{i,j}$ is calculated by:

$$\left(A_{DC}\right)_{i,j} = \left(H_c\right)_{i,j}\left[\left(C_{RU}\right)_{i,j}\begin{pmatrix}\frac{1}{\delta x}\\ \frac{1}{\delta y}\end{pmatrix} + \left(C_{LD}\right)_{i,j}\begin{pmatrix}\frac{1}{\delta x}\\ \frac{1}{\delta y}\end{pmatrix}\cdot\begin{pmatrix}1\\1\end{pmatrix}\right] \tag{5.38}$$

For each cell (i, j) the varying term $\{[B_{DC}]_{i,j}\}^k$ is calculated by:

$$\left[\left(B_{DC}\right)_{i,j}\right]^k = \left(H_c\right)_{i,j}\left[\left(C_{RU}\right)_{i,j}\begin{bmatrix}\frac{\left(H_{i+1,j}\right)^k}{2\delta x}\\ \frac{\left(H_{i,j+1}\right)^k}{2\delta y}\end{bmatrix} + \left(C_{LD}\right)_{i,j}\begin{bmatrix}\frac{\left(H_{i-1,j}\right)^k}{2\delta x}\\ \frac{\left(H_{i,j-1}\right)^k}{2\delta y}\end{bmatrix}\right]\begin{pmatrix}1\\1\end{pmatrix}\ldots$$

$$+\left[\left(H_c\right)_{i,j}\left(C_C\right)_{i,j}\begin{pmatrix}\frac{B_{i+1,j}-B_{i-1,j}}{2\delta x}\\ \frac{B_{i,j+1}-B_{i,j-1}}{2\,\delta y}\end{pmatrix}\begin{pmatrix}1\\1\end{pmatrix} + \Omega_{i,j}\,\delta x\delta y\right] \tag{5.39}$$

The coefficients $[C_{RU}]_{i,j}$, $[C_{LD}]_{i,j}$, and $[C_C]_{i,j}$ keep the same meaning as defined for unconfined flow.

5.4.2.2 3D algorithms

Assume the same premises as before. Consider now the complementary list of symbols:

- i, j, k: indices of the coordinate (x_i, y_j, z_k) of the middle cell (vertical z-axis pass through this point).
- $i-1$: index one step to the West of i.
- $i+1$: index one step to the East of i.
- $j-1$: index one step to the South of j.
- $j+1$: index one step to the North of j.
- $k-1$ (–): index one step to the Nadir of k.
- $k+1$ (–): index one step to the Zenith of k.
- [k] (L/T): anisotropic hydraulic conductivity, defined by the second order tensor:

$$k = \begin{pmatrix} k_{xx} & k_{xy} & k_{xz}\\ k_{yx} & k_{yy} & k_{yz}\\ k_{zx} & k_{zy} & k_{zz}\end{pmatrix} \tag{5.40}$$

- **J** (–): local hydraulic gradient, defined by the vector:

$$J = \begin{pmatrix} J_x \\ J_y \\ J_z \end{pmatrix} \tag{5.41}$$

- **q** (L/T): local specific discharge, defined by the vector:

$$\mathbf{q} = \begin{pmatrix} q_{xx} + q_{xy} + q_{xz} \\ q_{yx} + q_{yy} + q_{yz} \\ q_{zx} + q_{zy} + q_{zz} \end{pmatrix} \tag{5.42}$$

- q_r (L/T) and r=i or j or k: component of the specific discharge **q** along the r-direction resulting from the addition of the three parallel sub-components: $q_{ri}+ q_{rj}+ q_{rk}$.
- k_{rs} (L/T) and r or s=i or j or k: component of [k] that responds for the discharge vector q_{rs} parallel to the r-axis and induced by the hydraulic gradient component J_s. It is essential to remind that as $k_{rs}=k_{sr}$ this implies $q_{rs}=q_{sr}$.
- B (L): base elevation that may have a variable but a relatively gentle inclination, referred to an arbitrary horizontal datum H_{datum}.
- H (L): hydraulic head measured from the arbitrary horizontal datum H_{datum}.
- δx (L): the base dimension of the cuboid cell along the x-direction.
- δy (L): the base dimension of the cuboid cell along the y-direction.
- δz (L): the height of the cuboid cell, measured along the zenithal z-direction.
- Q_x (L^3/T): horizontal discharge across the vertical x-cross section: $Q_x=q_x \cdot \delta y \cdot \delta z$.
- Q_y (L^3/T): horizontal discharge across the vertical y-cross section: $Q_y=q_y \cdot \delta x \cdot \delta z$.
- Q_z (L^3/T): vertical discharge across the horizontal z-cross section: $Q_z=q_z \cdot \delta x \cdot \delta y$.
- $Q_{x,y,z}$ (L^3/T): sum of all discharge components: $Q_{x,y,z}=Q_x+Q_y+Q_z$.
- ω (L/T): vertical unit-area recharge at the phreatic surface.
- Θ (T^{-1}): unit-volume sink or source within the cuboid cell.

Modeling an accurate 3D simulation is more laborious than resorting to Dupuit's approach. Therefore, 3D models must only be used if necessary. Typically, 3D formations to be modeled have less thicknesses compared with their horizontal extensions; in these cases, Dupuit's solutions may be adequate, except if 3D contour plots near excavated foundations or nearby underground stopes of mines or underground power plants are needed. However, in these latter cases, attention must be paid to the possibility of having high hydraulic gradients, requiring a nonlinear relationship between **q** and **J**, as $\mathbf{q}=[k]\cdot \mathbf{J}^{\alpha}$, whose mathematical treatment is out of the scope of this book.[3]

Usually, an accurate 3D simulation requires splitting the system domain into N+1 slabs that usually are parallel to the slightly inclined base. These slabs are labeled from 0 to N. Consequently, adjacent slabs are labeled n and n+1. Assuming constant water density, it is possible to deduce for each slab, recognized by its label, its proper finite difference algorithm. Also, for the lower slab, labeled by 0, and for the higher slab,

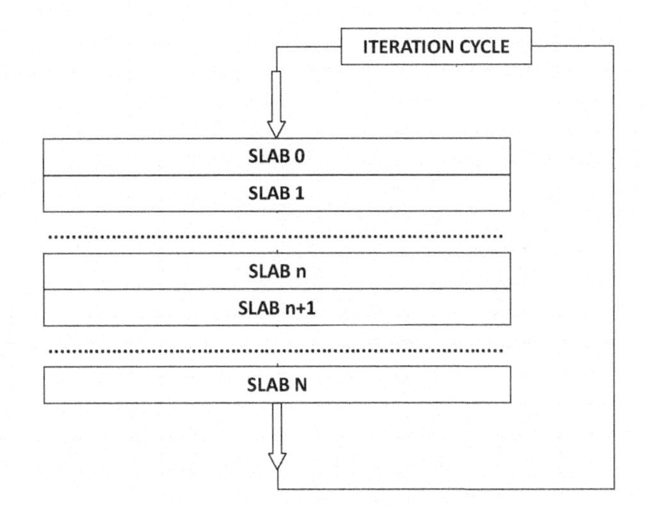

Figure 5.28 Iterative routine cycles are sequentially and orderly performed from the start to end of the slab pile until convergence. After the iteration routine for slab n has finished in every cycle, its output at the interface n/n + 1 enters as input for slab n + 1 before starting its iteration routine. Cycle order may be reversed.

labeled by N, it is possible to deduce the subsidiary algebraic equations for the boundary conditions. Iterative routine cycles are sequentially and orderly performed from the start to the limit of the slab pile until convergence. In every cycle, after finishing the iteration routine for slab n, its output at the interface n/n + 1 enters as input for slab n + 1 before starting the iteration routine for slab n + 1 (see Figure 5.28).

Example 5.4 helps to understand the use of the algorithm. Each problem may require an unusual approach, and the best way to understand 3D modeling principle is by doing this simple exercise.

Example 5.4

A slightly inclined layer of very weathered and very pervious calcareous-sandstone, 10 m thick on average, resting on a relative impervious and also inclined base, was contaminated upstream of a remnant of a weathered vertical fault, more pervious than the surrounding rock mass (see Figure 5.29 for more information).

To select the best pumping layout to pump the contaminated groundwater, many possibilities were simulated, varying the number of pumping wells, the pumping strength, and the well's location. This example shows the result of one of these simulations, not the one considered to be the better.

The model covered a volume of top area of 100×100 m and was divided into three slabs with the following initial arrangement before pumping:

- *Upper slab: thickness*: 2.5 m; elevation 95.0 m at the top and 92.5 m at the interface "upper-mid-slab"; initial thickness $\delta z = 2.5$ m.

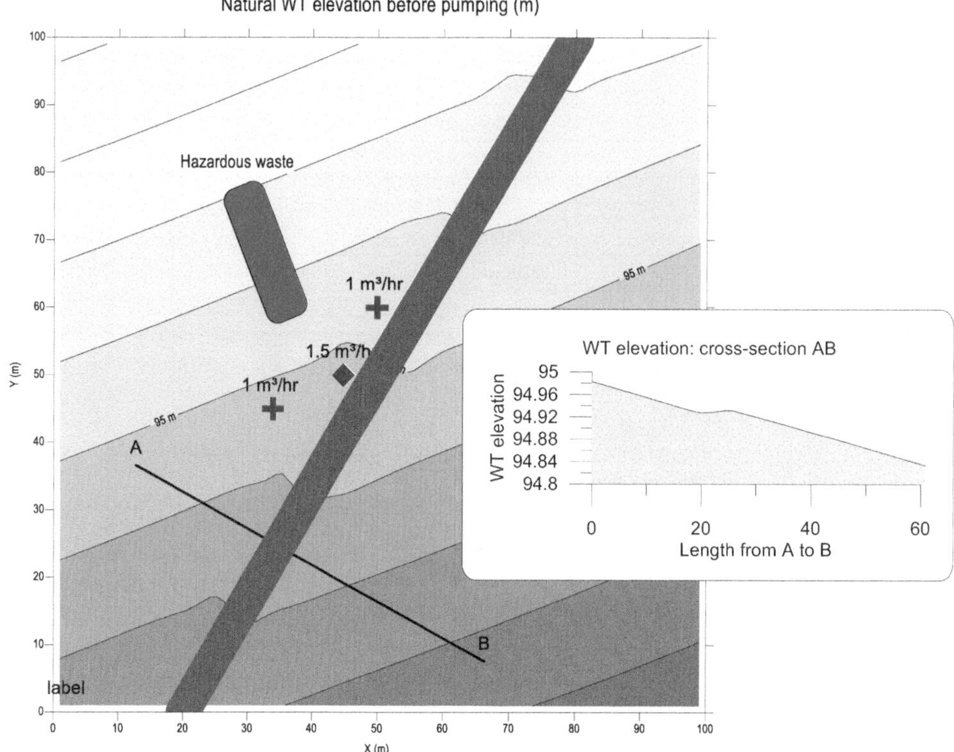

Figure 5.29 Groundwater flows from the top-left corner to the bottom-right corner. The trace of the remnant of a vertical fault, about 30 times more pervious than the calcareous-sandstone, was located downstream of the hazardous waste and allowed faster transport of contaminants along its trace besides the advective transport within the rock mass. In one of the layouts simulated, three pumps, with their filters just traversing the middle slab, were located as indicated in this figure. These pumping 5 m long filters were simulated by tube transmissivities having a vertical permeability ten times that of the rock mass. Their pumping rates are also indicated. Note the effect of the fault on the equipotentials contouring before pumping (slice AB). Note also the undesirable graphical distortion of the equipotential contouring at the neighborhood of highly contrasted cells. That distortion may be reduced using a denser grid.

- *Mid-slab: thickness*: 5.0 m; elevation 92.5 m at the interface "upper-mid-slab" and 87.5 m at the interface "mid-bottom slab"; initial thickness $\delta z = 5.0$ m.
- *Bottom slab: thickness*: 2.5 m; elevation at the interface "mid-bottom slab" and 82.5 and 85.0 m at the base; initial thickness $\delta z = 2.5$ m.

Each slab was divided into $20 \times 20 = 100$ cells. The quasi horizontal cross section of each cell measured 5×5 m, that is, $\delta x = 5$ m and $\delta y = 5$ m. However, to easily converge to the final WT configuration, each slab's thickness δz was progressively deformed as to best fit the changing WT boundary condition until convergence.

The estimated eigenvalues for hydraulic conductivity tensor of the weathered calcareous-sandstone, including minor discontinuities, were:

$$k_{\text{eigenvalues}} = \begin{pmatrix} 1.5\times10^{-5} \\ 1\times10^{-5} \\ 7.5\times10^{-5} \end{pmatrix} \text{m/s}$$

Referred to the xyz frame adopted, the transformed hydraulic conductivity was:

$$K_{CS} = \begin{pmatrix} 1.425\times10^{-5} & -2.234\times10^{-6} & -1.357\times10^{-5} \\ -2.234\times10^{-6} & 8.323\times10^{-6} & -4.054\times10^{-7} \\ -1.357\times10^{-7} & -4.054\times10^{-7} & 9.925\times10^{-6} \end{pmatrix} \text{m/s}$$

Corresponding eigenvalues and eigenvectors are shown in Figure 5.30.

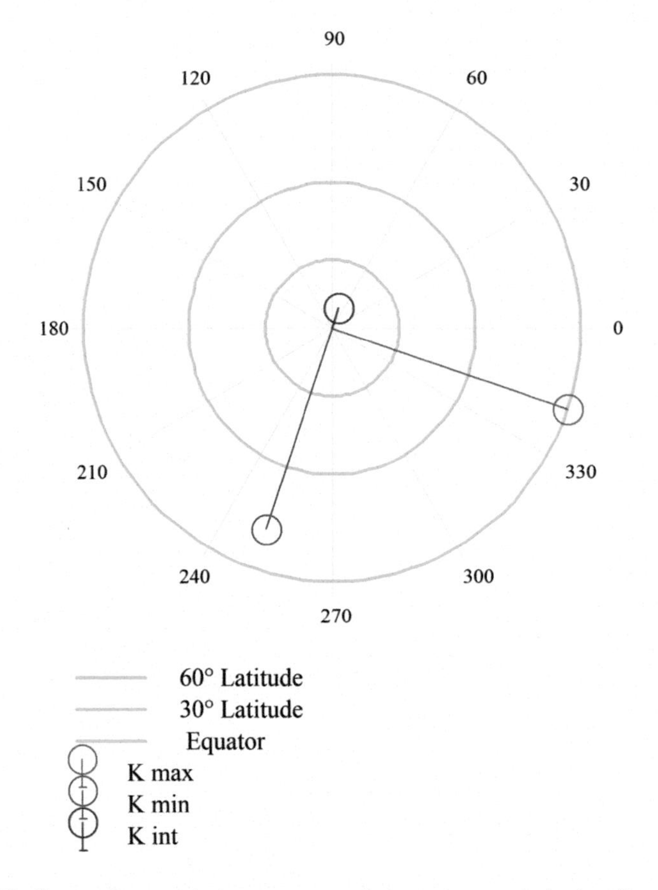

Figure 5.30 Eigenvalues and eigenvectors of the anisotropic hydraulic conductivity of the calcareous-sandstone slabs.

For the fault, the strike-transmissivity was taken as 30 times the maximum eigenvalue of the calcareous-sandstone. Its dip-transmissivity was taken as 80% of the strike transmissivity. Their values were:

$$T_{strike\_transmissivity} = 4.5 \times 10^{-4}\ m^2/s$$

$$T_{dip\_transmissivity} = 3.6 \times 10^{-4}\ m^2/s$$

Referred to the xyz frame adopted the transformed transmissivity for the vertical fault was:

$$T_{fault} = \begin{pmatrix} 1.191 \times 10^{-4} & -1.985 \times 10^{-4} & 0 \\ -1.985 \times 10^{-4} & 3.309 \times 10^{-4} & 0 \\ 0 & 0 & 3.6 \times 10^{-4} \end{pmatrix} m^2/s$$

The average vertical accretion ω simulated corresponded to infiltration of 3.5 mm/day.

To solve this problem more straightforwardly, one must deduce three finite difference algorithms for the continuity condition applied to the upper, middle, and bottom slab. Additionally, top and base boundary conditions require two more finite difference algorithms. There are many alternatives to do that, but the solution presented bellow gave satisfactory results. See Figure 5.31 to grasp the main points assumed.

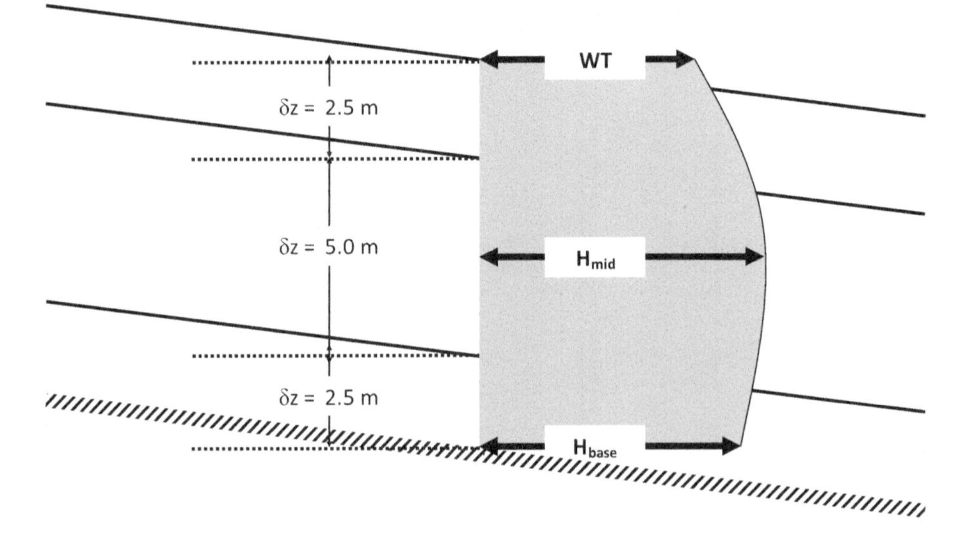

Figure 5.31 Initial thicknesses of slabs. Heads are calculated at the base, middle of the intermediate slab, and at the phreatic. Each slab's thickness δz at each cell was progressively contracted by the same %, aiming best fit the dropping WT boundary condition until convergence.

Continuity for the mid-slab (index k)

Consider the cuboid cell's central point referred to the coordinates (x_i, y_j, z_k) conveniently defined by the indices (i, j, k). The sum of the three discharges Q_i, Q_j, and Q_k traversing the three median cross sections is $Q_{i,j,k}=Q_i+Q_j+Q_k$. The total discharge $Q_{i,j,k}$ is calculated by:

$$Q_{i,j,k} = \begin{bmatrix} (k_{xx})_{i,j,k} & (k_{xy})_{i,j,k} & (k_{xz})_{i,j,k} \\ (k_{yx})_{i,j,k} & (k_{yy})_{i,j,k} & (k_{yz})_{i,j,k} \\ (k_{zx})_{i,j,k} & (k_{zy})_{i,j,k} & (k_{zz})_{i,j,k} \end{bmatrix} \begin{pmatrix} \delta y \delta z & 0 & 0 \\ 0 & \delta x \delta z & 0 \\ 0 & 0 & \delta x \delta y \end{pmatrix} \begin{bmatrix} \left(\frac{\partial}{\partial x} H\right)_{i,j,k} \\ \left(\frac{\partial}{\partial y} H\right)_{i,j,k} \\ \left(\frac{\partial}{\partial y} H\right)_{i,j,k} \end{bmatrix} \begin{pmatrix} 1 \\ 1 \\ 1 \end{pmatrix} \tag{5.43}$$

The immediate neighborhood of the cell cuboid at (i, j, k) are, respectively, located at (i – 1, j, k) and (i + 1, j, k) plus (i, j – 1, k) and (i, j + 1, k) plus (i, j, k – 1) and (i, j, k + 1). Then, its six adjacent faces are defined at mid-intervals (i + 1/2, j, k) and (i – 1/2, j, k) plus (i, j + 1/2, k) and (i, j – 1/2, k) plus (i, j, k – 1/2) and (i, j, k + 1/2). To approximate the discharges $Q_{i+1/2,j,k}$, $Q_{i-1/2,j,k}$, $Q_{i,\,j+1/2,k}$, $Q_{i,j-1/2}$, $Q_{i,j,k+1/2}$, and $Q_{i,j,k-1/2}$ that traverse these six faces by difference equations, one must again apply the rules discussed in Section 5.3 of this chapter.

As done before, continuity requires for each cell at (i, j, k) that:

$$\begin{aligned} &\left(Q_{i+\frac{1}{2},j,k} \quad -Q_{i-\frac{1}{2},j,k} \right) \ldots = 0 \\ &+\left(Q_{i,j+\frac{1}{2},k} \quad -Q_{i,j-\frac{1}{2},k} \right) \ldots \\ &+\left(Q_{i,j,k+\frac{1}{2}} \quad -Q_{i,j,k+\frac{1}{2}} \right) - \Theta i,j,k \cdot \delta x \cdot \delta y \cdot \delta z \end{aligned} \tag{5.44}$$

Approximating these discharges by corresponding finite difference equations in Equation 5.45, collecting and solving for $H_{i,j,k}$, one obtains the appropriate iterative algorithm:

$$\left(H_{i,j,k}\right)^{n+1} = \left[\left[\left(A_{3D}\right)_{i,j,k}\right]^{n}\right]^{-1}\left[\left(B_{3D}\right)_{i,j,k}\right]^{n} \quad (5.45)$$

In the above equation, n now stands for the iteration cycle.

For each cuboid cell (i, j, k), the term $(A_{3D})_{i,j,k}$ now changes for each n iteration because the thickness δz is continuously adjusted until the algorithm converges. This term is calculated by:

$$\left(A_{3D}\right)_{i,j,k} = \left[\left(C_{RUT}\right)_{ijk} + \left(C_{LDB}\right)_{ijk}\right]\left[\begin{pmatrix} \delta y\delta z & 0 & 0 \\ 0 & \delta x\delta z & 0 \\ 0 & 0 & \delta x\delta y \end{pmatrix}\begin{bmatrix} \frac{1}{\delta x} \\ \frac{1}{\delta y} \\ \frac{1}{\delta z} \end{bmatrix}\begin{pmatrix} 1 \\ 1 \\ 1 \end{pmatrix}\right] \quad (5.46)$$

The coefficients $[C_{RUT}]_{i,j,k}$ and $[C_{LDB})_{i,j,k}$ may result from arithmetic means involving, respectively, the adjacent cells:

$$\left(C_{RUT}\right)_{i,j,k}\cdots \quad (5.47)$$

$$= \begin{bmatrix} \frac{\left(k_{xx}\right)_{i+1,j,k} + \left(k_{xx}\right)_{i,j,k}}{2} & \frac{\left(k_{yx}\right)_{i,j+1,k} + \left(k_{yx}\right)_{i,j,k}}{2} & \frac{\left(k_{zx}\right)_{i,j,k+1} + \left(k_{zx}\right)_{i,j,k}}{2} \\ \frac{\left(k_{xy}\right)_{i+1,j,k} + \left(k_{xy}\right)_{i,j,k}}{2} & \frac{\left(k_{yy}\right)_{i,j+1,k} + \left(k_{yy}\right)_{i,j,k}}{2} & \frac{\left(k_{zy}\right)_{i,j,k+1} + \left(k_{zy}\right)_{i,j,k}}{2} \\ \frac{\left(k_{xz}\right)_{i+1,j,k} + \left(k_{xz}\right)_{i,j,k}}{2} & \frac{\left(k_{yz}\right)_{i,j+1,k} + \left(k_{yz}\right)_{i,j,k}}{2} & \frac{\left(k_{zz}\right)_{i,j,k+1} + \left(k_{zz}\right)_{i,j,k}}{2} \end{bmatrix}$$

$$(C_{LDB})_{i,j,k} \tag{5.48}$$

$$= \begin{bmatrix} \frac{(k_{xx})_{i-1,j,k}+(k_{xx})_{i,j,k}}{2} & \frac{(k_{yx})_{i,j-1,k}+(k_{yx})_{i,j,k}}{2} & \frac{(k_{zx})_{i,j,k-1}+(k_{zx})_{i,j,k}}{2} \\ \frac{(k_{xy})_{i-1,j,k}+(k_{xy})_{i,j,k}}{2} & \frac{(k_{yy})_{i,j-1,k}+(k_{yy})_{i,j,k}}{2} & \frac{(k_{zy})_{i,j-1,k}+(k_{zy})_{i,j,k}}{2} \\ \frac{(k_{xz})_{i-1,j,k}+(k_{xz})_{i,j,k}}{2} & \frac{(k_{yz})_{i,j,k-1}+(k_{yz})_{i,j,k}}{2} & \frac{(k_{zz})_{i,j,k-1}+(k_{zz})_{i,j,k}}{2} \end{bmatrix}$$

For each cell (i, j, k), the term $[(B_{3D})_{i,j,k}]^n$ also changes after each n iteration until the solution stabilizes. It is calculated by:

$$(B_{3D})_{i,j,k} = (C_{RUT})_{i,j,k} \left[\begin{pmatrix} \delta y \cdot \delta z_{i,j,k} & 0 & 0 \\ 0 & \delta x \cdot \delta z_{i,j,k} & 0 \\ 0 & 0 & \delta x \cdot \delta z \end{pmatrix} \begin{pmatrix} \frac{H_{i+1,j,k}}{\delta x} \\ \frac{H_{i,j+1,k}}{\delta y} \\ \frac{H_{i,j,k+1}}{\delta z_{i,j,k}} \end{pmatrix} \begin{pmatrix} 1 \\ 1 \\ 1 \end{pmatrix} \right] \ldots \tag{5.49}$$

$$+(C_{LDB})_{i,j,k} \left[\begin{pmatrix} \delta y \cdot \delta z_{i,j,k} & 0 & 0 \\ 0 & \delta x \cdot \delta z_{i,j,k} & 0 \\ 0 & 0 & \delta x \cdot \delta y \end{pmatrix} \begin{pmatrix} \frac{H_{i-1,j,k}}{\delta x} \\ \frac{H_{i,j-1,k}}{\delta y} \\ \frac{H_{i,j,k-1}}{\delta z_{i,j,k}} \end{pmatrix} \begin{pmatrix} 1 \\ 1 \\ 1 \end{pmatrix} \right]$$

$$+\Theta_{i,j,k} \cdot \delta x \cdot \delta y \cdot \delta z_{i,j,k}$$

Note that $\delta z_{i,j,k}$ has indices in both expressions to emphasize its changing character.

Continuity for the upper slab (index k + 1)

The algorithm for the upper slab was deduced by the same principles used for the mid-slab with two exceptions:

First: To get the head WT on the phreatic surface a virtual upper slab 2.5m thick was placed above the model. However, only half-thickness of 2.5m was effectively considered for computing.

Second: For obvious reasons, the vertical component of the total flow through the phreatic was null. Then, the corresponding hydraulic gradient was set equal to zero.

The appropriate iterative algorithm is:

$$\left(WT_{i,j,k+1}\right)^{n+1} = \left[\left[\left(A_{TOP}\right)_{i,j,k+1}\right]^{n}\right]^{-1}\left[\left(B_{TOP}\right)_{i,j,k+1}\right]^{n} \tag{5.50}$$

The terms $(A_{TOP})_{i,j,k+1}$ and $(B_{TOP})_{i,j,k+1}$, respectively, calculated are:

$$\begin{aligned}\left(A_{TOP}\right)_{i,j,k+1} &= \left(C_{RUT}\right)_{i,j,k+1} \left(\begin{pmatrix} \delta y\dfrac{\delta z_{i,j,k+1}}{2} & 0 & 0 \\ 0 & \delta x\dfrac{\delta z_{i,j,k+1}}{2} & 0 \\ 0 & 0 & \delta x\cdot\delta y \end{pmatrix}\begin{pmatrix} \dfrac{1}{\delta x} \\ \dfrac{1}{\delta y} \\ 0 \end{pmatrix}\right)\begin{pmatrix}1\\1\\1\end{pmatrix}\ldots \\ &+\left(C_{LDB}\right)_{i,j,k+1}\left(\begin{pmatrix} \delta y\dfrac{\delta z_{i,j,k+1}}{2} & 0 & 0 \\ 0 & \delta x\dfrac{\delta z_{i,j,k+1}}{2} & 0 \\ 0 & 0 & \delta x\cdot\delta y \end{pmatrix}\begin{pmatrix} \dfrac{1}{\delta x} \\ \dfrac{1}{\delta y} \\ \dfrac{1}{\delta z_{i,j,k+1}} \end{pmatrix}\right)\begin{pmatrix}1\\1\\1\end{pmatrix}\end{aligned} \tag{5.51}$$

$$\left(B_{TOP}\right)_{i,j,k+1} = \left(C_{RUT}\right)_{i,j,k+1}\left[\begin{pmatrix} \delta y\dfrac{\delta z_{i,j,k+1}}{2} & 0 & 0 \\ 0 & \delta x\dfrac{\delta z_{i,j,k+1}}{2} & 0 \\ 0 & 0 & \delta x\cdot\delta y \end{pmatrix} \begin{pmatrix} \dfrac{WT_{i+1,j,k+1}}{\delta x} \\ \dfrac{WT_{i,j+1,k+1}}{\delta y} \\ 0 \end{pmatrix}\begin{pmatrix}1\\1\\1\end{pmatrix}\right]\ldots \tag{5.52}$$

$$+\left(C_{LDB}\right)_{i,j,k+1}\left[\left(\begin{matrix}\delta y\dfrac{\delta z_{i,j,k+1}}{2} & 0 & 0\\ 0 & \delta x\dfrac{\delta z_{i,j,k+1}}{2} & 0\\ 0 & 0 & \delta x\cdot\delta y\end{matrix}\right)\right.$$

$$\left.\left(\begin{matrix}\dfrac{WT_{i-1,j+1,k+1}}{\delta x}\\ \dfrac{WT_{i,j-1,k+1}}{\delta y}\\ \dfrac{middle\_H_{i,j,k}}{\delta z_{i,j,k+1}}\end{matrix}\right)\left(\begin{matrix}1\\1\\1\end{matrix}\right)\right]\ldots$$

$$+\Theta_{i,j,k+1}\delta x\cdot\delta y\frac{\delta z_{i,j,k+1}}{2}+\omega_{i,j}\cdot\delta x\cdot\delta y$$

- Continuity for the bottom slab (index k –- 1):
  Continuity for the bottom slab was deduced as for the top slab with the difference that the no-flow boundary was the base itself. The appropriate iterative algorithm is:

$$\left(H_{i,j,k-1}\right)^{n+1}=\left[\left[\left(A_{BOTTOM}\right)_{i,j,k-1}\right]^{n}\right]^{-1}\left[\left(B_{BOTTOM}\right)_{i,j,k-1}\right]^{n} \tag{5.53}$$

- The terms $(A_{BOTTOM})_{i,j,k-1}$ and $(B_{BOTTOM})_{i,j,k-1}$ are, respectively, calculated by:

$$\left(A_{BOTTOM}\right)_{i,j,k-1}=\left(C_{RUT}\right)_{i,j,k-1}\left(\begin{matrix}\delta y\dfrac{\delta z_{i,j,k-1}}{2} & 0 & 0\\ 0 & \delta x\dfrac{\delta z_{i,j,k-1}}{2} & 0\\ 0 & 0 & \delta x\cdot\delta y\end{matrix}\right) \tag{5.54}$$

$$\left(\begin{matrix}\dfrac{1}{\delta x}\\ \dfrac{1}{\delta y}\\ \dfrac{1}{\delta z_{i,j,k}}\end{matrix}\right)\left(\begin{matrix}1\\1\\1\end{matrix}\right)\ldots$$

$$+\left(C_{LDB}\right)_{i,j,k-1}\begin{pmatrix}\delta y\dfrac{\delta z_{i,j,k-1}}{2} & 0 & 0\\ 0 & \delta x\dfrac{\delta z_{i,j,k-1}}{2} & 0\\ 0 & 0 & \delta x\cdot\delta y\end{pmatrix}\begin{pmatrix}\dfrac{1}{\delta x}\\ \dfrac{1}{\delta y}\\ 0\end{pmatrix}\begin{pmatrix}1\\1\\1\end{pmatrix}$$

$$\left(B_{BOTTOM}\right)_{i,j,k-1}=\left(C_{RUT}\right)_{i,j,k-1}\begin{pmatrix}\delta y\dfrac{\delta z_{i,j,k-1}}{2} & 0 & 0\\ 0 & \delta x\dfrac{\delta z_{i,j,k-1}}{2} & 0\\ 0 & 0 & \delta x\cdot\delta y\end{pmatrix}\begin{pmatrix}\dfrac{\left(H_{bottom}\right)_{i+1,j,k}}{\delta x}\\ \dfrac{\left(H_{bottom}\right)_{i,j+1,k}}{\delta y}\\ \dfrac{\left(H_{middle}\right)_{i,j,k+1}}{\delta z_{i,j,k}}\end{pmatrix}\begin{pmatrix}1\\1\\1\end{pmatrix}\ldots$$
$$+\left(C_{LDB}\right)_{i,j,k-1}\begin{pmatrix}\delta y\dfrac{\delta z_{i,j,k}}{2} & 0 & 0\\ 0 & \delta x\dfrac{\delta z_{i,j,k-1}}{2} & 0\\ 0 & 0 & \delta x\cdot\delta y\end{pmatrix} \quad (5.55)$$

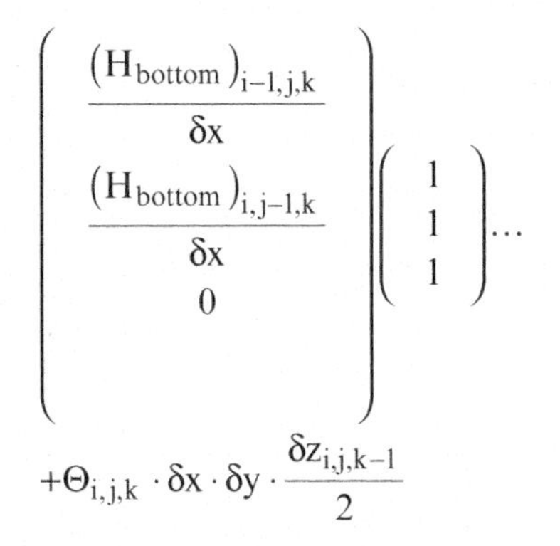

- Impervious base boundary condition (index $k-1$):

Phreatic and impervious boundary conditions are a bit more complicated. For the impervious base, the scalar product at point $(i, j, k-1)$ of its exterior normal by the specific discharge at this same point must be zero, as expected:

$$\begin{pmatrix} \dfrac{z_{i+1,j,k-1} - z_{i-1,j,k-1}}{2\,\delta x}\mathbf{u}_i \\ -\dfrac{z_{i,j+1,k-1} - z_{i,j-1,k-1}}{2\,\delta y}\mathbf{u}_j \\ \mathbf{u}_k \end{pmatrix} \begin{bmatrix} (q_x)_{i,j,k-1}\,\mathbf{u}_i \\ (q_y)_{i,j,k-1}\,\mathbf{u}_j \\ (q_z)_{i,j,k-1}\,\mathbf{u}_k \end{bmatrix} = 0 \tag{5.56}$$

In the above equation, the vectors $\mathbf{u}_i$, $\mathbf{u}_j$, and $\mathbf{u}_k$ stand for the unit vectors of the Cartesian axes xyz. Developing that condition gives the appropriate iterative algorithm:

$$\left(H_{i,j,k-1}\right)^{n+1} = \left[\left[\left(A_{NB}\right)_{i,j,k-1}\right]^{n}\right]^{n-1}\left[\left(B_{NB}\right)_{i,j,k-1}\right]^{n} \tag{5.57}$$

The terms $(A_{NB})_{i,j,k-1}$ and $(B_{NB})_{i,j,k-1}$ are, respectively, calculated by:

$$\left(A_{NB}\right)_{i,j,k-1} = \begin{pmatrix} \dfrac{z_{i+1,j,k-1} - z_{i-1,j,k-1}}{2\,\delta x} \\ \dfrac{z_{i,j+1,k-1} - z_{i,j-1,k-1}}{2\,\delta y} \\ 1 \end{pmatrix} \tag{5.58}$$

$$\begin{bmatrix} (k_{xx})_{i,j,k-1} & (k_{xy})_{i,j,k-1} & (k_{xz})_{i,j,k-1} \\ (k_{yx})_{i,j,k-1} & (k_{yy})_{i,j,k-1} & (k_{yz})_{i,j,k-1} \\ (k_{zx})_{i,j,k-1} & (k_{zy})_{i,j,k-1} & (k_{zz})_{i,j,k-1} \end{bmatrix} \begin{pmatrix} 0 \\ 0 \\ -\dfrac{3}{2\,\delta z_{i,j,k}} \end{pmatrix}$$

$$(A_{NB})_{i,j,k-1} = \begin{pmatrix} \dfrac{z_{i+1,j,k-1} - z_{i-1,j,k-1}}{2\,\delta x} \\ \dfrac{z_{i,j+1,k-1} - z_{i,j-1,k-1}}{2\,\delta y} \\ 1 \end{pmatrix} \tag{5.59}$$

$$\begin{bmatrix} (k_{xx})_{i,j,k-1} & (k_{xy})_{i,j,k-1} & (k_{xz})_{i,j,k-1} \\ (k_{yx})_{i,j,k-1} & (k_{yy})_{i,j,k-1} & (k_{yz})_{i,j,k-1} \\ (k_{zx})_{i,j,k-1} & (k_{zy})_{i,j,k-1} & (k_{zz})_{i,j,k-1} \end{bmatrix} \begin{pmatrix} -\dfrac{H_{i+1,j,k-1} - H_{i-1,j,k-1}}{2\,\delta x} \\ -\dfrac{H_{i,j+1,k-1} - H_{i,j-1,k-1}}{2\,\delta y} \\ -\dfrac{-H_{i,j,k+1} + 4H_{i,j,k}}{2\,\delta z_{i,j,k}} \end{pmatrix}$$

It is important to observe that the vertical gradient $J_{i,j,k-1}$ at the impervious base was calculated by the three-point formula, involving the heads $WT_{i,j,k+1}$, $H_{i,j,k}$, and $H_{i,j,k-1}$.

- Phreatic boundary condition (index $k+1$):

For the WT the scalar product at point (i, j, k − 1) of its exterior normal by the specific discharge at this same point must equate the vertical accretion $\omega_{i,j}$:

$$\begin{pmatrix} -\dfrac{WT_{i+1,j,k+1} - WT_{i-1,j,k+1}}{2\delta x}\mathbf{u}_i \\ -\dfrac{WT_{i,j+1,k+1} - WT_{i,j-1,k+1}}{2\delta y}\mathbf{u}_j \\ \mathbf{u}_k \end{pmatrix} \left[\begin{bmatrix} (q_x)_{i,j,k+1}\,\mathbf{u}_i \\ (q_y)_{i,j,k+1}\,\mathbf{u}_j \\ (q_z)_{i,j,k+1}\,\mathbf{u}_k \end{bmatrix} - \begin{pmatrix} 0\,\mathbf{u}_i \\ 0\,\mathbf{u}_j \\ w_{i,j}\,\mathbf{u}_k \end{pmatrix} \right] = 0 \tag{5.60}$$

Developing that condition gives the appropriate iterative algorithm:

$$\left(W_{i,j,k-1}\right)^{n+1} = \left[\left[(A_{NT})_{i,j,k-1}\right]^{n}\right]^{-1} \left[(B_{NT})_{i,j,k-1}\right]^{n} \tag{5.61}$$

The terms $(A_{NT})_{i,j,k+1}$ and $(B_{NT})_{i,j,k+1}$ are, respectively, calculated by:

$$(A_{NT})_{i,j,k+1} = \begin{pmatrix} -\dfrac{WT_{i+1,j,k+1} - WT_{i-1,j,k+1}}{2\,\delta x} \\ -\dfrac{WT_{i,j+1,k+1} - WT_{i,j-1,k+1}}{2\,\delta y} \\ 1 \end{pmatrix} \tag{5.62}$$

$$\left[\begin{bmatrix} (k_{xx})_{i,j,k+1} & (k_{xy})_{i,j,k+1} & (k_{xz})_{i,j,k+1} \\ (k_{yx})_{i,j,k+1} & (k_{yy})_{i,j,k+1} & (k_{yz})_{i,j,k+1} \\ (k_{zx})_{i,j,k+1} & (k_{zy})_{i,j,k+1} & (k_{zz})_{i,j,k+1} \end{bmatrix} \begin{pmatrix} 0 \\ 0 \\ \dfrac{3}{2\,\delta z_{i,j,k-1}} \end{pmatrix}\right]$$

$$(B_{NT})_{i,j,k+1} = \begin{pmatrix} -\dfrac{WT_{i+1,j,k+1} - WT_{i-1,j,k+1}}{2\,\delta x} \\ -\dfrac{WT_{i,j+1,k+1} - WT_{i,j-1,k+1}}{2\,\delta y} \\ 1 \end{pmatrix} \tag{5.63}$$

$$\left[\begin{bmatrix} (k_{xx})_{i,j,k+1} & (k_{xy})_{i,j,k+1} & (k_{xz})_{i,j,k+1} \\ (k_{yx})_{i,j,k+1} & (k_{yy})_{i,j,k+1} & (k_{yz})_{i,j,k+1} \\ (k_{zx})_{i,j,k+1} & (k_{zy})_{i,j,k+1} & (k_{zz})_{i,j,k+1} \end{bmatrix} \begin{pmatrix} -\dfrac{WT_{i+1,j,k+1} - WT_{i-1,j,k+1}}{2\,\delta x} \\ -\dfrac{WT_{i,j+1,k+1} - WT_{i,j-1,k+1}}{2\,\delta y} \\ \dfrac{-4(H)_{i,j,k} + (H)_{i,j,k-1}}{2\,\delta z_{i,j,k-1}} \end{pmatrix} + \begin{pmatrix} 0 \\ 0 \\ -\omega_{i,j} \end{pmatrix}\right]$$

It is essential to observe that the vertical gradient $J_{i,j,k+1}$ at WT was again calculated by the three-point formula, involving the heads $WT_{i,j,k+1}$, $H_{i,j,k}$, and $H_{i,j,k-1}$.

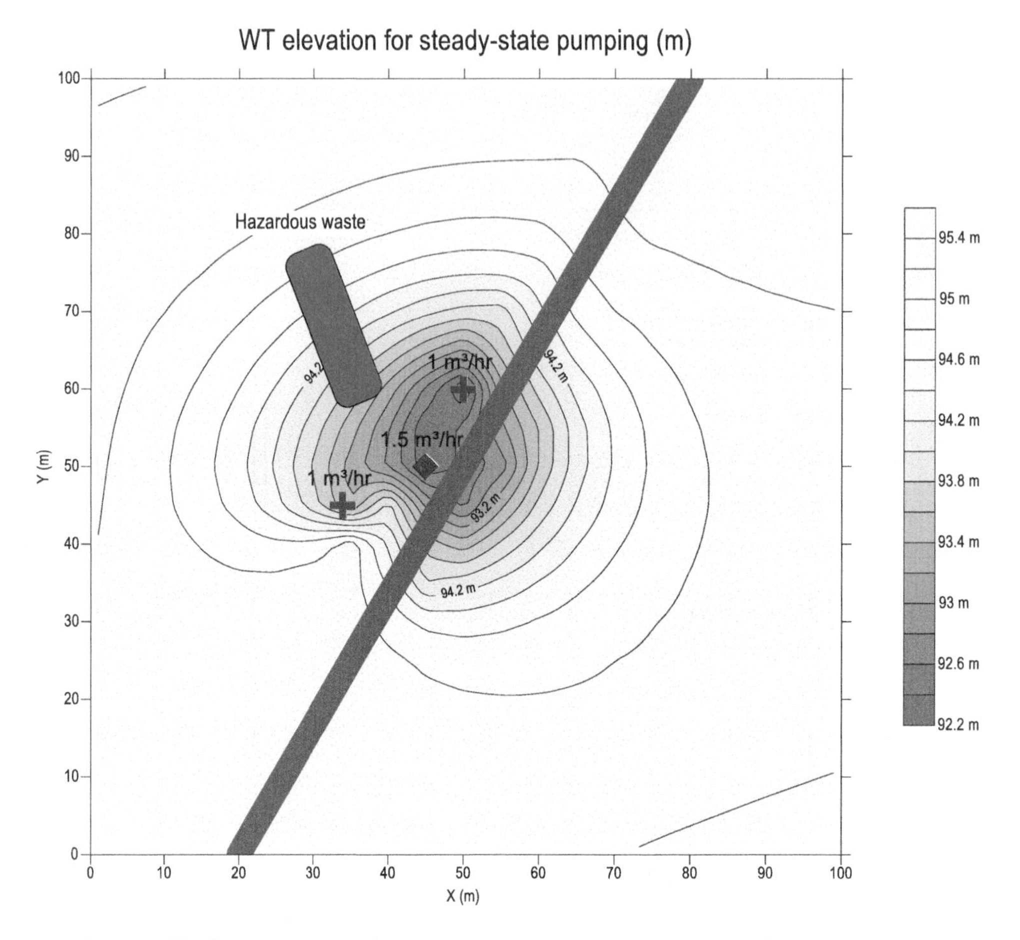

Figure 5.32 Contour map of the depressed WT showing the influence of the much pervious fault as a vertical and very narrow groundwater buffer.

These five conditions produce a system of simultaneous algebraic equations to be solved iteratively. It must be noted that more sophisticated algorithms can be deduced if judged convenient.

The results obtained for the case in hand are summarized in the following figures:

- Figure 5.30: Depressed WT for steady-state pumping.
- Figure 5.31: Contour map of equipotential at elevation 92 m.
- Figure 5.32: Contour map of equipotential at elevation 89 m.
- Figure 5.33: Contour map of equipotential at elevation 86 m.
- Figure 5.34: Contour map of total flow at the mid-slab.
- Figure 5.35: Contour map of total flow at the inclined base.
- Figure 5.36: 3D perspective of the pumping system (Figures 5.38 and 5.39).

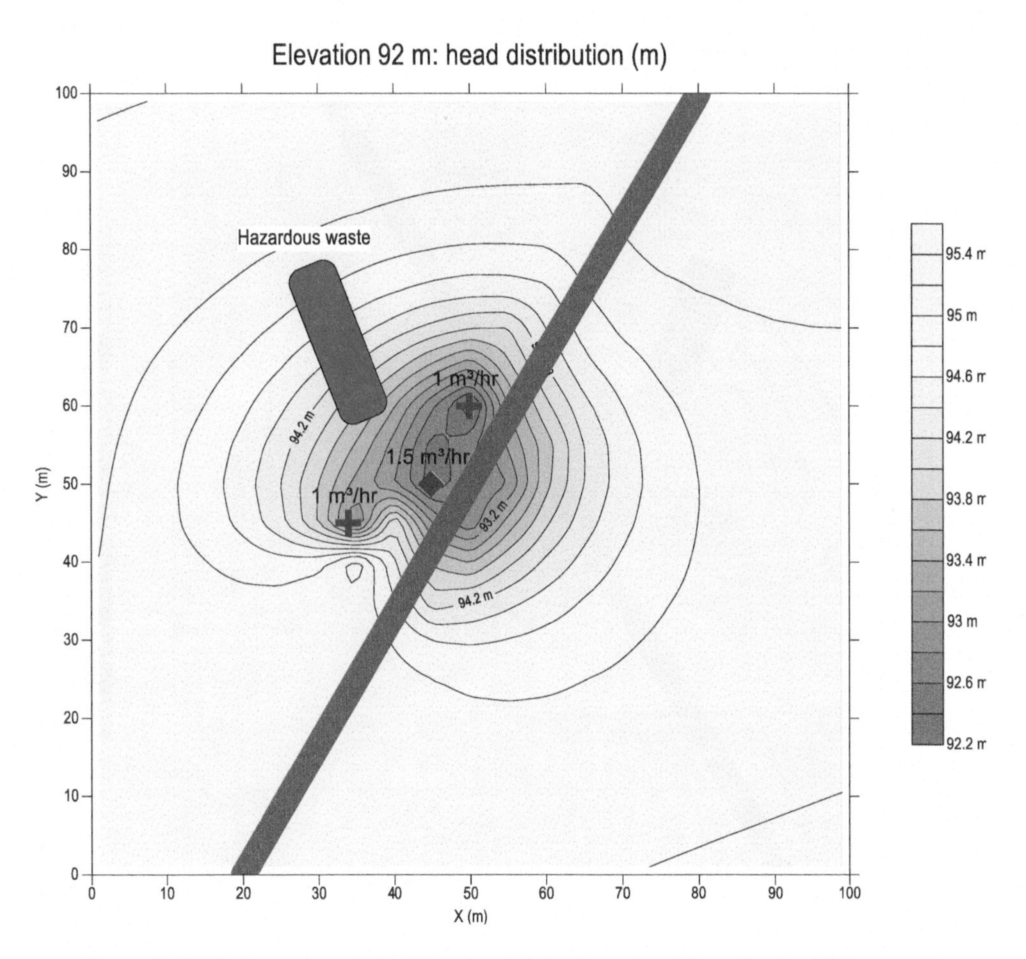

Figure 5.33 Contour map of equipotential at elevation 92 m. As the 3D xyz grid was progressively deformed until equilibrium, H-data at elevation 92 m was interpolated without difficulty from the grid (x, y, z, H) values.

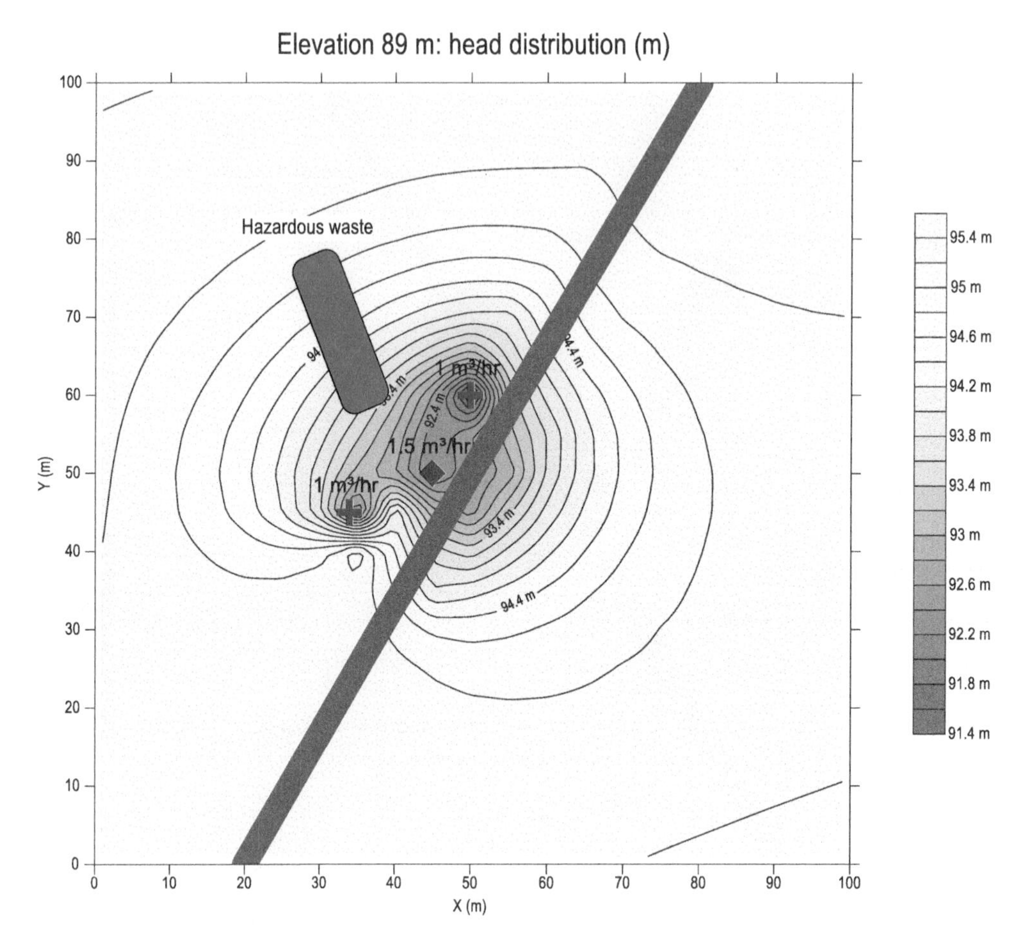

Figure 5.34 Contour map of equipotential at elevation 89 m. H-data interpolated without difficulty from the grid (x, y, z, H) values.

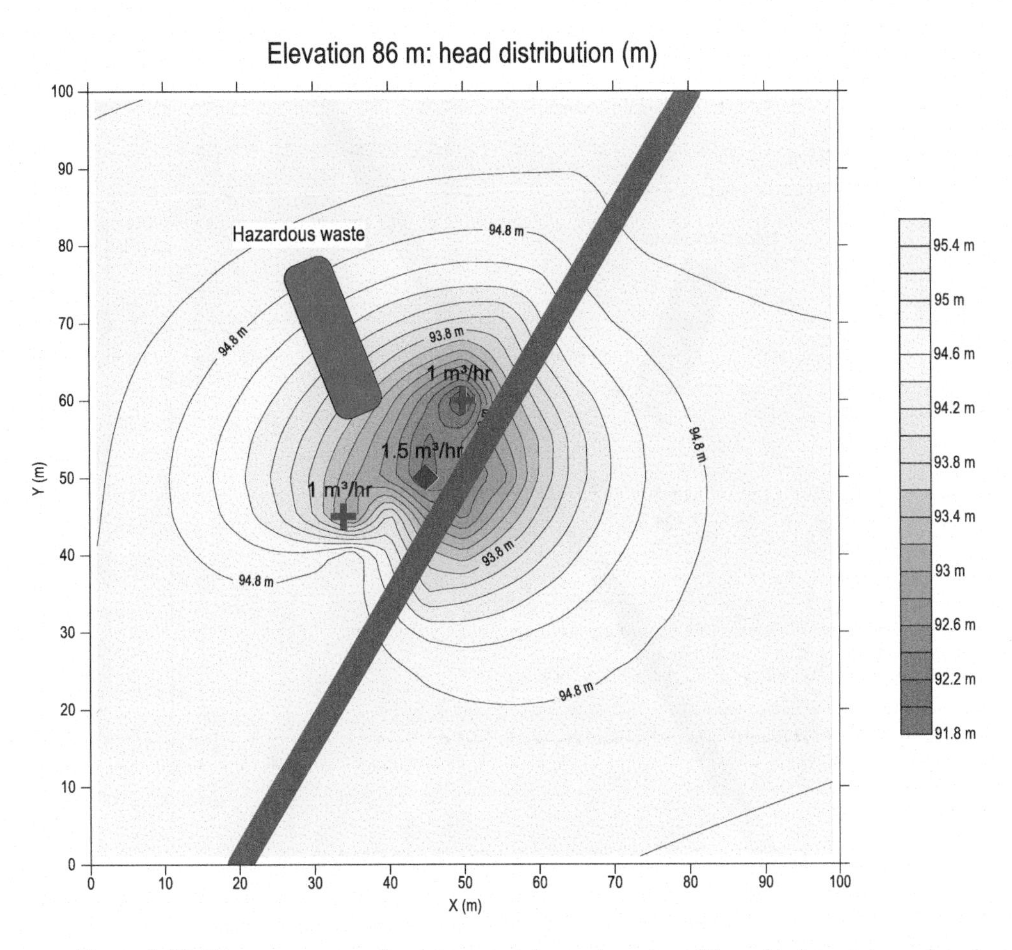

Figure 5.35 Contour map of equipotential at elevation 86 m. H-data interpolated without difficulty from the grid (x, y, z, H) values.

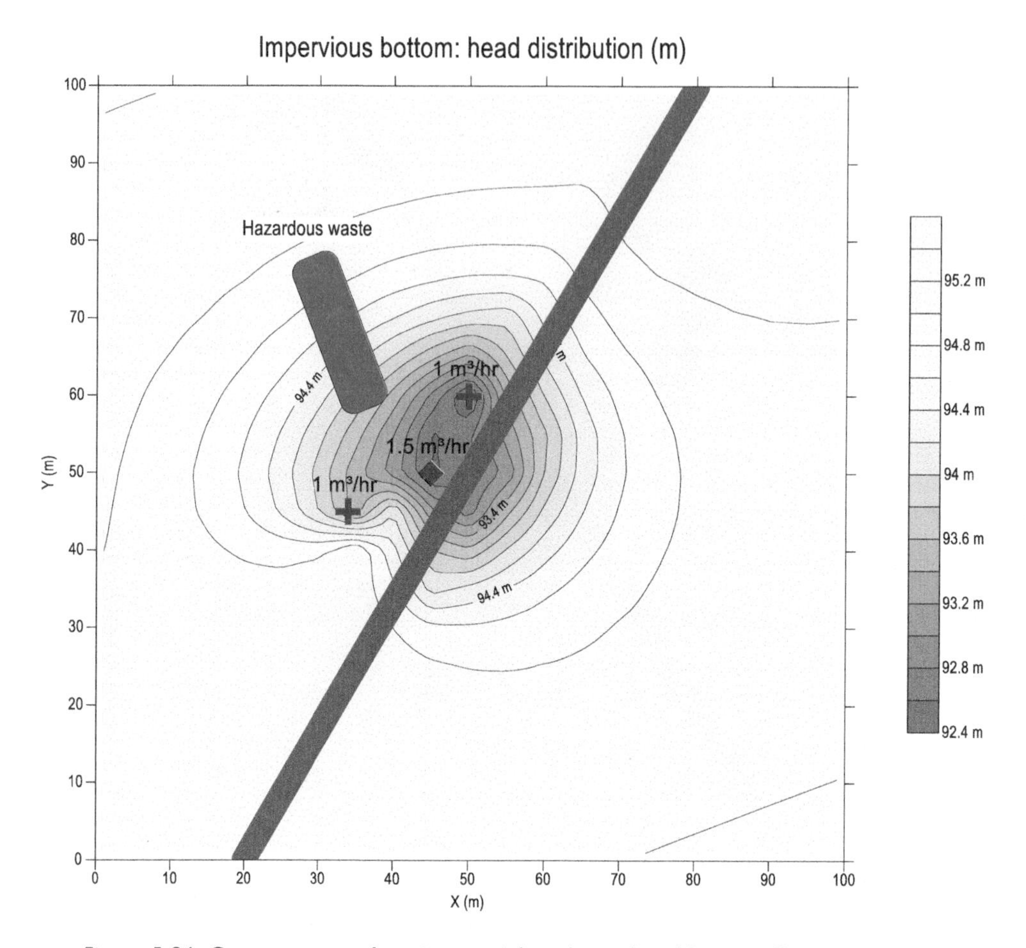

Figure 5.36 Contour map of equipotential at the inclined base, reflecting the proximity of the lower end of the 5 m pump filter.

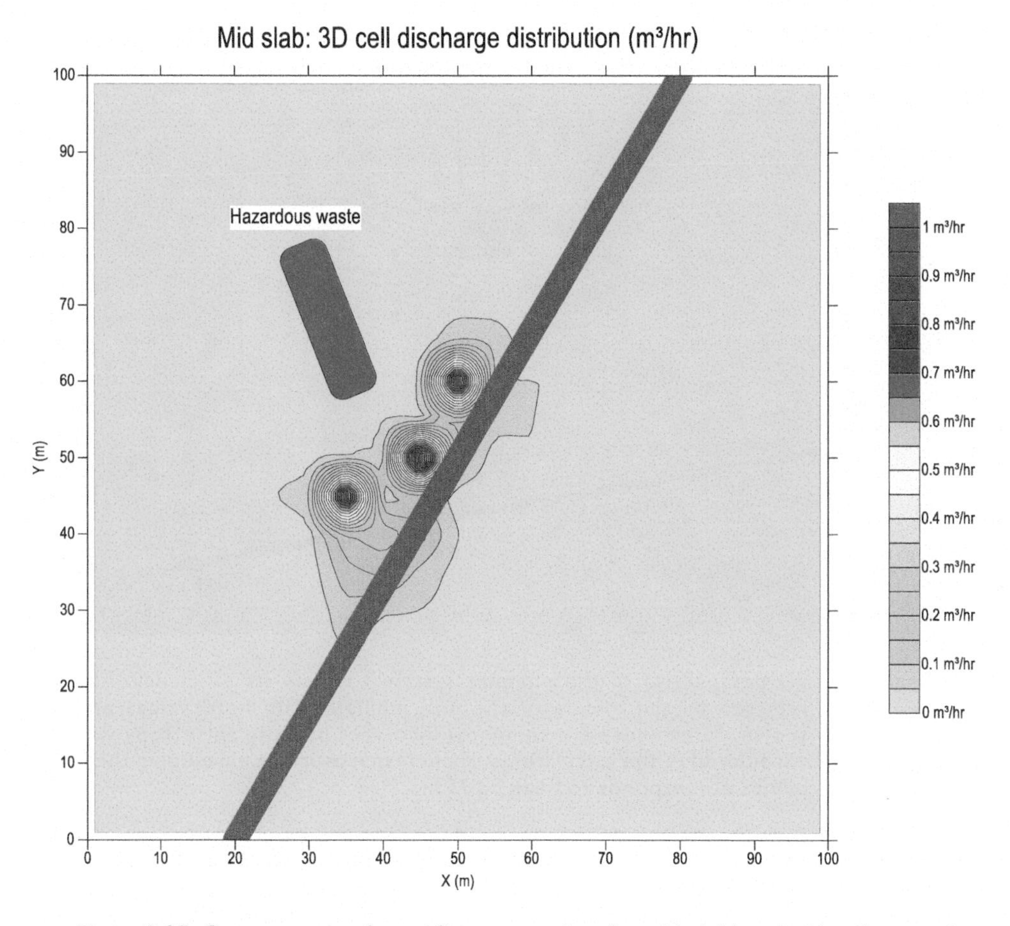

Figure 5.37 Contour map of total flow traversing the mid-slab's cuboid cells, entering and leaving by its six faces. Note the pumping efficiency near the fault by reversing its natural flow and contributing to mitigate and reverse the contaminant transport.

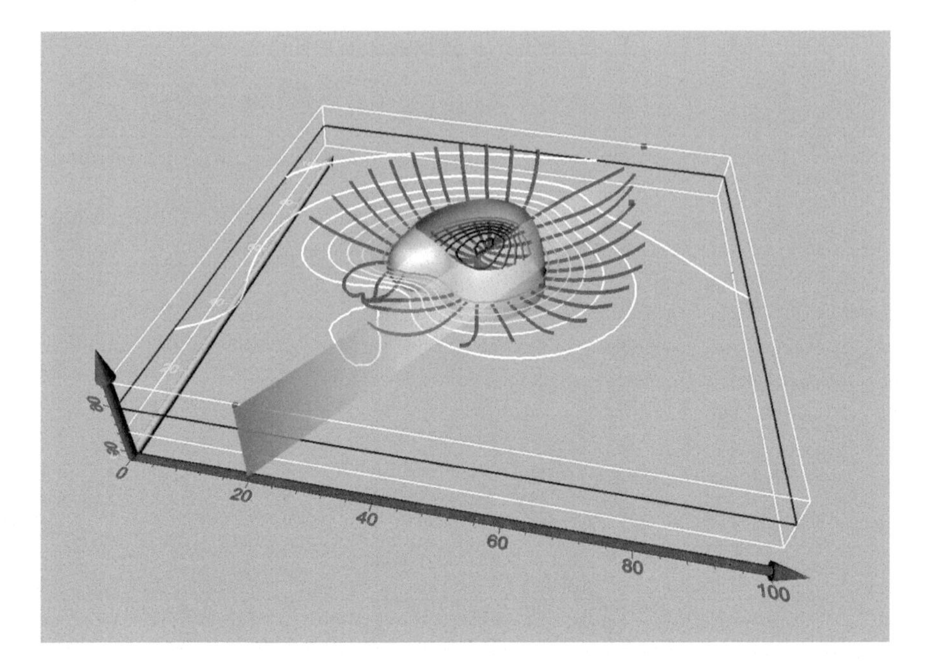

Figure 5.38 3D perspective of the pumping system to show its efficiency. All data referred to the middle of the intermediate slab: head values are not shown. Streamlines are somehow incorrect because anisotropy was not considered in the calculations. Half close isosurface enveloping the three pumps corresponds to head 93.75 m.

As pointed out, the 3D grid was successively adapted to correspond to all requirements (see Figure 5.39).

Finally, it must be told that instabilities did not occur in this exercise. If it happens, the cure is the same recommended for the Dupuit ' s approach in Example 5.3.

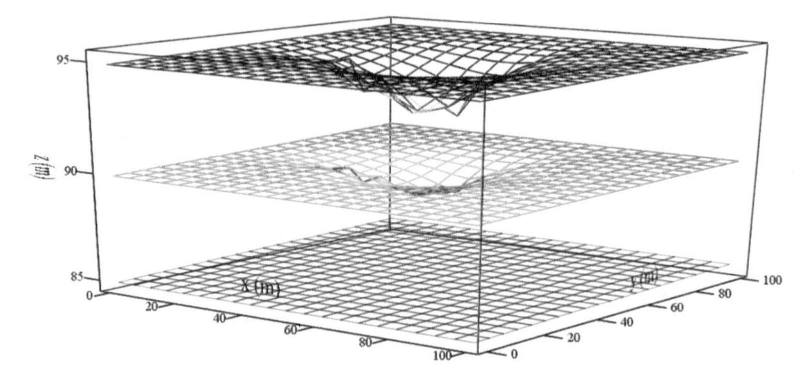

Figure 5.39 Perspective of the self-adjusted grid as to satisfy all continuity and boundaries requirements (deformed scales).

5.4.3 Transient solutions

Transient solutions follow traditional procedures, either explicit or implicit. For example, the explicit algorithm for a confined aquifer modeled by Dupuit's approach is:

$$
\left[(H_c)_{i,j}\right]^{n+1} = \left[(H_c)_{i,j}\right]^{n} + \frac{1}{(S_C)_{i,j}}\,\frac{\delta t}{\delta x\,\delta y}\left[\left[(H_c)_{i,j}\right]^{n}\begin{bmatrix} (C_{RU})_{i,j}\begin{bmatrix}\dfrac{(H_{i+1,j})^n-(H_{i,j})^n}{\delta x}\\ \dfrac{(H_{i,j+1})^n-(H_{i,j})^n}{\delta y}\end{bmatrix} - (C_{LD})_{i,j}\begin{bmatrix}\dfrac{(H_{i,j})^n-(H_{i-1,j})^n}{\delta x}\\ \dfrac{(H_{i,j})^n-(H_{i,j-1})^n}{\delta y}\end{bmatrix} \\ +(C_C)_{i,j}\begin{bmatrix}\dfrac{B_{i+1,j}-B_{i-1,j}}{2\,\delta x}\\ \dfrac{B_{i,j+1}-B_{i,j-1}}{2\,\delta y}\end{bmatrix}\end{bmatrix}\cdot\begin{pmatrix}1\\1\end{pmatrix} + \Omega_{i,j}\,\delta x\,\delta y\right]
\tag{5.64}
$$

As a first tentative to get a stable explicit algorithm, the Fourier coefficient $FN_{i,j}$ must be restricted to:

$$
0 < FN_{i,j} = \frac{1}{(S_C)_{i,j}}\cdot\frac{\delta t}{\delta x\cdot\delta y} < \frac{1}{2}
\tag{5.65}
$$

In any case, a suitable Crank-Nicolson scheme may also be used, as explained formerly.

But now, a different and variable Fourier coefficients $FN_{i,j}$ must translate the complexity of inhomogeneous and anisotropic systems, and the available experience on that matter, specifically for fractured rocks, is scarce. On the other hand, calibration resorting to observed temporal data series, besides being sometimes tricky if there are too many conflicting data, may not be reliable for time-spans other than the monitored. In practice, if time predictions are really needed, a simple time-dependent model may approximately answer current questions for a restricted time-span (see Figure 5.40). For real 3D simulation, the problem is still more complicated.

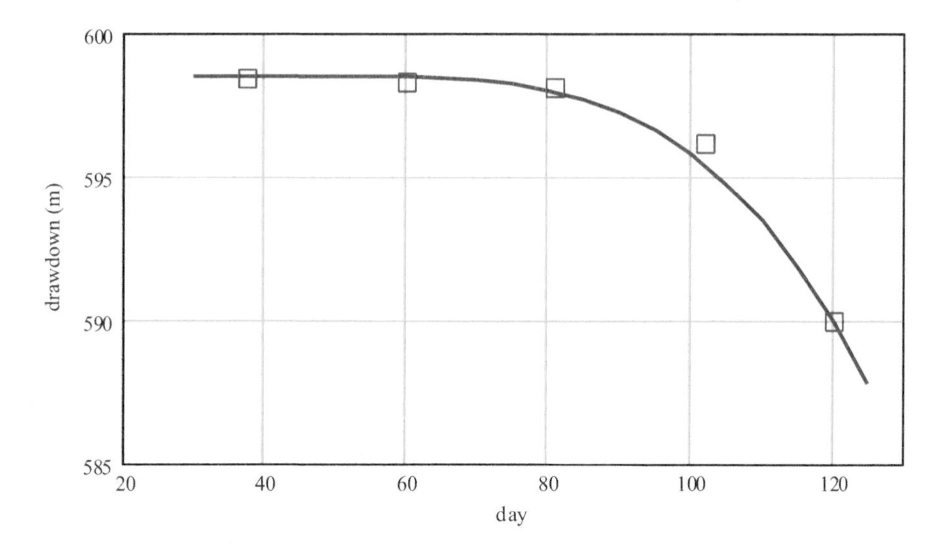

Figure 5.40 The theoretical transient underground mine dewatering in an anisotropic pervious media was roughly approximated by a closed solution for a finite linear drain calibrated by five monitored points.

Notes

1 A well organized and commented anthology on the subject can be found in "Rock Fractures and Fluid Flow", National Academic Press, Washington, 1966.
2 Field tests, as those proposed by Hshieh et al., considers the hidden contributions of all unseen discontinuities, with no exception, and implicitly integrate the effect of the intact rock matrix.
3 A non-linear solution applied to the analysis of design envelope for the uplift pressure under concrete gravity dams, is found in Franciss, F. O., "Practical comments derived from theoretical consideration on the effect of grouting and drainage on uplift", *XIV Large Dams Congress*, Q. 52, R. 70, Rio de Janeiro, 1982.

References

Bear, J., 1972, *Dynamics of Fluids in Porous Media*, American Elsevier, New York.

Lam, C. Y., 1994, *Applied Numerical Methods for Partial Differential Equations*, Prentice-Hall, Upper Saddle River.

Press, W. H., Teukolsky, S. A., Vetterling, W. T. and Flannery, B. P., 2007, *Numerical Recipes: The Art if Scientific Computing*, Cambridge University Press, Cambridge, Third Ed.

Varga, R. S., 1962, *Matrix Iterative Analysis*, Prentice-Hall, Upper Saddle River.

Wang, H. F. and Anderson, M. P., 1982, *Introduction to Groundwater Modeling: Finite Difference and Finite Element Methods*, W. H. Freeman and Company, San Francisco.

Chapter 6

Applications

6.1 Preliminaries

This last chapter contains three examples of groundwater flow through large net of fractures. The first and second examples concern its high capacity to transmit pressure or transport groundwater over exceptionally long distances. The third example exemplifies how pervious megafractures redistribute heat continually stored at very deep horizons around the periphery of residual reliefs giving rise to mild hot springs.

6.2 Fracture interconnectivity

6.2.1 *Interconnection between reservoirs*

Figure 6.1 shows a dam reservoir over basalt flat strata and some piezometers around it. Due to the basalt strata's horizontal pervious characteristics, local environmental agencies' constraints require long-term checking of the reservoir filling effects.

Figure 6.2 shows the gradual water table rise at the control piezometers after the filling-up, from level 480m to level 647m. With only one exception, all piezometers in the lake's proximity but located at higher altitudes did not show a significant impact. However, piezometer ANG11, about 10km away from the dam axis, probably connected to the reservoir by a pervious basalt horizontal contact, reacted with some delay to the filling of the lake. Even at less distance from the reservoir, other piezometers did not show similar behavior, certainly because its final depths were above this high pervious contact.

6.2.2 *Pollution bypassing a wide river*

6.2.2.1 The problem

Located at the right bank of a wide river, around 500m downstream of a dam axis, an industrial plant dedicated to Zn concentration from raw mine pellets faced an unexpected problem in 2012 (see Figure 6.3).

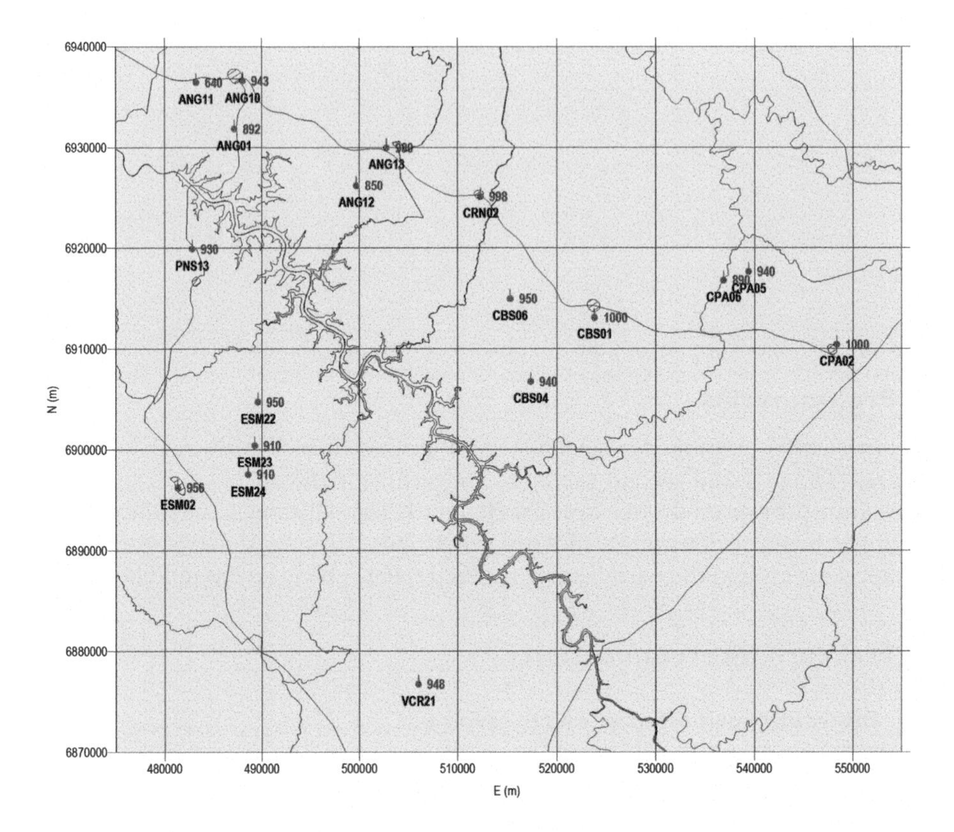

Figure 6.1 Dam location, dam reservoir boundaries, and location of some piezometers around the reservoir over extensive basalt strata, some of them marked by pervious contacts. This map shows the elevations of these piezometers near its locations. Piezometers' final depths are around 100 ± 40 m.

In the early 1960s, at the start of the plant operations, some hauling activities generated some Zn contamination spots but rapidly alleviated. On the other hand, as the whole area occupied by the industry was separated from the left bank by a large river, the possibility that the spread of Zn could reach the left bank was considered unfeasible. Indeed, this river, facing the industrial plant, is about 200–250 m large $\times$ 6–9 m depth, continuously discharging 300–1,500 m^3/s of water daily. All occasional temporary small spots of Zn dispersion were immediately held under control for environmental safety (see Figure 6.4).

Unexpectedly, some reliable reports in 2011 claimed that the water from four wells located at the river's left bank seemed polluted with dissolved $ZnSO_4$. Then, the industrial plant manager started an urgent campaign of field investigation to see what was happening. Additional field investigations include electro-resistivity surveys and continuous downhole scanning in addition to the systematic quality control of water samples from the suspected wells. The suspected wells were promptly deactivated (see Figure 6.5), and since then, daily uninterrupted freshwater supply was assured to the left bank consumers by the industry.

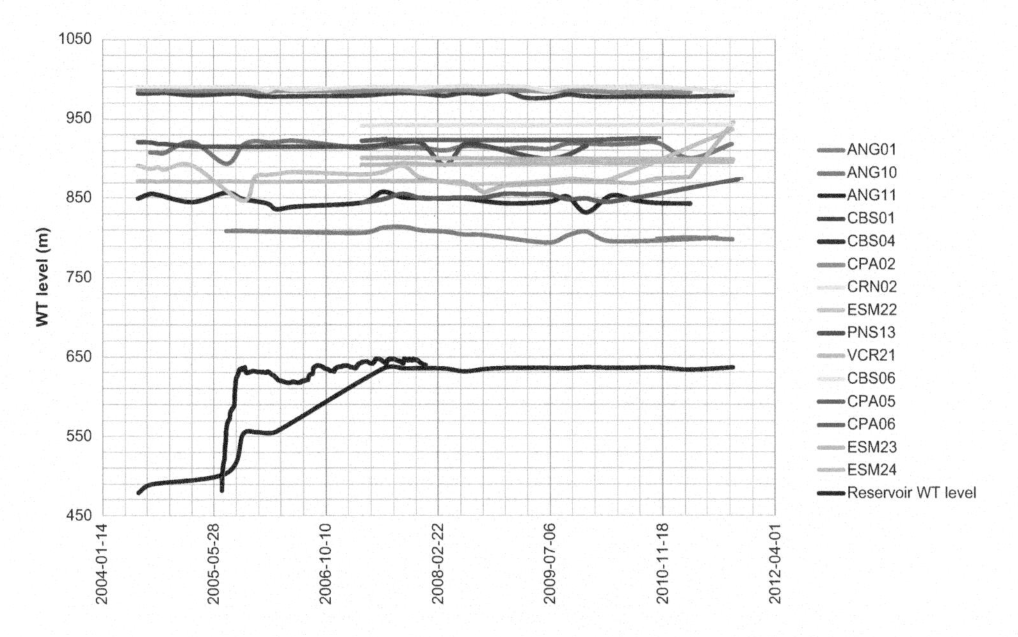

Figure 6.2 Water table rise in 15 piezometers around the dam reservoir during its first filling up. Only the piezometer ANG11, located ≈10 km away from the reservoir lake, reacted sensibly, probably because of its connection to the lake by a pervious contact between basalt layers. The other piezometers did not show a similar behavior because its final depths were above this contact.

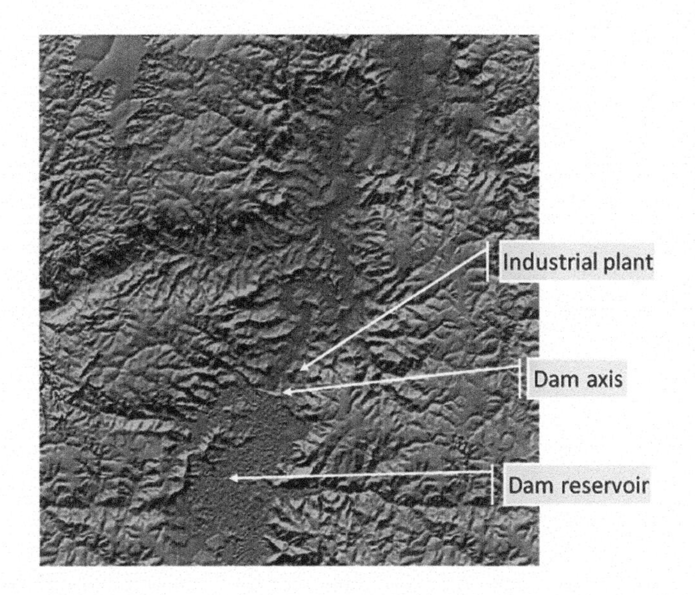

Figure 6.3 Satellite image.

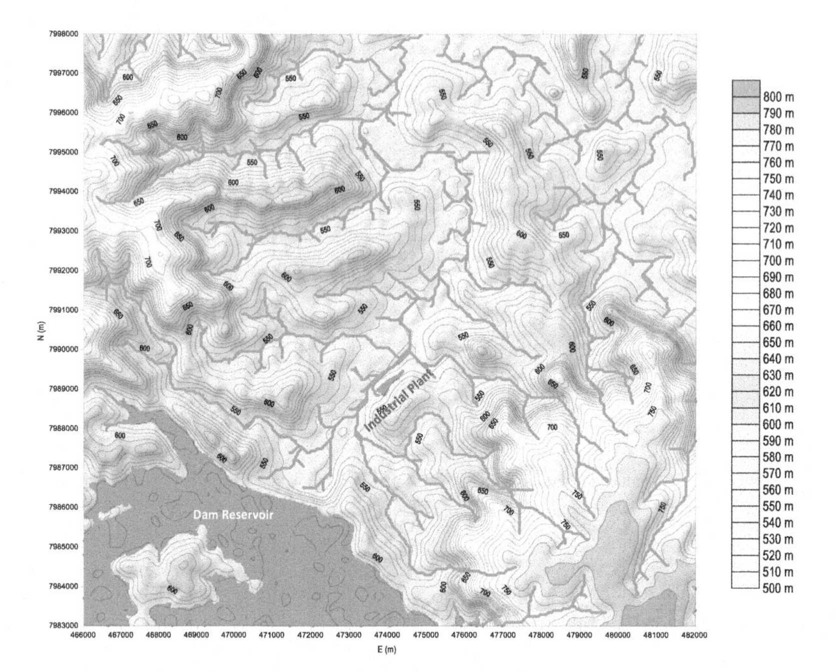

Figure 6.4 Digital elevation model includes the dam axis, part of its reservoir, and the general location of the industrial plant. It shows the river and its main tributaries. This river is about 200–250 m wide × 6.9 m depth in front of the industrial area, continuously discharging 300–1,500 m^3/s of water daily (model generated at 100 × 100 m square grids centers).

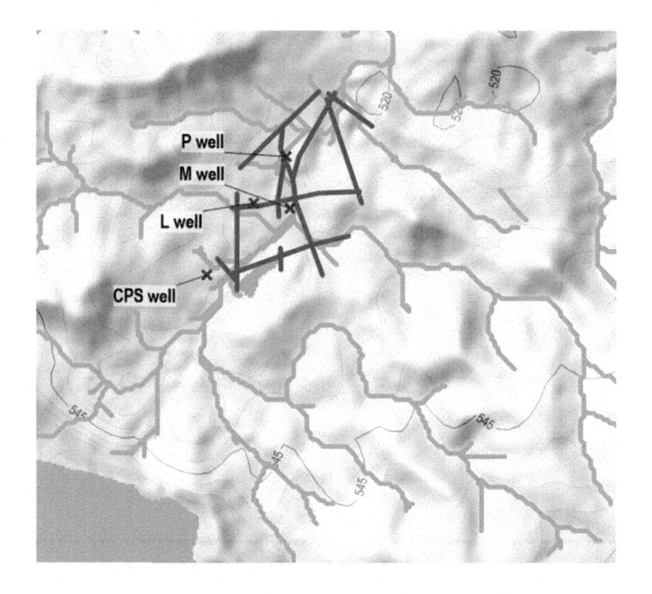

Figure 6.5 This figure shows the location of the electro-resistivity surveys profiles and the suspected wells.

6.2.2.2 The anomalies

After its completion, the geophysical survey disclosed electrical resistivity anomalies compliant with two structural directions, NS/N15E and N30W, and, more discreetly, to EW and N45W. Moreover, two incongruities stood out, one subparallel to the river, NS to N5W, and other to N30W.[1] They were so strong as to obliterate other structures, particularly near the river (see Figure 6.6).

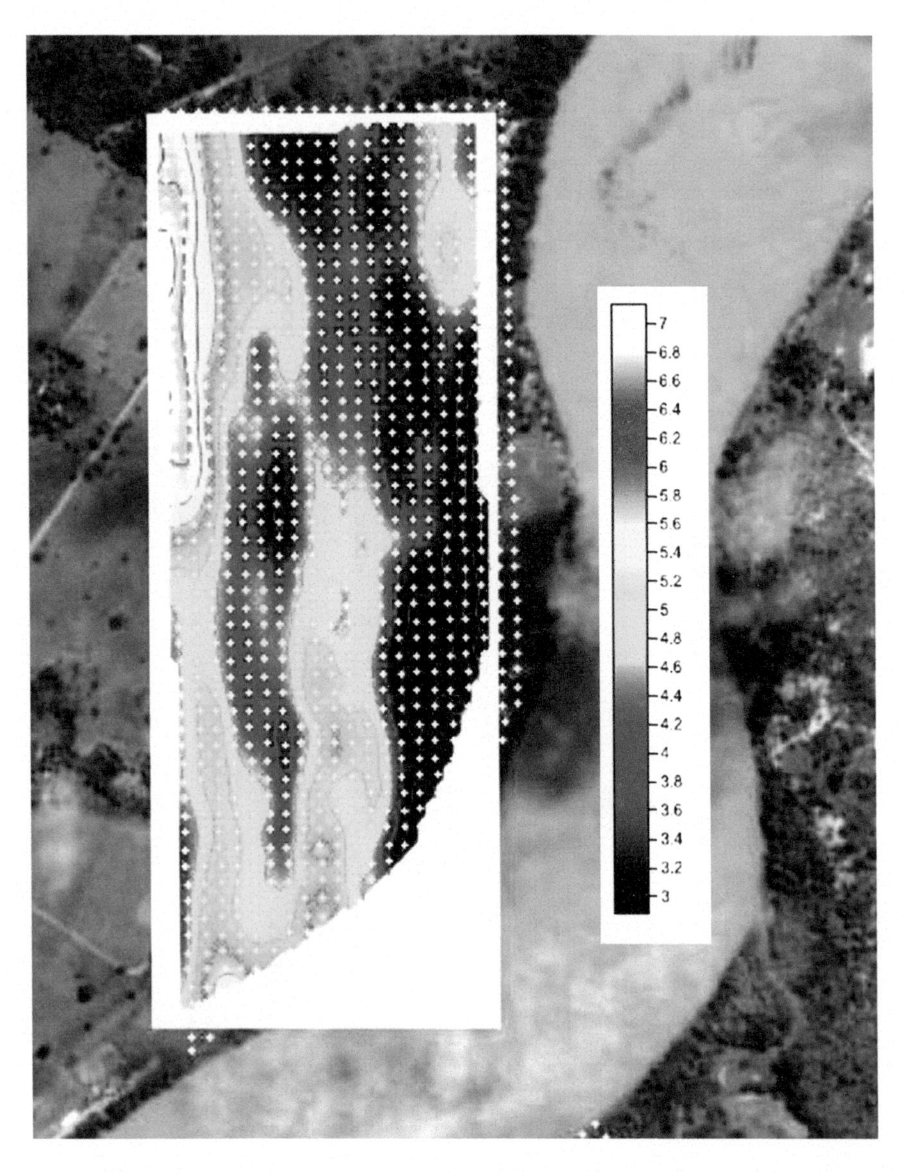

Figure 6.6 This figure summarizes the electro resistivity survey results near the P well. The lower the resistivity value (dark, deep blue, and blue; color scale in ohm.m), the higher the probable concentration of dissolved Zn.

6.2.2.3 *The scanned images pieces of evidence*

Figures 6.7 and 6.8 show the digitally scanned P well profiles located on the river's left bank, stand facing the industrial plant (see Figure 6.5). These profiles show the depth scale P (m), the ascending or descending water velocities v (m/s), the Reynolds numbers Re (–), the variations of the hydraulic gradient J (–), and the computed variations of the hydraulic head H (m). All these data aided modeling the groundwater flow through the fractured rock basement.

The main results of these profiles: heads, and entry gradients, indicated in the table below, aided calibrating the model parameters.

Well	*E (m)*	*N (m)*	*WT elev. (m)*	*Z base (m)*	*H base (m)*	*J escape gradient (–)*
P	474,440.37	7,990,904.91	515.85	463.00	515.85	–0.000038
M	474,501.00	7,990,135.00	516.07	446.00	516.05	–0.000285
L	473,957.00	7,990,216.00	518.19	456.00	518.17	–0.000322
BA	473,241.00	7,989,151.00	517.64	463.00	517.62	–0.000366

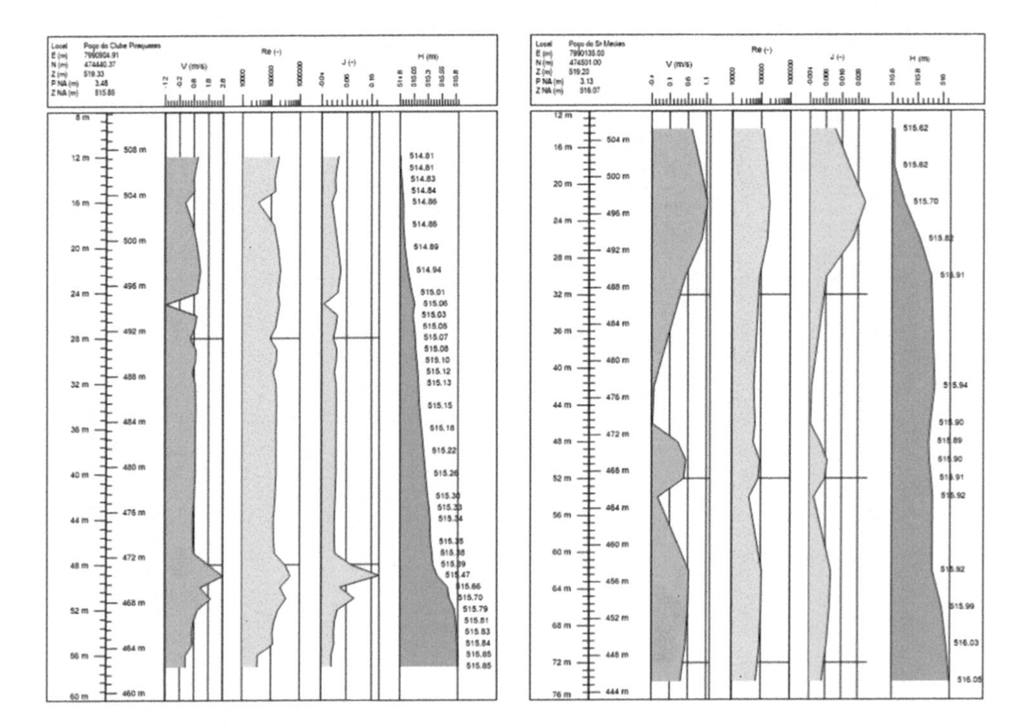

Figure 6.7 Hydraulic analysis results for two wells.

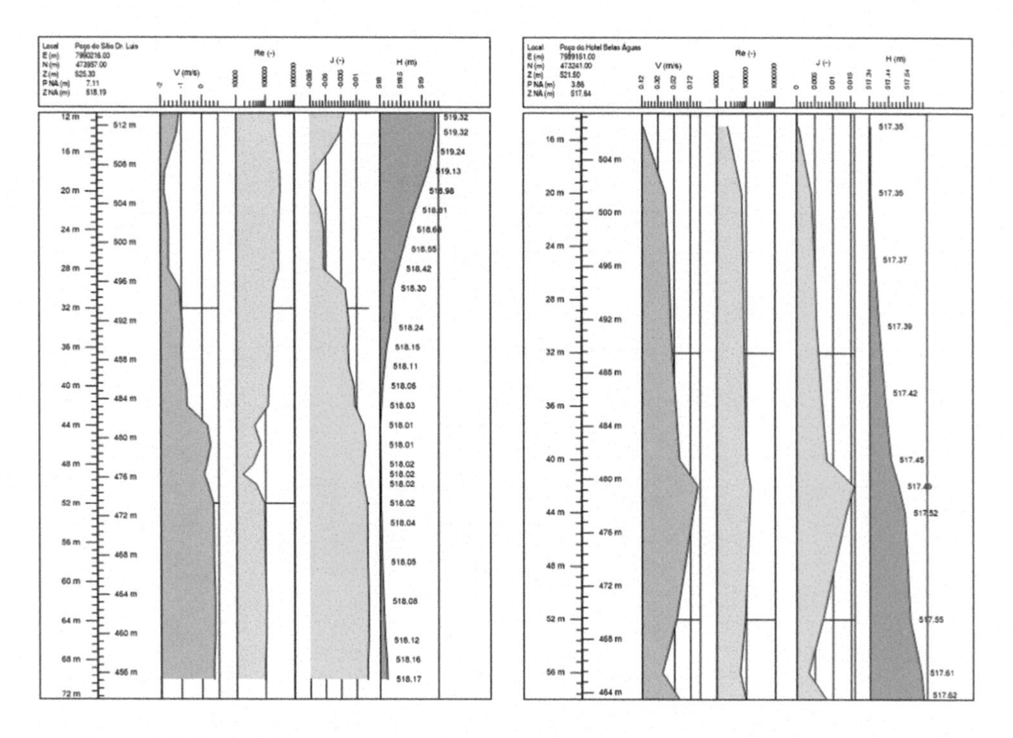

Figure 6.8 Hydraulic analysis results for more two wells.

Pieces of evidence of Zn contamination were observed in all full 360° scan logs, as exemplified in Figures 6.9–6.14.

These investigations revealed that the pollution at the left bank, despite the river barrier, came from advective groundwater transport of Zn at deeper depths through the fractures in rock basement. Probably the pollution sources were remains of dissolved Zn at the right bank.

6.2.2.4 The hydraulic properties of the megafractures net

The modeled 3D volume has three 100 m thick reverse slabs, piled from 300 to 600 m altitudes, covering an area of 16 × 15 km. It integrated the rain infiltration and the groundwater percolation under the plant area and under the dam foundation but assuming that the grouting effect extent under the dam foundation did not reach more than 30–50 m deep. Moreover, hydraulic boundary conditions were defined very far from the plant area to reduce the inevitable inaccuracies in their prescription, mainly for the deep groundwater percolation through the megafractures. Boundary conditions included the dam reservoir level, the relief morphology, the complete fluvial net, and known groundwater head levels in deep wells.

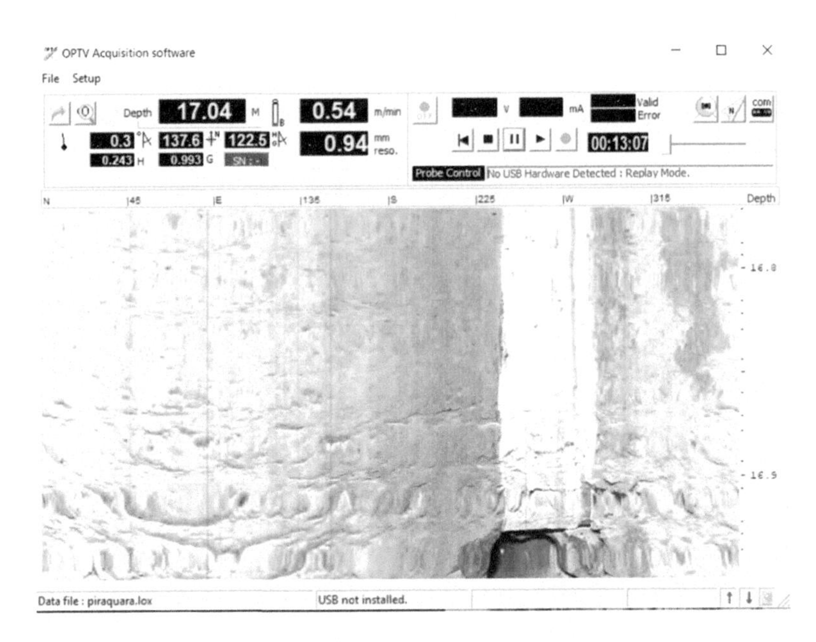

Figure 6.9 Full 360° lateral color image a long-term deactivated P well (depth and geographic orientation indicated). The white stains denote a slow *upward* movement of the dissolved $ZnSO_4$.

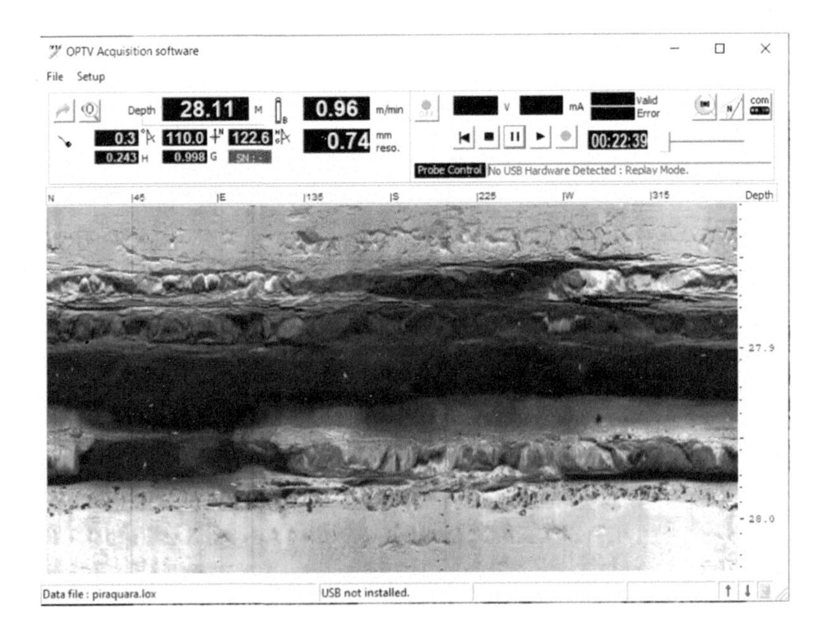

Figure 6.10 Full 360° lateral color image of a deactivated P well (depth and geographic orientation indicated). Horizontal open fracture partly eroded during the well development.

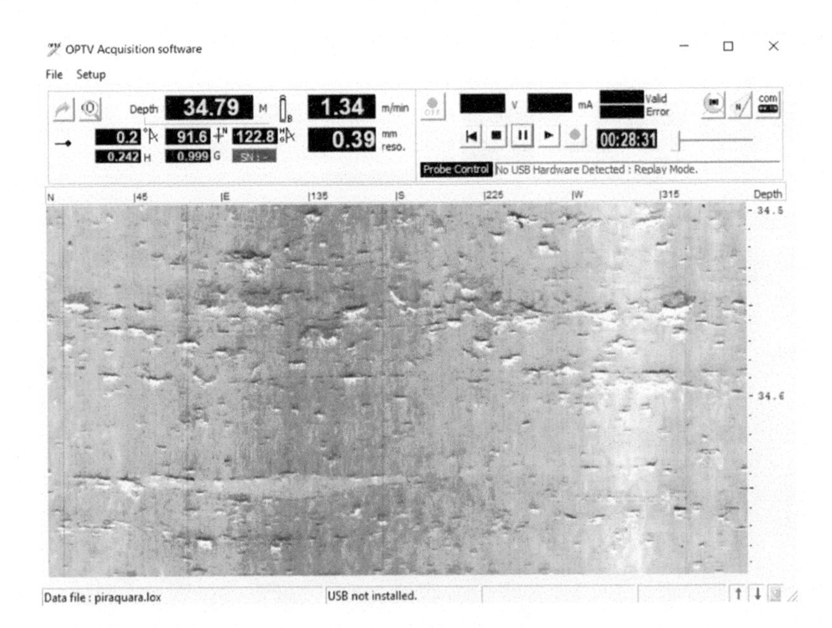

Figure 6.11 Full 360° lateral color image of long-term deactivated P well (depth and geographic orientation indicated). The white stains in this image denote, as opposed to a previous one, a slow *downward* movement of dissolved $ZnSO_4$. Other wells show similar direction reversal, translating the groundwater percolation's intricate pattern through the crisscrossed fracture 3D net.

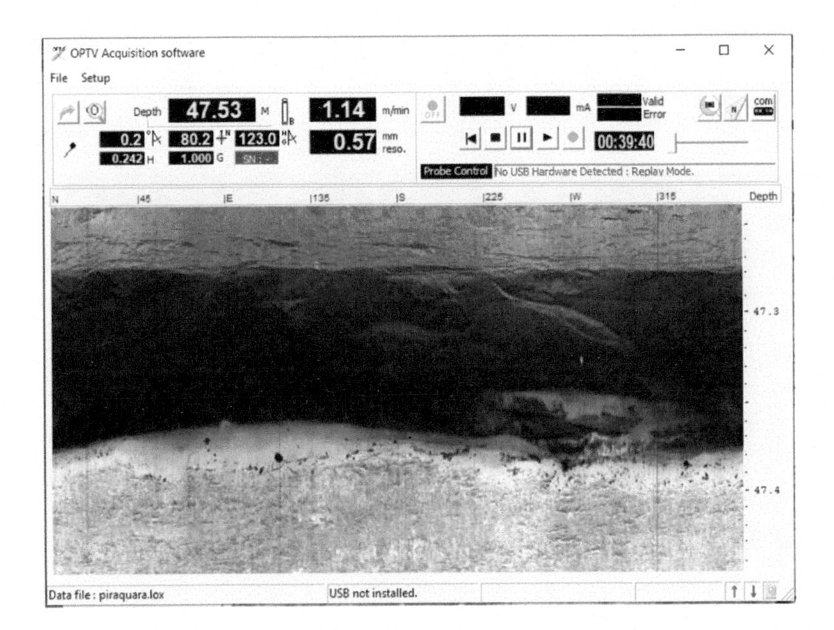

Figure 6.12 Full 360° lateral color image of deactivated P well (depth and geographic orientation indicated). Horizontal open fracture partly eroded during the well development at the contact between weathered and sound formations.

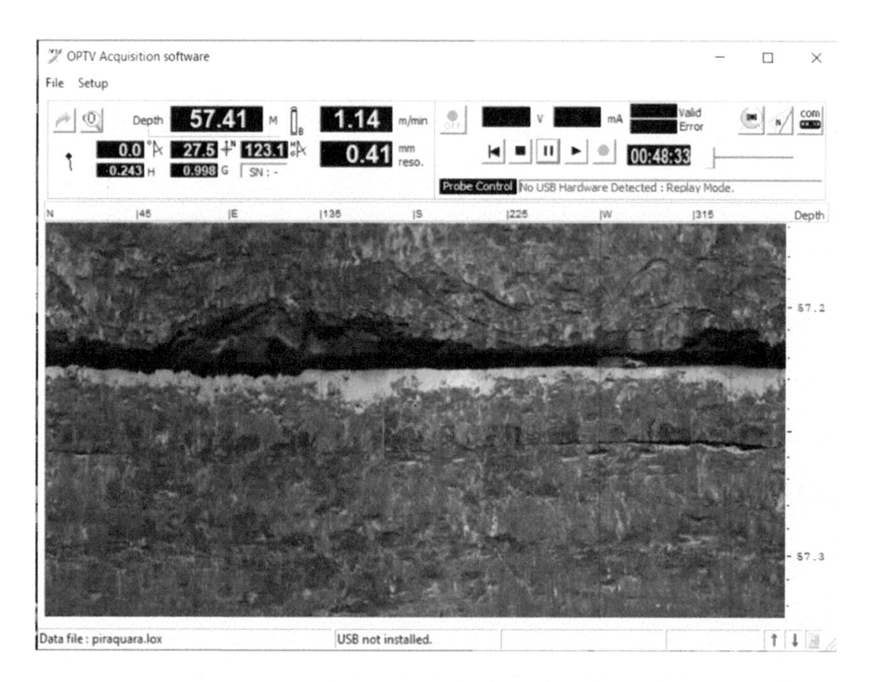

Figure 6.13 Full 360° lateral color image of deactivated P well (depth and geographic orientation indicated). Horizontal open fracture in the lower formation.

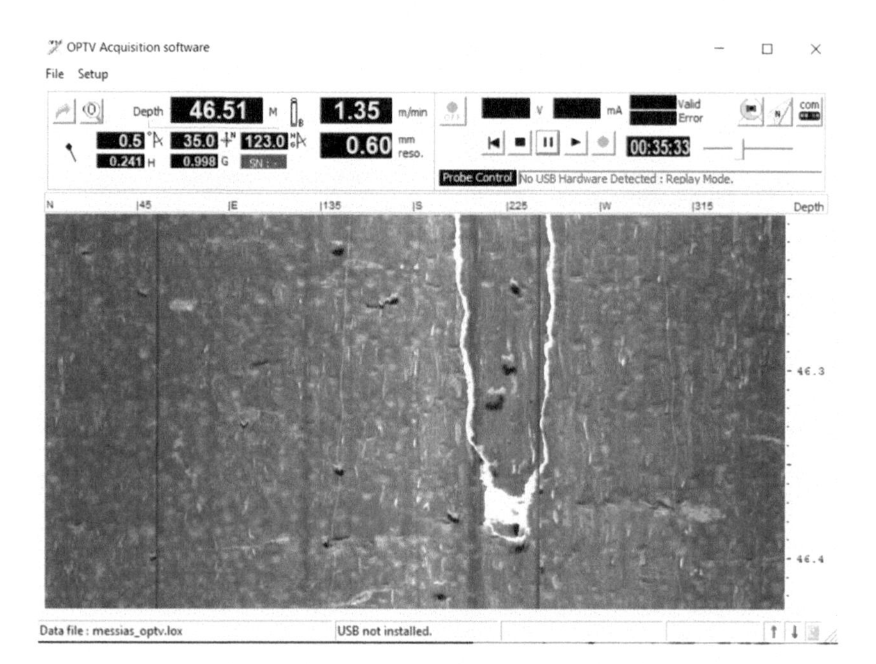

Figure 6.14 Full 360° lateral color image of long-term deactivated M well (depth and geographic orientation indicated). Subvertical fracture with $ZnSO_4$ stains, followed from 40.99 m to 46.4 m extents.

Regional rocks[2] are low-grade metasiltstones immersed in a sericite-chlorite dark gray matrix (or yellow matrix, if weathered). Intermingled arkoses and metasandstones also occur. Bare rock outcrops are visible at road cuts of a near highway and secondary motorways. They show incipient metamorphism and its sedimentary structure gentle dives to E-NE. Sometimes, ripple marks and cross-stratification are present.

Figure 6.15 shows a radial graph combining "fracture direction × fracture apparent length." That information allowed a good approximation for the average regional anisotropic hydraulic conductivities of the fractured rock basement.

In this preliminary model, the almost flat metasediments' average horizontal permeability was taken as the geometric average of corresponding known values at three near similar geological areas, reaching 1.039×10^{-5} m/s. The permeability tensor of the smaller discontinuities crossing the metasediments was taken as 4.976×10^{-7} assuming tetragonal symmetry, average spacing of 3 m, and average hydraulic opening of 500 μ (0.0005 m). The resulting equivalent hydraulic conductivity for the metasediments and associated smaller discontinuities were estimated as follows (see Chapter 2):

$$K_{\text{metasediment\_matrix}} = \begin{pmatrix} 1.039 \times 10^{-5} & 0 & 0 \\ 0 & 1.039 \times 10^{-5} & 0 \\ 0 & 0 & 4.976 \times 10^{-7} \end{pmatrix} \text{m/s}$$

However, for the subvertical megafractures, mostly accountable for the groundwater advective transport of the dissolved metals at great depths and at a considerable distance from the industrial site, their "equivalent" anisotropic transmissivity was considered proportional to the average continuous apparent length revealed by regional photo analysis (see Chapter 2). The second-order tensor shown below, corresponding to the indexed point (5, 31), exemplifies one of the 3 × 2,787 second-order tensors estimated for the whole model.

$$k_{\text{megafractures}_{5,31}} = \begin{pmatrix} 1.373 \times 10^{-5} & 0 & 0 \\ 0 & 3.363 \times 10^{-4} & 0 \\ 0 & 0 & 6.24 \times 10^{-4} \end{pmatrix} \text{m/s}$$

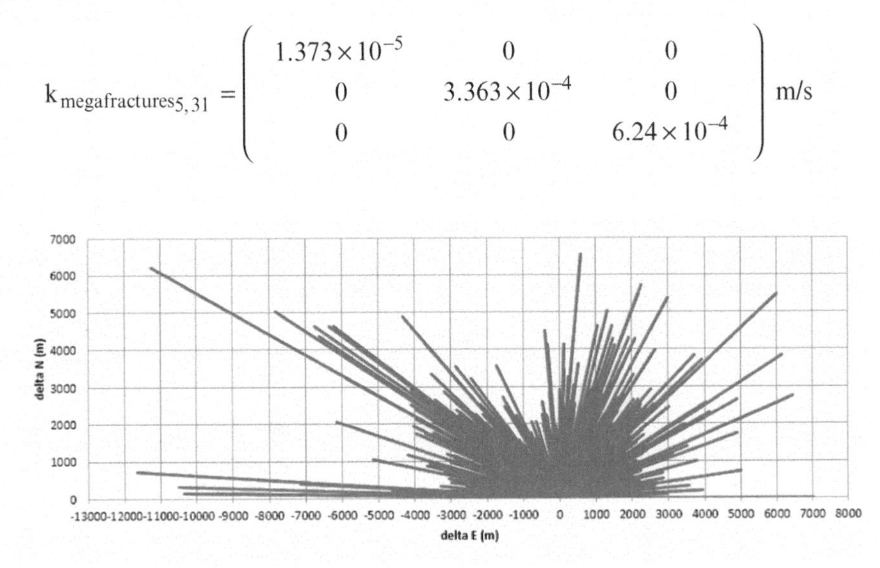

Figure 6.15 Radial graph "fracture direction × fracture apparent length."

Figure 6.16 shows megafractures lineaments in the modeled area but with lineation lengths beyond their limits.[3]

6.2.2.5 What happens during heavy rains

The mathematical simulation was made for a typical weather condition that occurs during the rainy season. Its results show the unequal hydraulic behavior of the fissured and laminated metasediments compared with the highly permeable megafractures at contacts between different permeable strata. Figure 6.17 shows typical rainy infiltration mode during heavy and prolonged rains.

These megafractures, even if slowly diminishing their transmissivities with deepness, may remain permeable far more than 1,000 m depth in the Brazilian Central Region. The deactivated P well (see above), marked by upward slow water seepage mostly

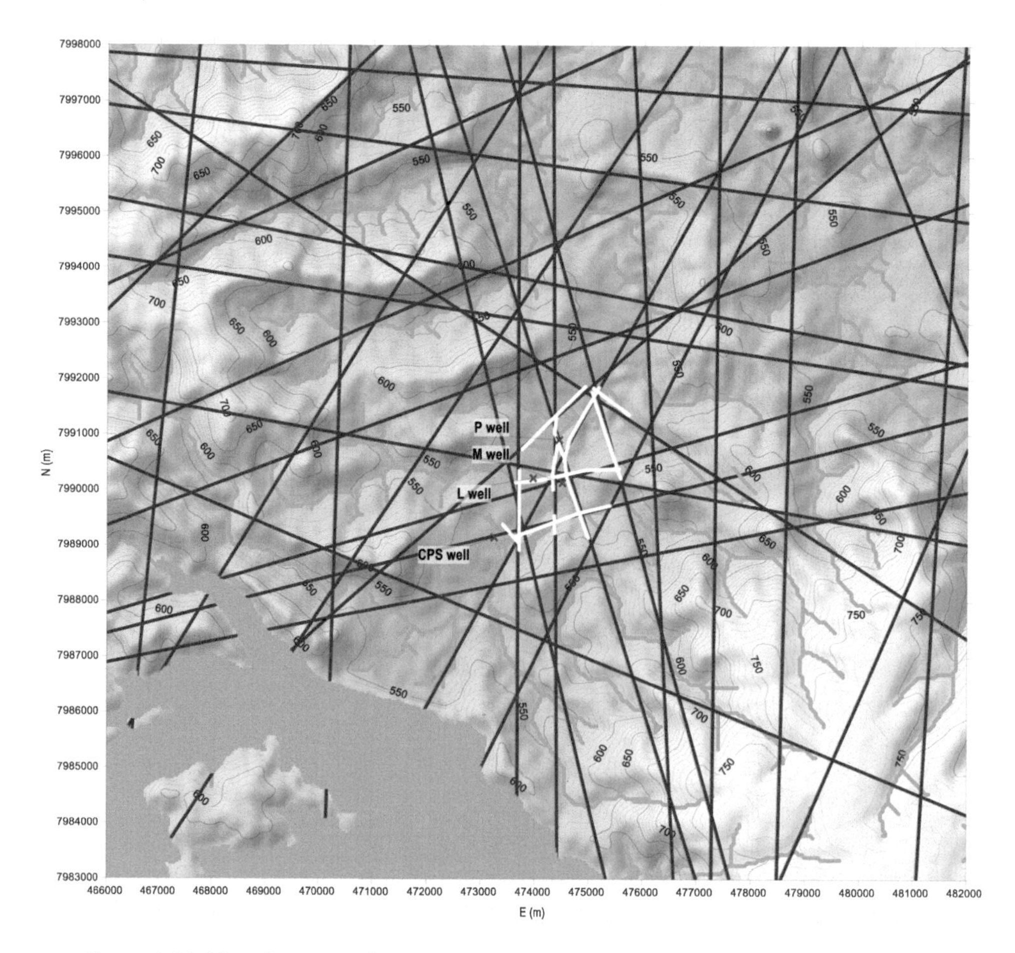

Figure 6.16 Megafractures lineaments over the modeled area based on photo analysis. The white lineaments correspond to the base-lines of the electro resistive survey.

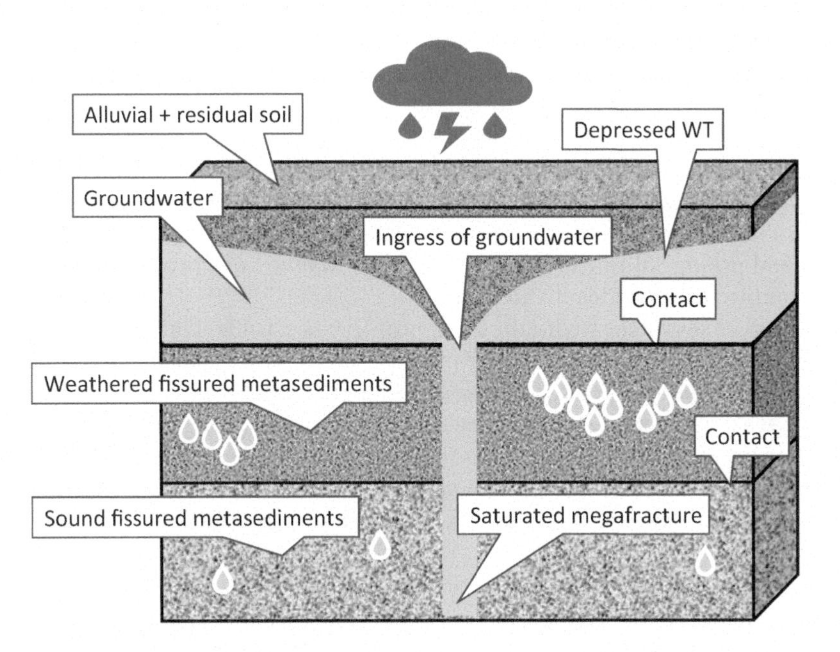

Figure 6.17 Schematic hydrogeologic behavior during heavy and prolonged rains occurring after the dry season. In this case, the almost flat contacts behave as horizontal drains, and the megafractures perform as high-capacity downward drains of the groundwater of top alluvial and residual soils.

coming from megafractures and pervious flat contacts, is an example of that singular behavior. In this case, the groundwater seeps through the interstitial porosity of all types of incoherent pervious top formations but percolates very deep through the very pervious megafractures traversing the coherent rocky basement. Therefore, as in the present example, deep percolation can bypass a wide river. In the concerned industrial plant, that process caused spreading over the left bank of the old-concentrated Zn pollution occurring in the right margin.

As a second conclusion, it was made clear that any new well located within or in the vicinity of the already polluted areas could spread more Zn pollution around that area. Therefore, at the start of the problem, the plant manager has to intuitively take the best decision by deactivating the existing wells, not allowing new ones, and assuring the delivery of good quality water to the plant neighbors in agreement with the environmental agencies.

6.2.2.6 The 3D model results

The 3D model had three superposed reverse thin slabs, as exemplified in Section 5.4.2.2, Chapter 5. Each slab was subdivided by $100 \times 100 \times 100$ m subsystems, up to 24,311 cells. These slabs' shape was like a thin reverse sheet of square cardboard, having a thickness/size ratio equal to 1/400. The model boundary conditions were calibrated by Dirichlet (known hydraulic heads at some points) and by Neumann conditions at its top (rain infiltration), at its bottom (downward groundwater flow), and its four sides (groundwater flow transfer). Figure 6.18 shows the graphical distribution of the Neumann condition at the model top.

Figure 6.19 shows a perspective of the 3D model output at its mid-slab after heavy and prolonged rains. In this condition, the WT inside the megafractures is higher than in the metasediments.

Figure 6.20 shows the hydraulic head vertical profile near the P well. The reversal of these heads near the top of the profile signals the start of the artesian behavior.

Figure 6.21 shows a typical top view of the 3D model output at its mid-slab after heavy and prolonged rains. This figure also shows the head profile along an indicated megafracture near the deactivated wells.

Figure 6.22 shows the hydraulic head profile, but estimated at an elevation of 450 m, along the indicated megafracture near the deactivated wells in Figure 6.18.

Figure 6.23 shows a typical top view of the 3D model results at its mid-slab during the dry season. This figure also shows the head profile along an indicated megafracture near the deactivated wells.

Figure 6.24 shows the hydraulic head profile along the indicated megafracture, but estimated at an elevation of 450 m, near the deactivated wells in Figure 6.23.

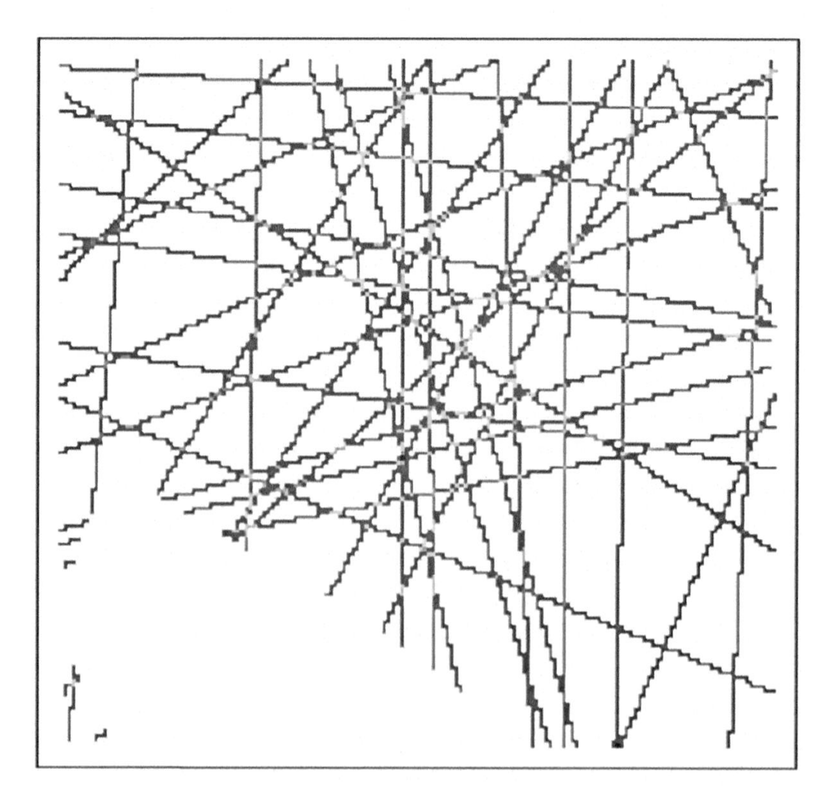

Figure 6.18 Graphical distribution of the Neumann condition values at the model top: rain infiltration is higher at the megafractures crisscrossing and smaller for the remaining area. Unequal numerical values of the rain infiltration are not indicated to avoid overcrowded drawing.

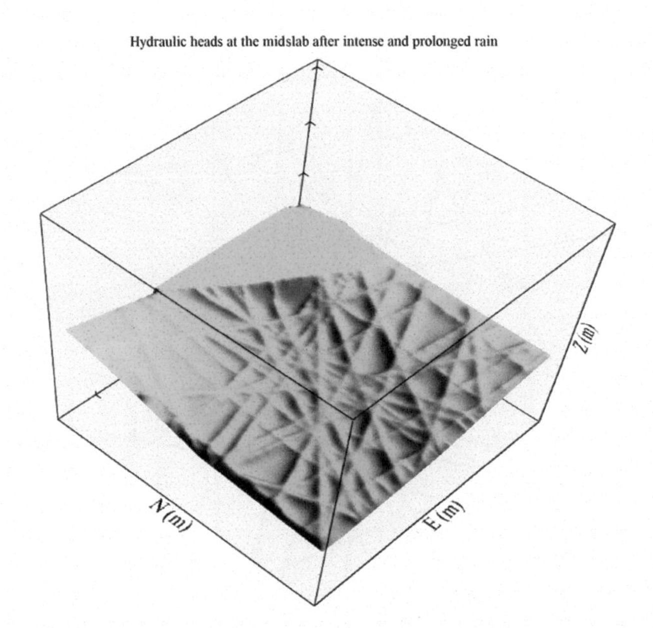

Figure 6.19 Perspective of the 3D model results at its mid-slab after heavy and prolonged rains. In this condition, the WT inside the megafractures is higher than in the metasediments. It is easy to understand why the nearby deactivated wells may accuse a weak artesian behavior. The water curtains percolating through the megafractures feed the groundwater percolating the rock matrix (concave surface). Note that the Z vertical scale is 400 times the E and N horizontal scales. In undistorted scales, the "beehive" pattern is unobserved.

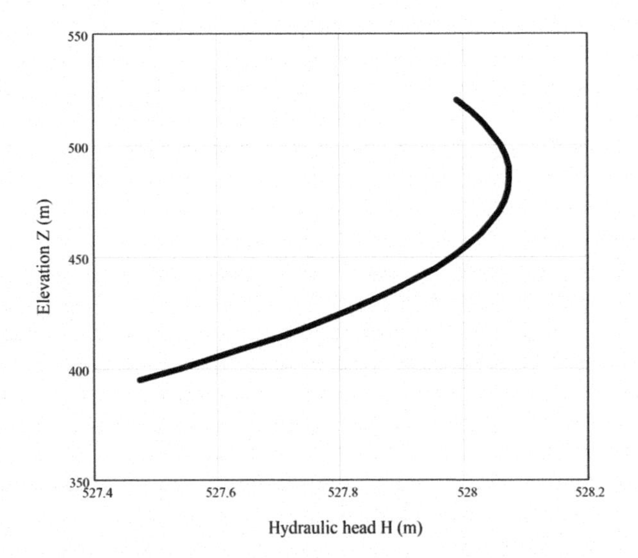

Figure 6.20 Hydraulic head vertical profile close to the P well. The reversal of these heads near the top of the profile signals the start of the artesian behavior.

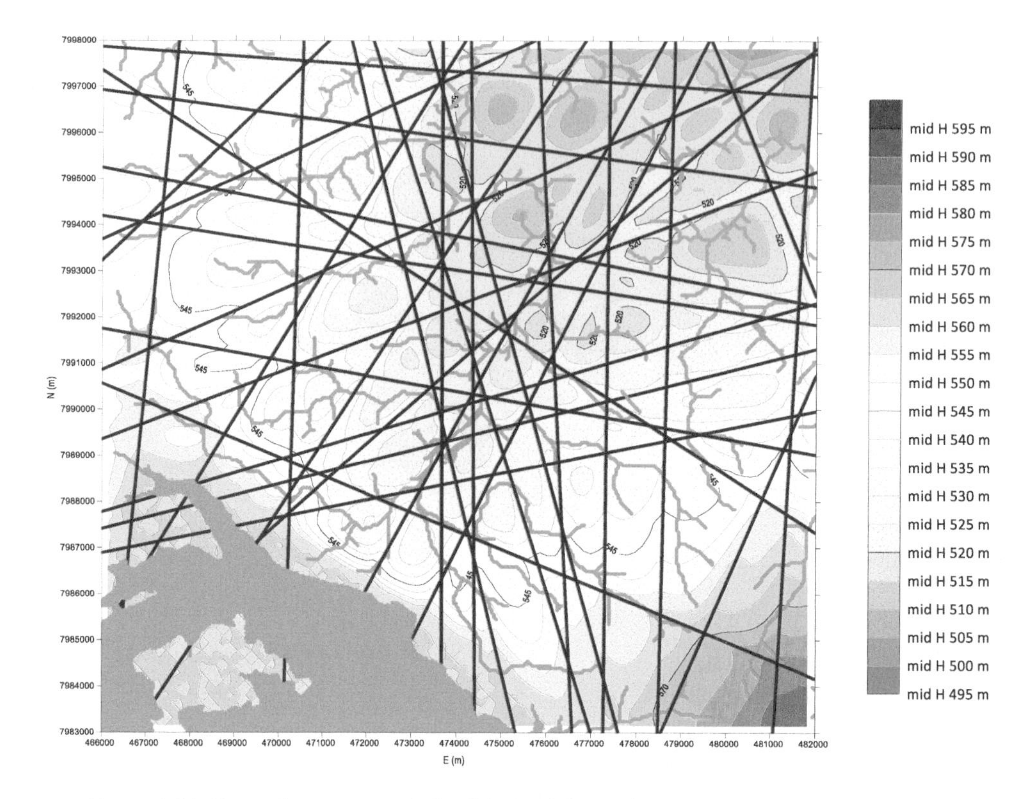

Figure 6.21 Typical top view of the 3D model results at its mid-slab after heavy and prolonged rains. In this condition, the WT within the megafractures is higher than in the metasediments.

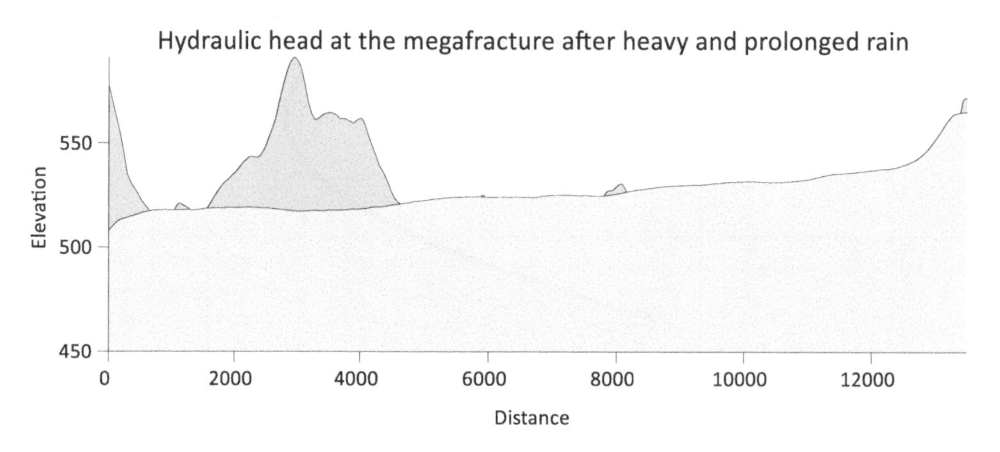

Figure 6.22 Hydraulic head profile, but estimated at an elevation of 450 m, along an indicated megafracture near the deactivated wells (see Figure 6.21). In this condition, the WT inside the megafractures is higher than in the metasediments.

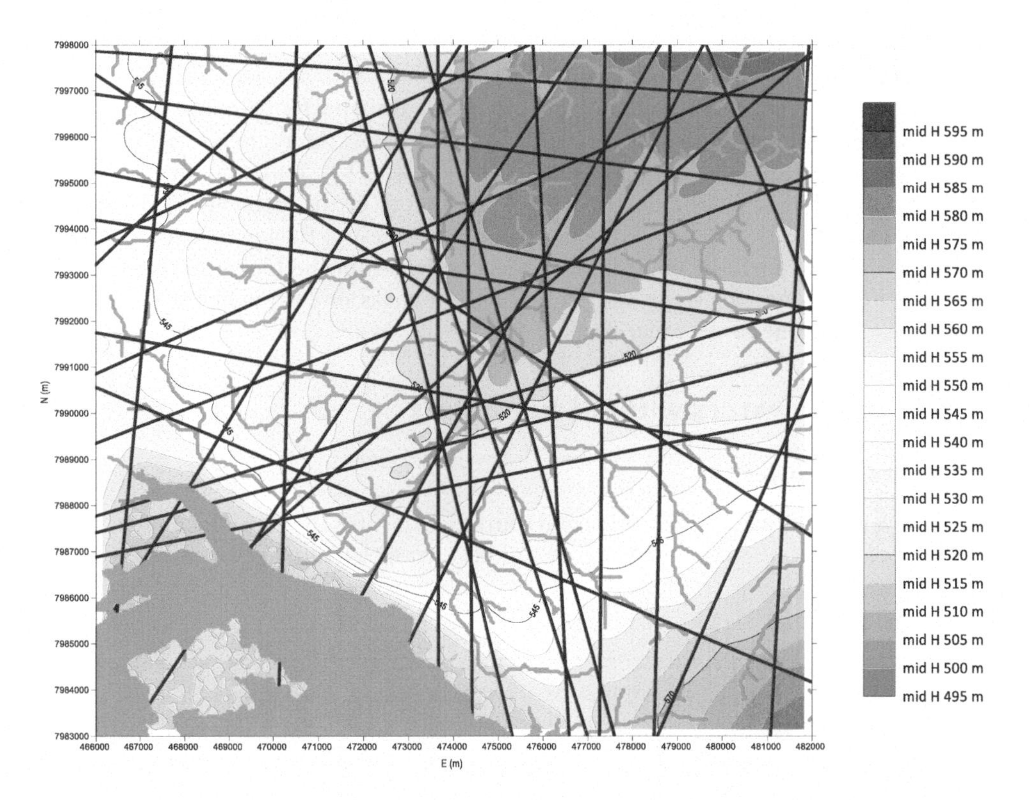

Figure 6.23 Typical top view of the 3D model results at its mid-slab during the dry season. In this condition, the WT inside the megafractures is less high than in the metasediments.

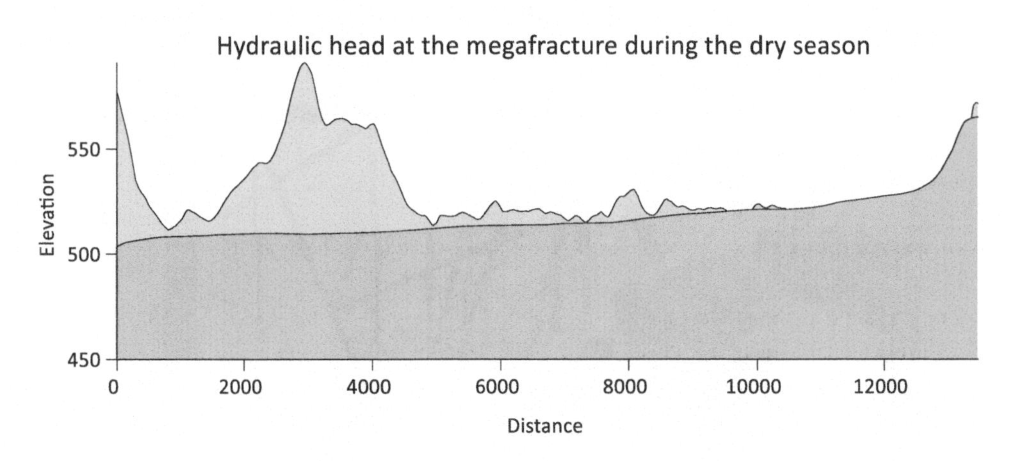

Figure 6.24 Hydraulic head profile, but estimated at an elevation of 450 m, along the indicated megafracture near the deactivated wells in Figure 6.20. In this condition, the WT inside the megafractures is not so higher than in the metasediments.

As the vertical hydraulic conductivities and the hydraulic heads are simulated, the net ascending groundwater discharges at each cell model can be easily computed and map its distribution, as shown in Figure 6.25.

As a general conclusion, the risk of advective transport of dissolved metals at great depths depends on the local relief's characteristics, the regional pervious megafractures, and geological contacts' net characteristics. These interconnecting singular discontinuities favor the deep advective transport of the elements in solution in groundwater. This transport takes place from the highest to the lowest equipotentials, whose values are partially conditioned by the relief, from the higher altitudes to the lower altitudes, and can, as in the present case, bypass under the channel base of a wide river, from one to the other bank.

6.3 Thermal waters without hot points

6.3.1 Preliminaries

Many low-grade Brazilian hydrothermal sources, from 30°C to 50°C, occur at points at the base of the tabular relief of Serra de Caldas (mean elevation: 960 m; area: 100 km^2;

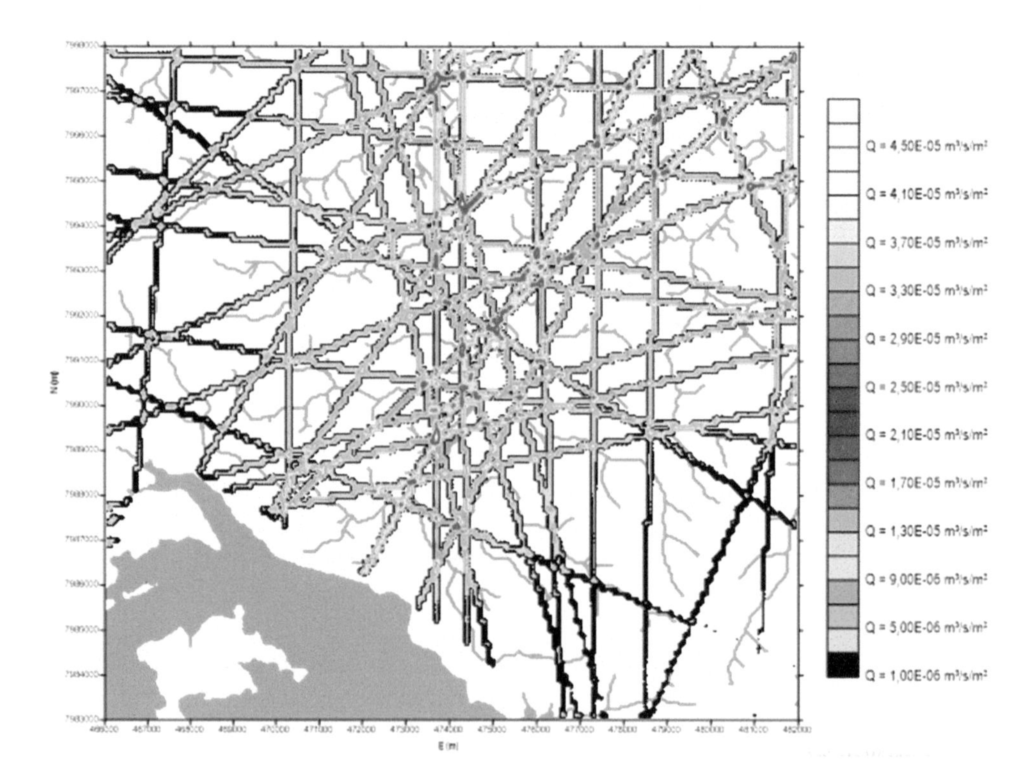

Figure 6.25 Map of the net ascending groundwater discharge after the rainy season at each mid-slab's grid cell.

thick weathered top: 30 m). Its thermalism includes the famous Rio Quente, the Caldas Novas municipality, and the Pirapitinga Pond, as shown in Figures 6.26–6.28.
The Rio Quente tourist complex exploits several thermal springs located at the NW of the Serra de Caldas; whose coalescence (partly human-made) originates the Rio Quente hot stream having average annual flow of around 1,400 L/s. All thermal sources are almost aligned with presumed structural lines.

Pumping wells, drilled before 1990, have their filters down to 400 m elevation. New wells (more than 900 m deep) drilled since 1990 had already captured thermal waters down to −500 m elevation.

Among other natural anomalies so far disappeared, it is worth mentioning the natural thermal springs that in the old days arose at the banks of the Ribeirão de Caldas and the famous Pirapitinga Hot Pond, nowadays amounting only to 20 L/s. The old Poço da Porca and Ribeirão do Bagre hot springs have disappeared or were inadvertently filled with compacted soils.

The construction of the Corumbá hydroelectric plant was preceded by many types of investigations and inquiries to assess that the impact of its future cold-water

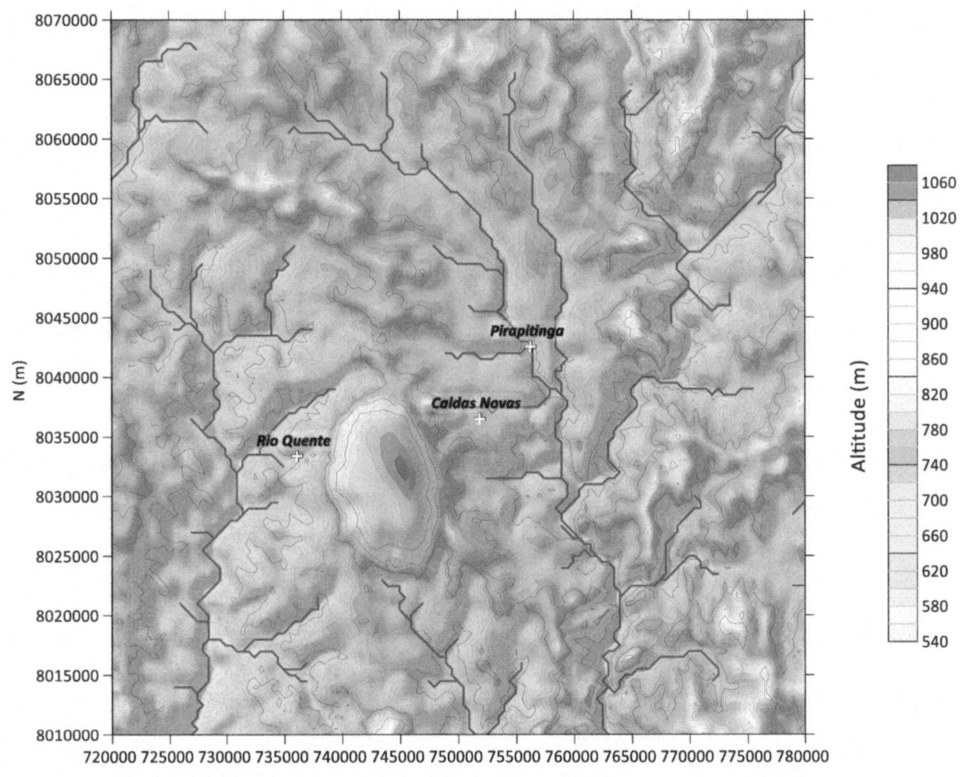

Figure 6.26 Plan view including the Serra de Caldas, the municipality of Caldas Novas, and the Pirapitinga Pond.

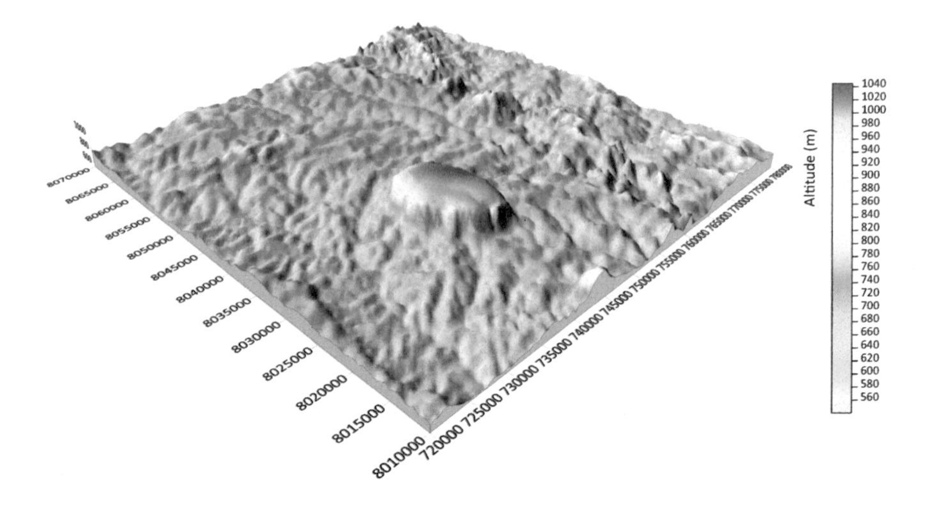

Figure 6.27 Perspective of the region of Caldas Novas. The tabular relief of the Serra de Caldas is a remnant of the basal surface of an old erosion cycle. Its weathered top horizon is nearly 30 m thick. Near to the E-edge of this perspective, the Serra da Matinha mountain belt, averaging 850 m high, runs NNW.

Figure 6.28 Aerial view of the Serra de Caldas. Note the location of five monitoring piezometers. At its extreme right, this image shows part of the reservoir of the hydroelectric plant of Corumbá.

reservoir would not compromise the hydraulic behavior of the thermal springs. All routine checks carried out systematically, from 1994 to 2000, indicated that the reservoir would not compromise the hydro-thermalism phenomena.

6.3.2 *Abridged geologic data*

Figure 6.29 shows a schematic and simplified map of the main geological features of the Serra de Caldas region.[4] The rock basement, typified by gneisses, and associated metavolcanic and metasedimentary rocks are well structured.

Figure 6.30 shows the radial histogram of total apparent-length × strike-direction inferred by air-photo-analysis. Fracture lengths dominate in the NW and NNW directions, as well as NE and SEN. The NW and NNW directions correspond to the most pronounced fractures, probably reactivated during Neotectonics (Higher-Quaternary-Tertiary). Fracture crossings may undoubtedly serve as subvertical escapes for heated waters in deep depth.[5]

Figure 6.31 shows part of the 3D model for deep groundwater percolation and the geothermal advective heat transport. Proper hydraulic and thermal parameters characterize each lithological units.

6.3.3 *Model premises*

6.3.3.1 Generalities

The general hydrothermal model of the Caldas Novas region simulates the natural phenomena prior to the pumping-induced thermal waters exploitation. Its main objective was to assess, as a first approximation, its hydrothermal potential and set its limits.

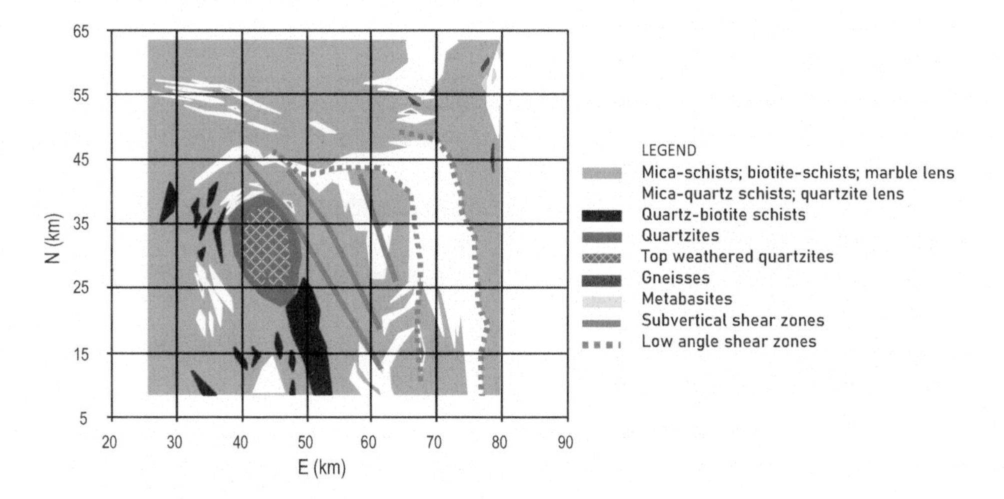

Figure 6.29 Simplified geological map of the Serra de Caldas region's main structural features.

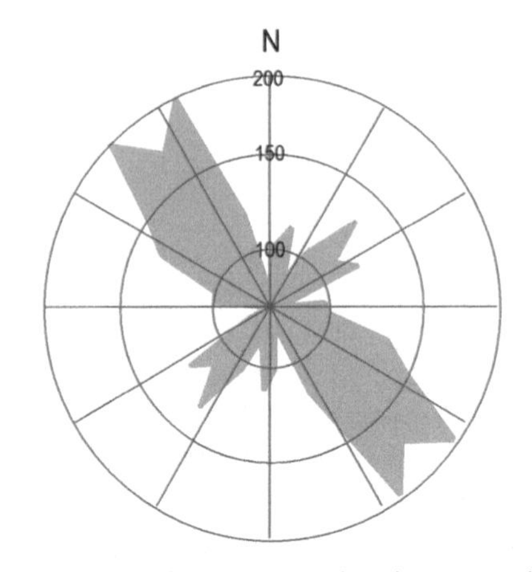

Total apparent-length x strike-direction (km)

Radial histogram (area: 2,800 km^2)

Figure 6.30 Radial histogram of the "accumulated apparent length × strike direction" concerning subvertical fractures at the Caldas Novas region. Its general pattern is well defined. It seems that "extension fractures" due to compressive stresses dominate over typical "shear fractures." These subvertical megafractures allow deep runoff infiltration and subsequent percolation across the fracture net. However, subvertical and low-angle shear zones are less efficient in this aspect.

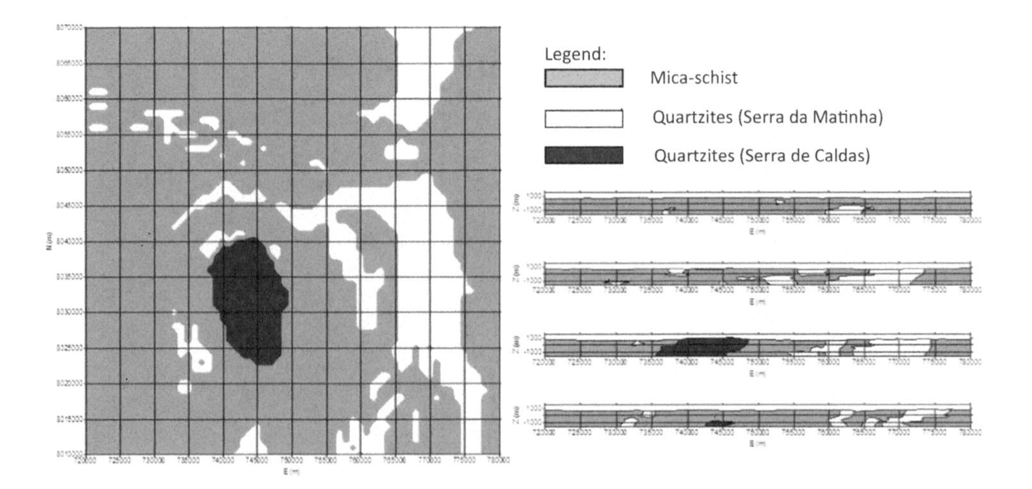

Figure 6.31 Distribution of the mapped and inferred lithological units. Some WE undeformed profiles stand facing the corresponding N coordinate.

This model, made with hydraulic and thermal parameters adjusted to the geological conditions and the monitored data, allowed the following conclusions:

- The Caldas Novas hydrothermal anomalies did not come from any geological "hot points" or other types of geological heat sources.[6] This thermalism, as well at other similar geomorphological provinces, derives from the advective redistribution, by the groundwater percolation, of the excess heat – *relative to the regional average* – continuously accumulated under the base of the Serra Caldas tabular relief.
- The Serra de Caldas, as the central recharge zone of the regional hydrothermalism, favors deep groundwater infiltration, attaining low altitudes ≈–1,500 m, and subsequent spread and upward movement of the heated waters to higher elevations.
- During the rainy season, a substantial part of the runoff infiltrates and remains perched within the weathered top of the Serra de Caldas quartzites, immediately below its colluvial cover. This perched aquifer, stored in a total volume around 30 m × 100 km^2, has an "effective" volume of around 1% of this bulk volume, that is, 3×10^{12} L, but continuously dropping.
- The isotopic and chemical signatures of the Caldas Novas' groundwater suggest other residual reliefs as recharging areas (600—750 m altitude), as exemplified by the Serra da Matinha. Their weathered rock top operates as porous and transient groundwater reservoirs that can stock and gradually release a substantial water amount.
- The average mild temperature of the anomalies, particularly at Rio Quente, Caldas Novas, and Pirapitinga, exceeds the regional average by more than 10°C, forming a relatively modest hydrothermal system if compared with other hydrothermal systems, famous by much higher temperatures (somewhat less than 100°C at small depths).
- Chemical analysis of hot and cold groundwater samples from natural springs revealed incipient mineralization, implying short percolation pathways. Cl, Br, and I are absent, typical of heated pluvial infiltration.
- Figure 6.32 shows maps of the thermal water temperature at 500, 550, and 600 m elevations, below the Caldas Novas urban perimeter (20 wells in 1985–1986). The temperature profile also included in this figure suggests a process of advective heat transport characterized by a high-localized concentration of heat in-depth, followed by the rise of deep heated and pressurized water, probably through preferential percolation paths at the intersection of subvertical megafractures.

6.3.3.2 *Effective mean regional porosity*

Figure 6.33 shows the 180 days recession curve for the 1997 dry-period at a fluviometric station located in the Rio Corumba. That fluviometric station comprises 28,400 km^2, covers all fluvial net around the Caldas Novas region, and gauged a total volume of 2.09×10^9 m^3 during the 1997 dry period. Then, assuming the 1997' dry-season depletion could be equated to the corresponding previous rainy-period total recharge, it was possible to estimate an average recharge of 4.758×10^{-9} m/s for the Caldas Novas region. For practical applications, that average infiltration rate can be considered a downward Darcy's velocity v_D driven by a gravitational gradient $J_v \approx 1$.

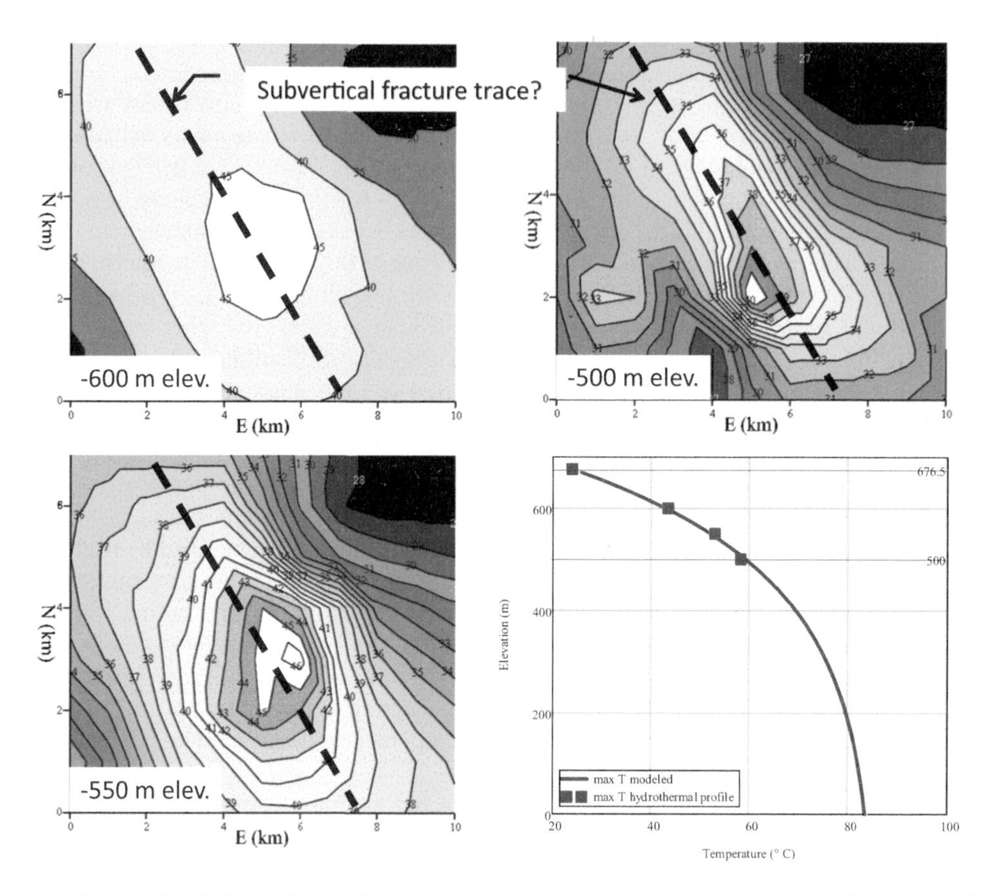

Figure 6.32 Caldas Novas thermal water temperatures at elevation 500, 550, and 600 m. Thermal profile under a very hot dome (see adjoining maps).

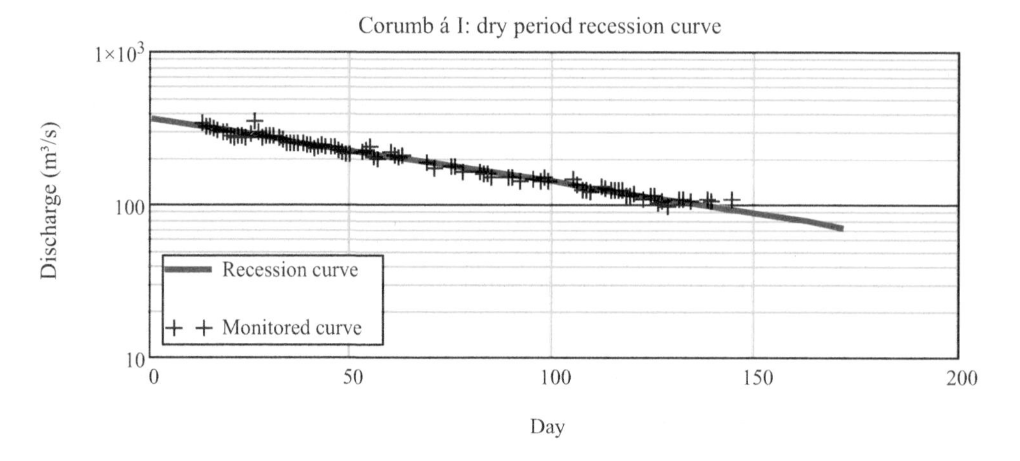

Figure 6.33 Recession curve at Corumbá I fluviometric station in 1997.

The mean "effective" drawdown velocity v_e of the falling WT at a transversal cross section at the eroded top of the Serra de Caldas, dully monitored by five equidistant piezometers during the 1997 dry period, was calculated as $v_e = 5.73 \times 10^{-7}$ m/s. The mean effective regional porosity n_e of the rock matrix was estimated as $v_e/v_D = 0.83\%$. This value was then taken as a weighted average for all regional fractured lithologies' effective porosity, basically shales and quartzites.

6.3.3.3 Permeability

The inference of the characteristics of the hard rock basement's hydraulic transmissivities at the Caldas Novas region was based on their geometric characteristics, as revealed by photo-analysis, and calibrated from all available pumping-test results carried out for wells at depths from 100 to 1,000 m.

Figure 6.34 shows the values of the main transmissivities T at various elevations Z and their trend $T = f(Z)$. Notice that the increase of the transmissivities T values with the decrease of the productive elevations Z is in line with the regional experience at

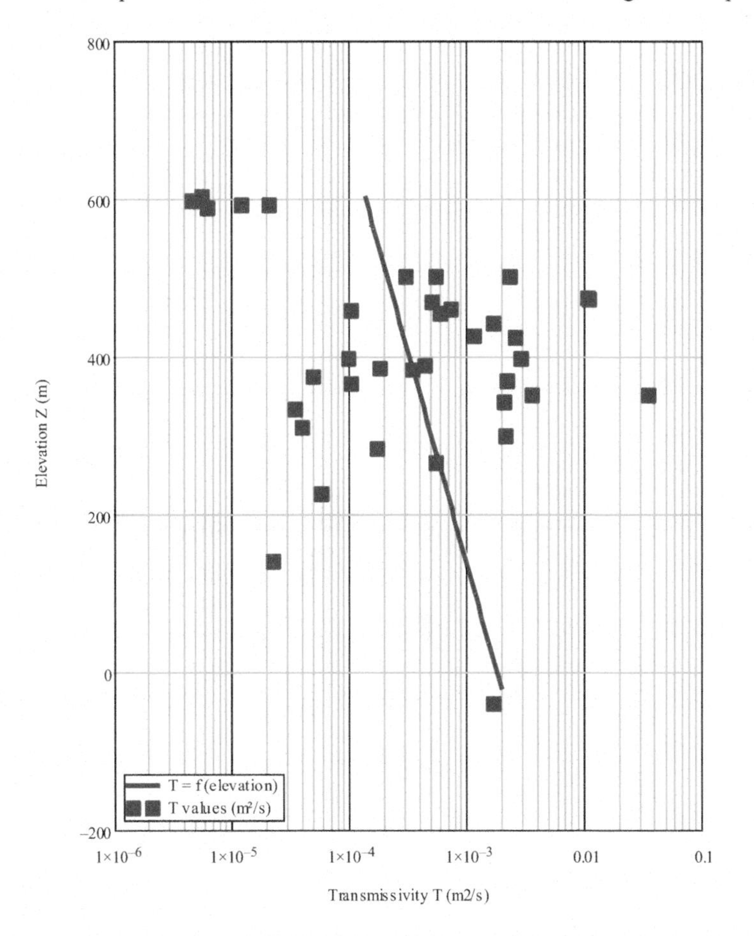

Figure 6.34 "Transmissivity T" × "Productive elevations Z." That observed trend is in line with the regional experience.

Caldas Novas. Reasons linked to the lithological differentiation between the upper levels of shales and the lower levels of quartzites, possibly with enrichment of dolomitic bodies, and the progressive reduction of the rocky basement's weathering with increasing depths, may explain that observed fact. Similar findings may also occur in the Scandinavian and African basements. However, that general rule, down to more than 1,000 m, contradicts local opposite trends down to 100 m, commonly observed in the intertropical climate.

6.3.3.4 The groundwater recharge discussion

The role of the groundwater recharge at the top of Serra de Caldas has always been controverting. The reduced effective porosity of its quartzites, around 1%, excluding its 30 m thick weathered top, were arguments against the contribution of the Serra de Caldas to the regional thermalism. However, their meteoric character, low mineralization, and high-water pressure of the marginal thermal springs surrounding the Serra de Caldas tabular relief disprove these arguments.

On the other hand, from August 1994 to September 1998, statistical tests confirmed a positive time-lagging correlation between the time-series seasonal variation of the WT levels monitored at deep wells at the top Serra de Caldas and the corresponding time-series monitored at the thermal wells at Caldas Novas region, thus sanctioning the active role of the Serra de Caldas tabular relief. Also, the conclusions drawn from the observations by of Professor Hans Zojer in 1986[7] played a decisive role in supporting the importance of the Serra de Caldas in the regional thermalism (see Figure 4.30, Chapter 4, Section 4.6.1).

The spatial distributions of the concentration gradients of the isotopic signatures of D and O18 defined two groups, A and B, of peculiar thermal waters, as shown in Figure 6.35.

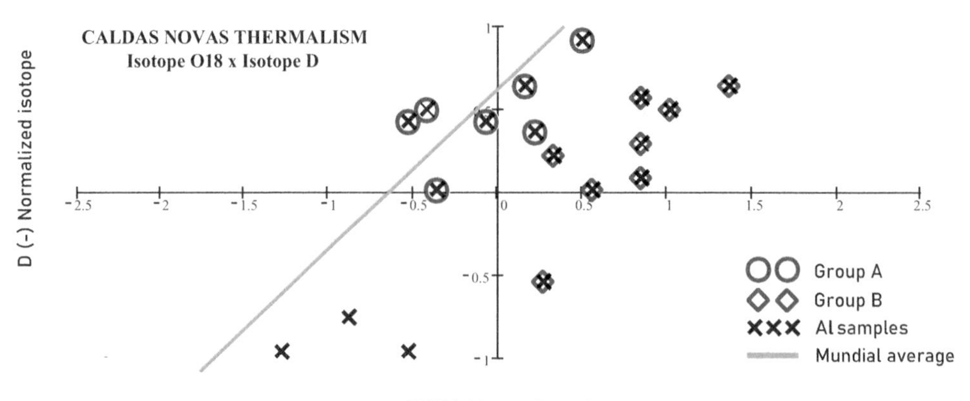

Figure 6.35 Normalized isotopic signatures of two different groups, A and B, of thermal waters, identified by appropriate mathematical technique. The presence of these two groups implies different origins and paths.

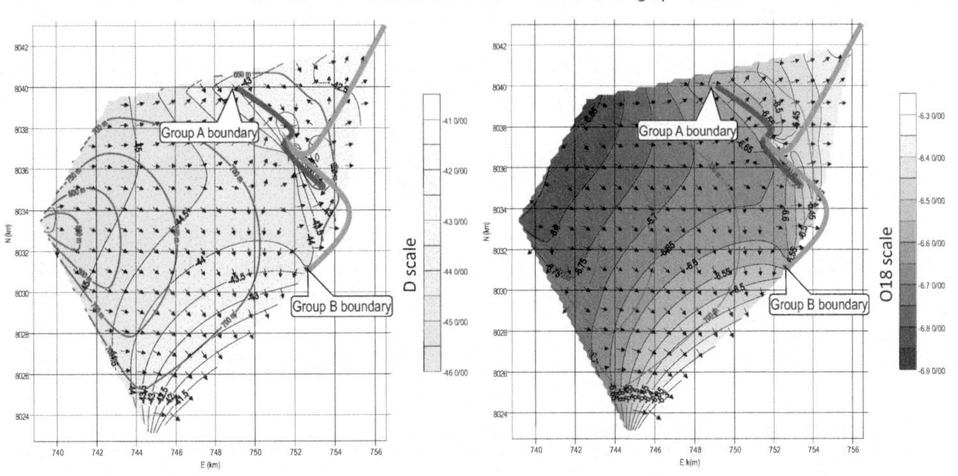

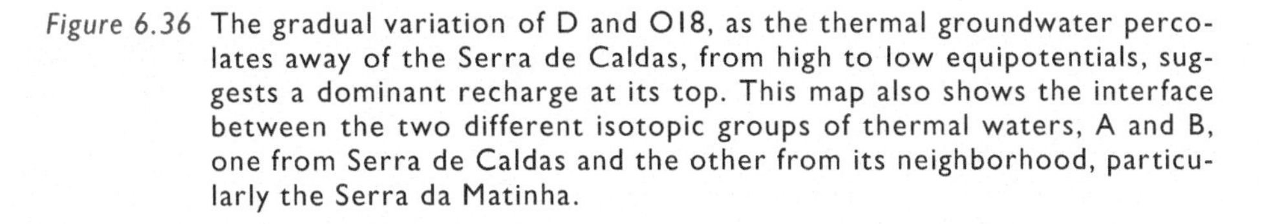

Figure 6.36 The gradual variation of D and O18, as the thermal groundwater percolates away of the Serra de Caldas, from high to low equipotentials, suggests a dominant recharge at its top. This map also shows the interface between the two different isotopic groups of thermal waters, A and B, one from Serra de Caldas and the other from its neighborhood, particularly the Serra da Matinha.

The location of the sampling points defined by D and O28 isotopic signatures imply groundwater circulation, from high altitudes to low altitudes, as shown in Figure 6.36.

The above-summarized results, including detailed geochemical analyses by Petrobras (Section 6.3.3.1, endnote 6), validated the hypothesis that the Serra de Caldas is the dominant regional hydrothermalism recharge zone. Its tabular relief favors deep water infiltration at high altitudes ≈1,000 m, and subsequent heating at low altitudes ≈–1,500 m, followed by the radial spreading and hot groundwater rising to higher elevations.

6.3.3.5 Fundamentals of heat transport

Thermal *advective transport* is the process of heat transport carried by percolating heated water across pervious rocks. The amount q_A of the heat advectively transferred by percolation groundwater is:

$$q_A = mcT\mathbf{q}_H$$

where:

q_A denotes the specific heat flowrate *convectively* transferred.

m_w, c_w, and T, respectively, denote the specific mass, the specific heat, and the temperature of the groundwater.

$\mathbf{q}_H$ denotes the specific flowrate of the groundwater, equal to $-K_H(\partial H/\partial \mathbf{r})$, where K_H is the hydraulic conductivity, $\partial H/\partial \mathbf{r}$ is the hydraulic gradient, and $\mathbf{r}$ is the position vector.

For a typical 3D finite cell δx·δy·δz centered at the point (i, j, k), the finite difference expression for the component Q_{Ax} of the total *advective* heat transfer Q_A through the cross section δx·δy is:

$$Q_{A_{i,j,k}} = m_W \cdot c_W \cdot \left[-k_H \cdot \delta y \cdot \delta x \cdot \left(\frac{H_{i+1,j,k} - 2H_{i,j,k} + H_{i-1,j,k}}{2\delta x} \right) \right] \cdot T_{i,j,k}$$

The components Q_{Ay} and Q_{Az} have similar expressions, and the total heat transfer is $Q_A = Q_{Ax} + Q_{Ay} + Q_{Az}$.

Thermal *conductive transport* is the process of heat transport carried by the electrons in motion or through vibrations of the mineralogical mesh of the pervious rocks itself. The amount q_{KT} of the heat *conductively* transferred obeys Fourier's law, analogous to Darcy's law:

$$\mathbf{Q}_C = -K_C (\partial T / \partial \mathbf{r})$$

where:

$\mathbf{Q}_C$ denotes the specific heat flowrate *conductively* transferred;
K_C denotes the thermal conductivity of the rock;
T denotes the temperature;
$\partial T / \partial \mathbf{r}$ denotes the rock thermal gradient;
$\mathbf{r}$ denotes the position vector.

The basic steady-state differential that equation simultaneously govern both advective and conductive transport is:$\text{div}[K_C \cdot \mathbf{grad}(T)] - m_w c_w \cdot \mathbf{q}_A \cdot \mathbf{grad}(T) = 0$

where, besides the symbols defined above:

grad(T) denotes the gradiente operator applied to the (scalar) temperature T;
div denotes the divergence operator applied to the (vector) $K_C \cdot \mathbf{grad}(T)$.

For a typical 3D finite cell δx·δy·δz centered at the point (i, j, k), the finite difference expression for the component Q_{Cx} of the total *conductive* heat transfer $\mathbf{Q}_C$ through the cross section δx·δy is:

$$Q_{C_{i,j,k}} = -K_C \cdot \delta y \cdot \delta x \cdot \left(\frac{T_{i+1,j,k} - 2T_{i,j,k} + T_{i-1,j,k}}{2\delta x} \right)$$

The components Q_{Cy} and Q_{Cz} have similar expressions, and the total heat transfer is $Q_C = Q_{Cx} + Q_{Cy} + Q_{Cz}$.

The complete finite difference equation simultaneously applied to the conductive and advective heat transport is:

$$\begin{gathered} \left[\left(Q_{C_{i+1}} - Q_{C_{i-1}} \right) + \left(Q_{C_{j+1}} - Q_{C_{j-1}} \right) + \left(Q_{C_{k+1}} - Q_{C_{k-1}} \right) \right] \ldots = 0 \\ + \left[\left(Q_{A_{i+1}} - Q_{A_{i-1}} \right) + \left(Q_{A_{j+1}} - Q_{A_{j-1}} \right) + \left(Q_{A_{k+1}} - Q_{A_{k-1}} \right) \right] \end{gathered}$$

where the subscripts C and A, respectively, stand for *conductive* and *advective*.

The complete finite difference equations for the *conductive* transport (Fourier's law) are analogous to those presented for the *groundwater flow* (Darcy's law). However, the finite difference equations for *advective* transport is a little more laborious. For example, the first term in the continuity equation, $Q_{A(i+1)}$-$Q_{A(i-1)}$, is now:

$$Q_{A_{i+1,j,k}} - Q_{A_{i-1,j,k}} = -\left(m_W \cdot c_W \cdot \delta y \cdot \delta z\right) \cdot$$

$$\left[\left(K_{H_{i+1,j,k}} \cdot \frac{H_{i+1,j,k} - 4H_{i,j,k} + 3H_{i-1,j,k}}{2\delta x} \cdot T_{i+1,j,k} - K_{H_{i-1,j,k}} \cdot \frac{3H_{i-1,j,k} + 4H_{i,j,k} - H_{i+1,j,k}}{2\delta x} \cdot T_{i-1,j,k} \right) \ldots + \left(K_{H_{i,j+1,k}} \cdot \frac{H_{i,j+1,k} - 4H_{i,j,k} + 3H_{i,j-1,k}}{2\delta y} \cdot T_{i+1,j,k} - K_{H_{i,j-1,k}} \cdot \frac{3H_{i,j-1,k} + 4H_{i,j,k} - H_{i,j+1,k}}{2\delta y} \cdot T_{i-1,j,k} \right) \ldots + \left(K_{H_{i,j,k+1}} \cdot \frac{H_{i,j,k+1} - 4H_{i,j,k} + 3H_{i,j,k-1}}{2\delta z} \cdot T_{i+1,j,k} - K_{H_{i,j,k-1}} \cdot \frac{3H_{i,j,k-1} + 4H_{i,j,k} - H_{i,j,k+1}}{2\delta z} \cdot T_{i-1,j,k} \right) \right]$$

Note that the hydraulic gradients at points $i+1$, $i-1$ ... $k+1$ are off-centered in the second-order derivatives' numerical expression. The remaining terms for the j and k directions follow the same way.

6.3.4 Modeling results

6.3.4.1 Dimensions and parameters

The following list summarizes the principal dimensions and parameters for the heat advective transfer 3D models at the Caldas Novas region.

- Five horizontal slabs of 3,600 km^2 total area, bounded by coordinates E720–780 km and N8,010–8,070 km.
- 1.772×10^6 cubic finite difference elements measuring $500 \times 500 \times 500$ m.
- Average thermal *conductivities*:
  - Schists: 3.0 W/m/K
  - Serra da Matinha: 3.5 W/m/K
  - Serra de Caldas: 4.0/W/m/K
- Thermal boundary conditions:
  - Top morphological relief, Dirichlet condition = 23°C.
  - Bottom horizontal boundary, Neumann condition ascending geothermal flux = 0.065 W/m^2.

- Model vertical sides: Neumann condition applied after truncating the neighborhood at the edge of the grid model.
- Water specific heat: cal/gm/K.
- Peclet number: 1,000 kg/m^3.

6.3.4.2 The heat transfer processes

The conductive transfer of the geothermal flow does not yet simulate the Caldas Novas thermalism's ultimate reality because the first group of results only contemplates the ascension of the terrestrial heat flow by *conductive* transfer through the crustal rocks, from the altitude −1,500 to 500 m. This first model's high interest lies in the localization of the heat concentrations, emphasizing the importance of Serra de Caldas in the whole process.

Figures 6.37–6.39 resume this conductive heat ascension process.

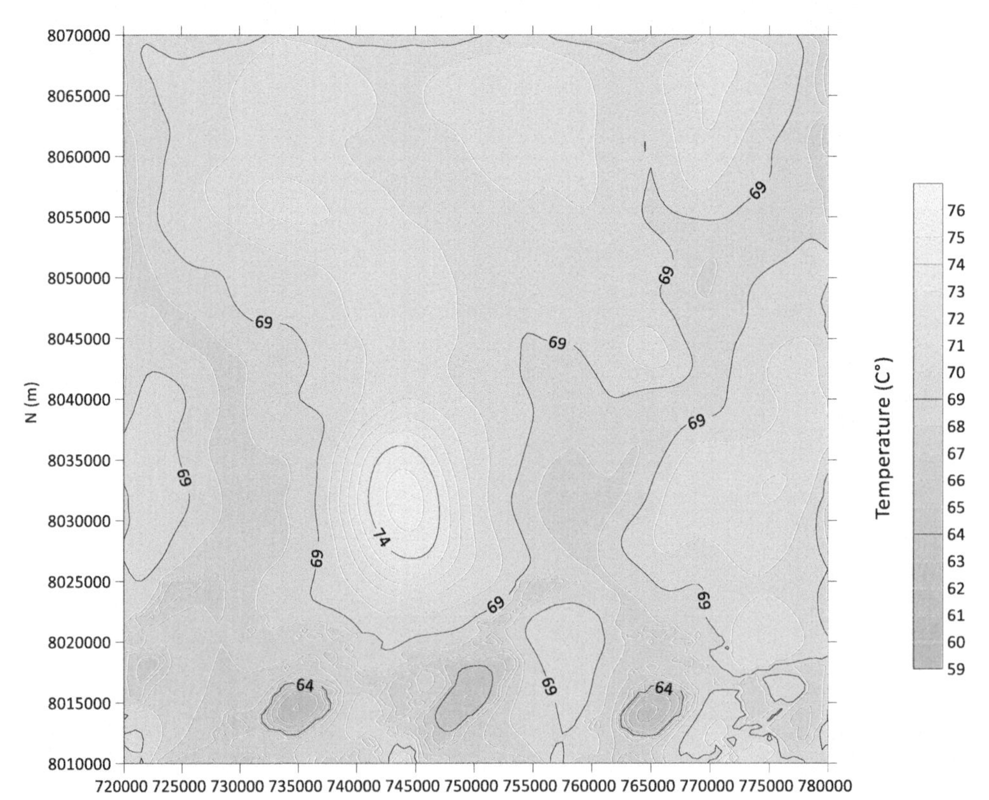

Figure 6.37 Map of the temperature distribution only due to the conductive geothermal flow at altitude −1,500 m at the Caldas Novas region. Thermal water temperatures vary from 59°C to 76°C.

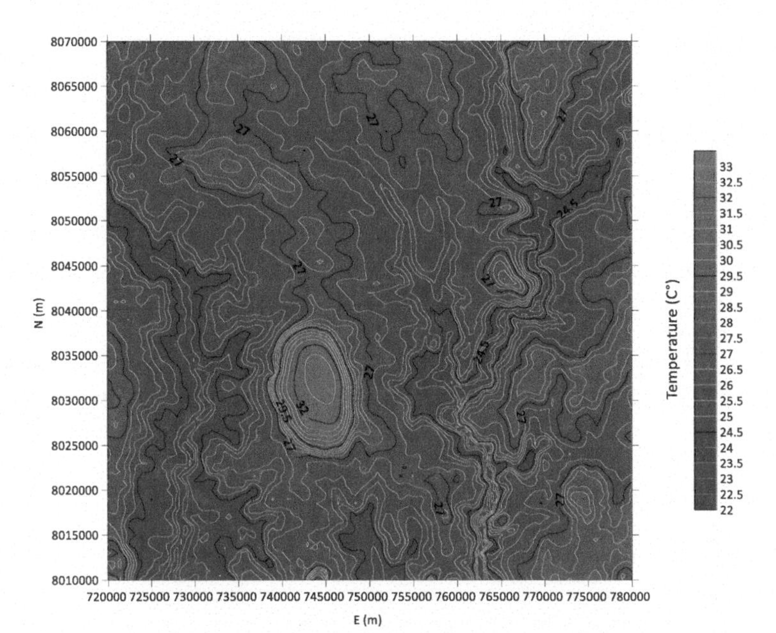

Figure 6.38 Map of the temperature distribution only due to the conductive geothermal flow at altitude 500 m at the Caldas Novas region. Thermal water's much lower temperatures, than in the preceding map, vary from 22°C to 33°C.

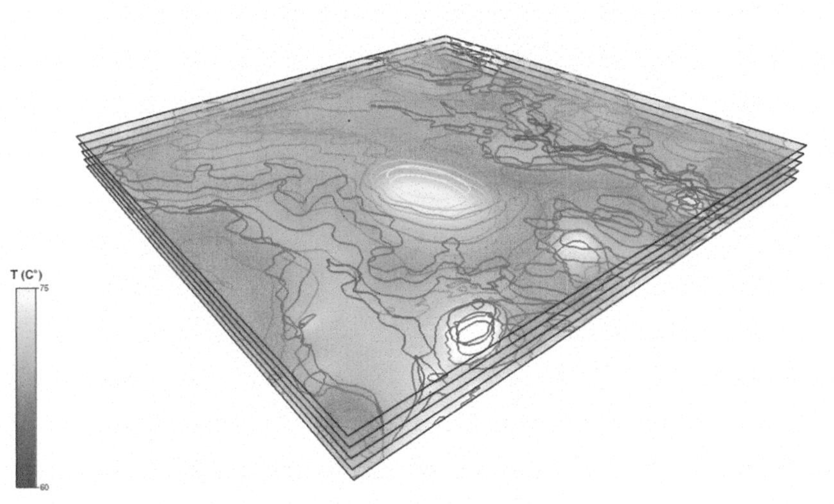

Figure 6.39 Perspective of the distribution of temperatures only due to conductive geothermal flow within the middle of each of the 3D model five horizontal slabs, located at altitudes −1,500, −1,000, −500, 0, and 500 m. This result only concerns the geothermal flow transferred by conductive processes. The groundwater temperatures decrease from the bottom to the top slab. The high heat concentration under the tabular relief of the Serra de Caldas is conspicuous.

Figures 6.40–6.42 show the groundwater equipotentials at the Caldas Novas region. That percolation implies the advection transfer of the geothermal heat.

Figures 6.43–6.45 show the final redistribution of the heat that is advectively carried by the groundwater as it spreads and rises around the Serra de Caldas tabular relief.

6.3.4.3 Modeling the thermalism gradual decay

The gradual temperature decay of the Caldas Novas thermalism, after 25 and 100 years of permanent exploitation, but without environment control measures, was simulated by a non-linear electro-analog circuitry, as shown in Figure 6.46.

The circuit was calibrated with discharges data monitored since August 1986 and temperature variations measurements from October 1994 to September 2001. Tentative prediction for the thermal profile gradual temperature decay at elevation 600 m after 100 years of exploitation without environment control measures is shown in Figures 6.47–6.49.

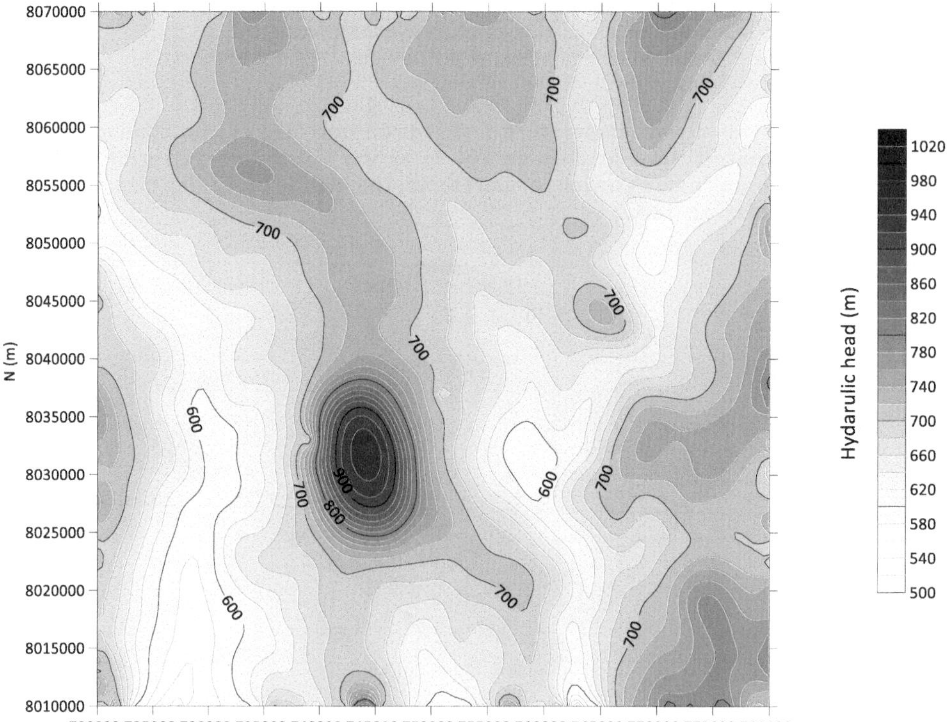

Figure 6.40 Map of the groundwater equipotentials at altitude −1,500 m at the Caldas Novas region. These equipotentials vary from 500 to 1020 m. Higher potentials concentrate under the Caldas tabular relief. This configuration favors the continuous radial and upward advective motion of the high-temperature groundwater under that relief to the neighborhood of the Serra de Caldas at high altitudes.

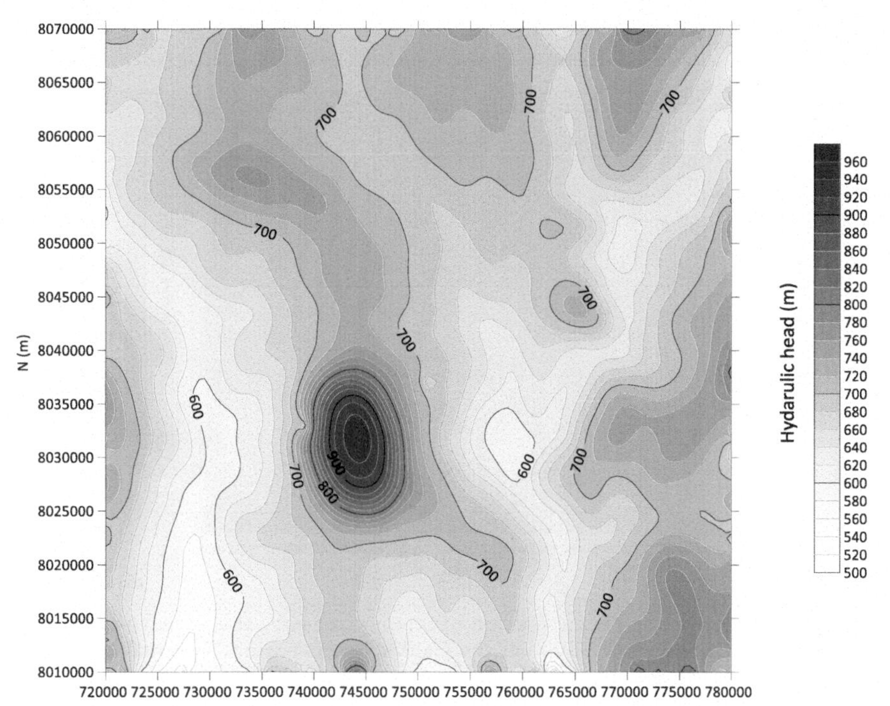

Figure 6.41 Map of the groundwater equipotentials at altitude 500 m. Now, the hydraulic heads vary from 500 to 1,020 m. As expected, higher potentials concentrate under the Caldas tabular relief. As in the precedent figure, that configuration favors the continuous radial and upward advective motion of the hot groundwater.

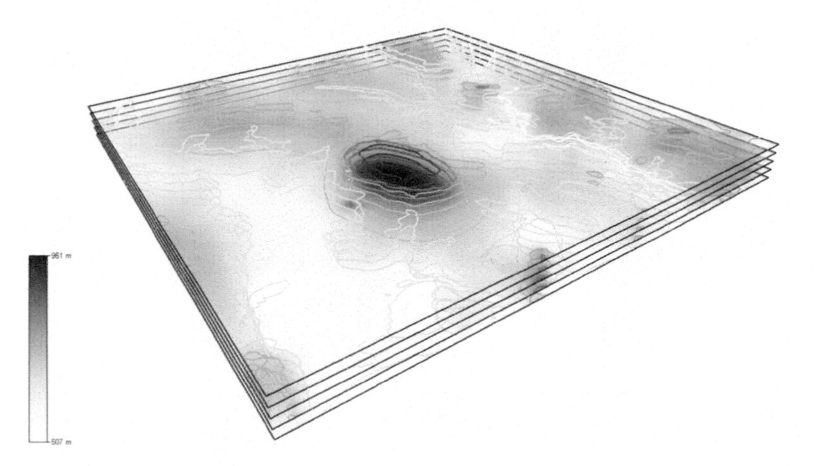

Figure 6.42 Perspective of the distribution of the hydraulic heads within the middle of each of the 3D model five horizontal slabs, located at altitudes −1,500, −1,000, −500, 0, and 500 m. The groundwater percolates from the dark to the light areas, favoring artesian springs and wells.

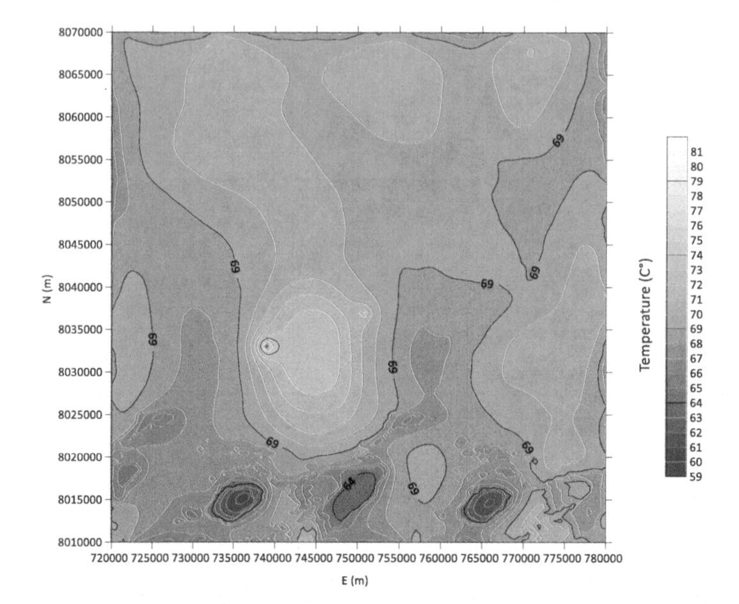

Figure 6.43 Map of the redistribution of temperatures by the advective heat transfer at altitude −1,500 m at the Caldas Novas region. Thermal water temperatures are much lower and vary only from 59°C to 81°C, a bit smaller than the original conductive transfer, as expected.

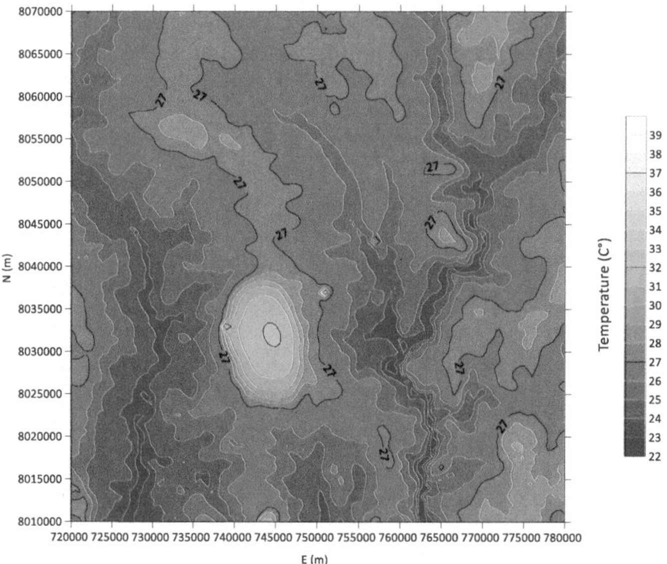

Figure 6.44 Map of the redistribution of temperatures by the advective heat transfer at altitude 500 m at the Caldas Novas region. Thermal water temperatures vary only from 22°C to 39°C. Note the temperature concentrations near the Hot River, Caldas Novas hot springs, and Pirapitinga Pool, respectively, located at the west and east side, Serra de Caldas.

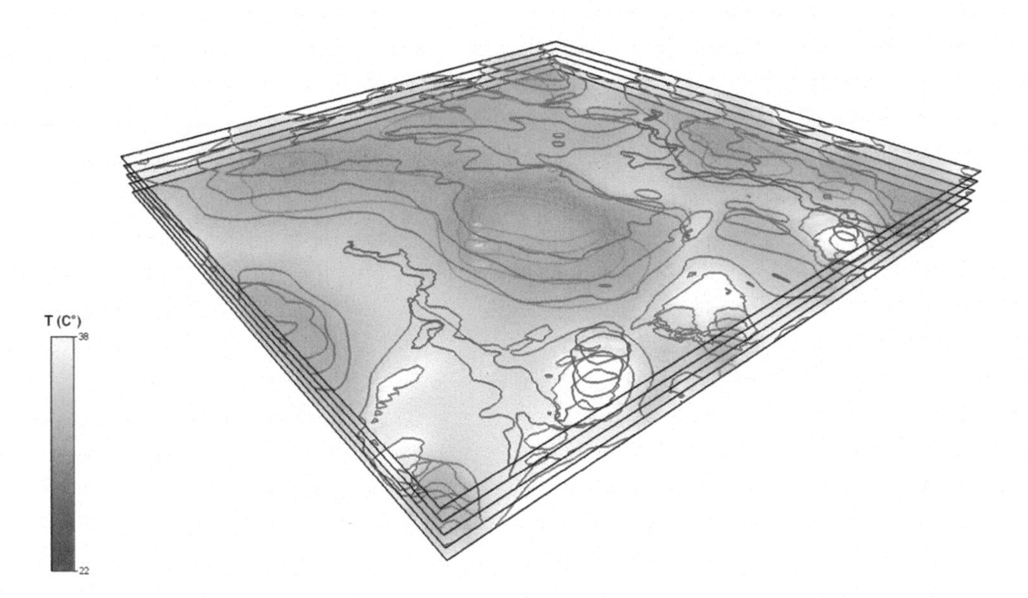

Figure 6.45 Perspective of the redistribution of temperatures by the groundwater advective heat transfer at altitudes −1,500, −1,000, −500, 0, and 500 m at the Caldas Novas region. Thermal water temperatures decrease from the bottom to the top slab. The dispersion of the heat concentration under the tabular relief of the Serra de Caldas is evident. Legend corresponds to the top slab.

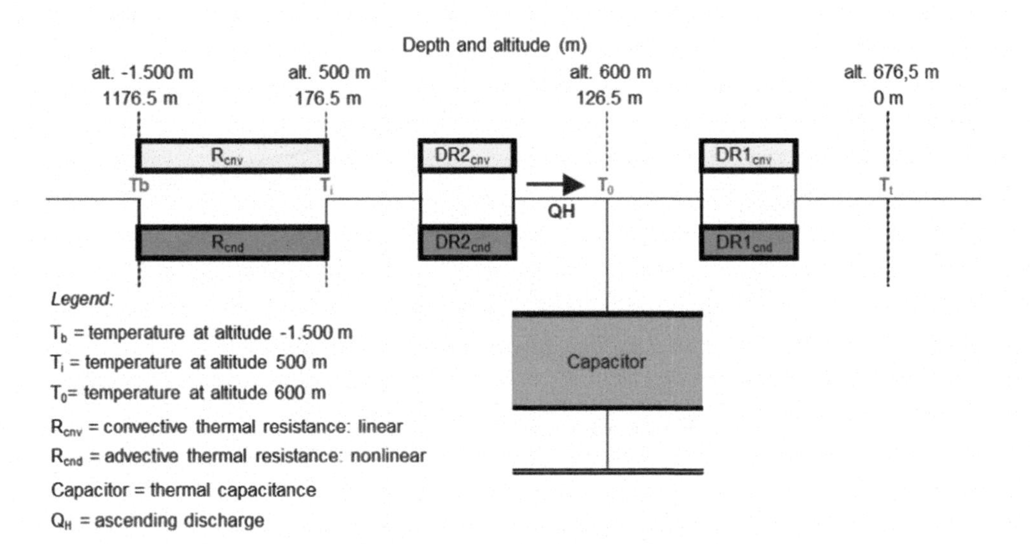

Figure 6.46 Electro-analog for predicting the thermalism gradual decay of its resources after 25 and 100 years of continuous exploration.

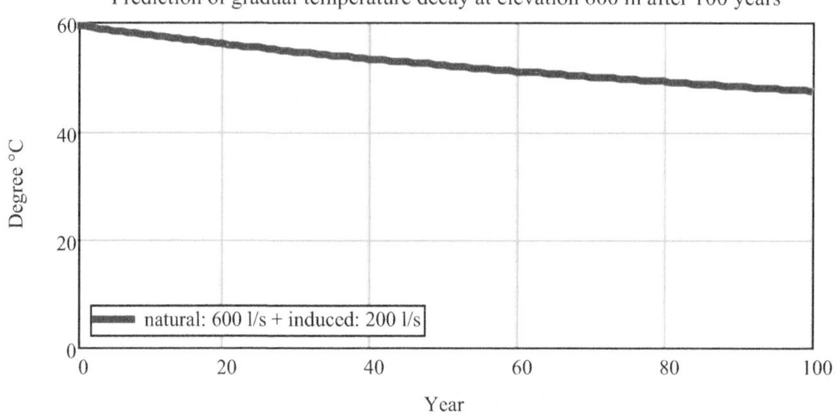

Figure 6.47 Prediction of the gradual decay of the maximum temperature of the thermal profile at altitude 600 m after a century of exploration of the hydrothermal springs under Caldas Novas, but without environment control measures.

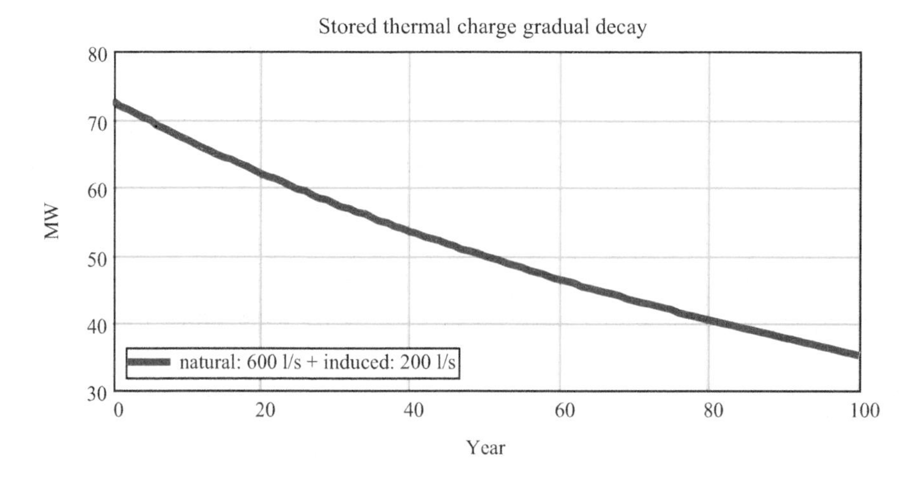

Figure 6.48 Prediction of the gradual decay of the stored thermal energy under the Caldas Novas Town, above 500 m altitudes after a century of exploration of the hydrothermal springs under Caldas Novas, but without environment control measures.

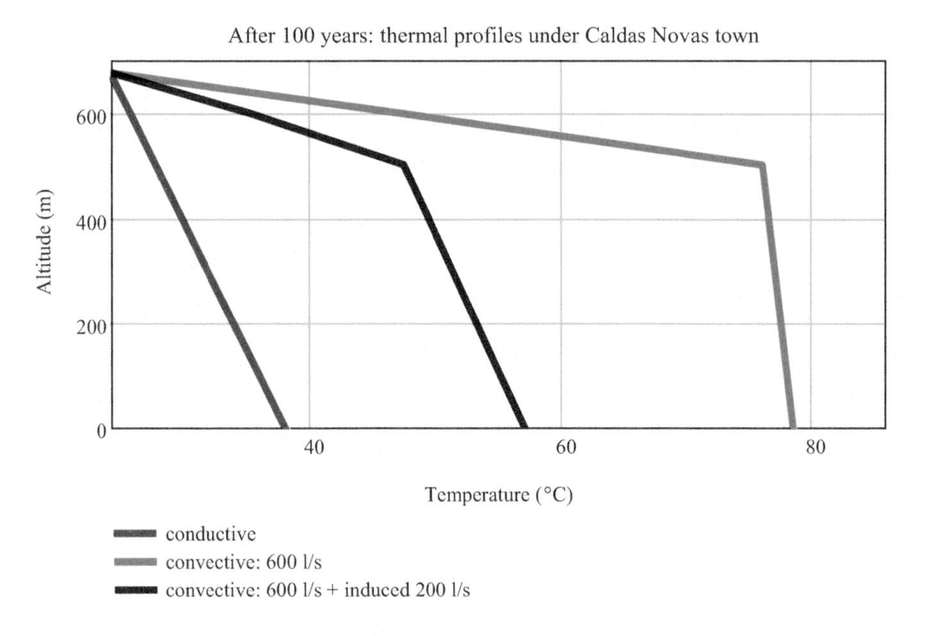

Figure 6.49 Prediction of the gradual change of thermal profiles under the Caldas Novas Town, above 500 m altitudes after a century of exploration of the hydrothermal springs under Caldas Novas, but without environment control measures.

Notes

1 Information source: IPT, SP, Brazil, Technical Report No. 127 334-205, 2012/03/20.
2 Source of information: "Regional Hydrogeology – Conceptual Model", Technical Report CM05-RT-32, item 4, Regional Geology, R0, July, Geoconsultoria, 2007.
3 Information source: IPT, SP, Brazil, Technical Report No. 127 334-205, 2012/03/20.
4 DNPM/CPRM: "Projeto Estudo Hidrogeológico da Região de Caldas Novas", Relatório Final, Departamento Nacional da Produção Mineral – DNPM/Companhia de Pesquisa de Recursos Minerais – CPRM, Goiânia, 1980.
5 IPT: "Geologia, Geomorfologia e Prospecção Geofísica na Região de Caldas Novas – Síntese dos Relatórios IPT.31858 e IPT.31843" e IPT, "Geologia e geomorfologia da região de Caldas Novas – GO", REL 11250 – IPT.31858 (94), "Prospecção geofísica na região de Caldas Novas – GO", REL 11 247 – IPT.31843 (53).
6 Prof. Hans Zojer, "Features of Subsurface Drainage Systems in the Caldas Novas Thermal District", 11/86; PETROBRÁS, "Estudo das Exsudações Gasosas de Rio Quente e Caldas Novas", SEGEQ 29/96; FURNAS, "Radioatividade das Águas de Caldas Novas", DEC.T.067.95; CPRM, "Estudo Hidrogeológico da Região de Caldas Novas", 1980.
7 "Features of Subsurface Drainage Systems in the Caldas Novas Thermal District", Prof. Hans Zojer, Institute for Geothermics and Hydrogeology, Research Center Joanneum, Graz, Áustria, 1986.

Bibliography

Abu Dhabi City Municipality (2014) Dewatering Guidelines. Spatial Data Division, Geotechnical, Geophysical, and Hydrogeological Investigation Project, (GGHIP), Project No. 13-5015, Revision 1.

Anderson, M.P., Ward, D.S., Lappala, E.G. and Prickett, T.A. (1992) Computer models for subsurface water. In *Handbook of Hydrology*, ed. D.R. Maidment, pp. 22.1–22.34. McGraw-Hill, Inc., New York.

Anderson, M.P. and Woessner, W.W. (1992) *Applied Groundwater Modeling*. Academic Press, San Diego.

Anon (1993) *Guidelines for Drinking Water Quality*. World Health Organization, Geneva.

Assaad, F., LaMoreaux, P.E. and Hughes, T. (2004) *Field Methods for Geologists and Hydrogeologists*. Springer, Berlin.

Atkinson, B.K. (1987) *Fracture Mechanics of Rock*, 534 pp. Academic Press, Orlando.

Barnett, B., Townley, L.R., Post, V., Evans, R.E., Hunt, R.J., Peeters, L., Richardson, S., Werner, A.D., Knapton, A. and Boronkay, A. (2012) *Australian Groundwater Modelling Guidelines*. National Water Commission, Canberra.

Barenblatt, G.I., Entov, V.M., Ryzhik, V.M. (1990) *Theory of Fluid Flows through Natural Rocks*. Kluwer Academic Publishers, Dordrecht.

Bear, J. (1972) *Dynamics of Fluids in Porous Media*, American Elsevier, New York.

Bear, J. (1979) *Hydraulics of Groundwater*. McGraw-Hill, New York.

Bear, J. and Verruijt, A. (1987) *Modeling Groundwater Flow and Pollution*. Reidel, Dordrecht.

Bear, J., Zaslavsky, D. and Irmay, S. (1966) *Physical Principles of Water Percolation and Seepage*. Technion, Haifa.

Bisson, R.A. and Lehr, J.H. (2004) *Modern Groundwater Exploration*. Wiley, New York.

Brown, R.H., Konoplyantsev, A.A., Ineson, J. and Kovalevsky, V.S. (Eds.) (1972) *Groundwater Studies*. UNESCO, Paris.

Brown, S.R. (1987) Fluid flow through rock joints: the effect of surface roughness, *J. Geophys. Res.*, Vol. 92 No. B2, pp. 1337–1347.

Busch, K.-F., Luckner, L. and Thiemer, K. (1993) Geohydraulik. In *Lehrbuch der Hydrogeologie*, ed. Matthess, G., Band 3. Gebrüder Borntraeger Verlag, Berlin-Stuttgart.

Cedergren, H. R. (1967) *Seepage, Drainage, and Flow Nets*. John Wiley & Sons, New York.

Couland, O., Morel, P. and Caltagirone, J.-P. (1986) Effects non linéaires dans les écoulements en milieu poreux, *C.R. Acad. Sci.*, Vol. 302, pp. 263–266.

Cripps, J.C., Bell, F.G. and Culshaw, M.G. (Eds.) (1986) *Groundwater in Engineering Geology*. Geological Society, London.

Dagan, G. (1989) *Flow and Transport in Porous Media*. Springer-Verlag, Berlin.

Darcy, H. (1856) *Les fontaines publiques de la ville de Dijon*. Dalmont, Paris.

Davis, S.N. and De Weist, R.J.H. (1966) *Hydrogeology*. Wiley, New York.

Delleur, J. (Ed.) (2007) *The Handbook of Groundwater Engineering.* Co-published by CRC Press LLC, and by Springer-Verlag GmbH & Co. Boca Raton, Heidelberg.

de Marsily, G. (1986) *Quantitative Hydrogeology – Groundwater Hydrology for Engineers.* Academic Press, Orlando.

Diersch, H.-J.G. (1985) *Modellierung und numerische Simulation geohydrodynamischer Transportprozesse, Habilitationsschrift.* Akademie der Wissenschaften der DDR, Berlin.

Domenico, P.A. and Schwartz, F.W. (1990) *Physical and Chemical Hydrogeology.* John Wiley & Sons, New York.

Dunnicliff, J. (1988) *Geotechnical Instrumentation for Monitoring Field Performance.* John Wiley & Sons, Canada, Ltd., Etobicoke.

Engelder, T., Fischer, M.P. and Gross, M.R. (1993) *Geological Aspects of Fracture Mechanics, A Short Course Manual,* 281 pp. Geological Society of America, Boulder.

Fetter, C.W. (1999) *Contaminant Hydrogeology.* 2nd ed. Prentice-Hall, Upper Saddle River.

Fetter, C.W. (2001) *Applied Hydrogeology.* 4th ed. Prentice-Hall, Upper Saddle River.

Fetter, C.W. (2004) *Applied Hydrogeology.* 4th ed. Prentice-Hall, Upper Saddle River.

Fitts, C.R. (2002) *Groundwater Science.* Academic Press, New York.

Freeze, R. A. and Cherry, J.A. (1979) *Groundwater.* Prentice-Hall, Englewood Cliffs.

Haitjema, H.M. (1995) *Analytic Element Modeling of Groundwater Flow.* Academic Press, San Diego.

Hamill, L. and Bell, F.G. (1986) *Groundwater Resource Development.* Butterworth, London.

Harr, M. E. (1962) *Groundwater and Seepage.* McGraw-Hill Inc., New York.

Heath, R.C. (1995) *Basic Groundwater Hydrology.* Water-Supply Paper 2200, 7th printing. US Geological Survey, Denver.

Hem, J.D. (1985) *Study and Interpretation of the Chemical Characteristics of Natural Water.* Water Supply Paper 2254. 3rd ed. US Geological Survey, Washington.

Hill, M.C. (1998) Methods and Guidelines for Effective Model Calibration, US Geological Survey Water-Resources Investigation Report 98-4005.

Hiscock, K.M. (2002) *Sustainable Groundwater Development.* Special Publication 193, Geological Society, London.

Huyakorn, P.S. and Pinder, G.P. (1983) *Computational Methods in Subsurface Flow.* Academic Press, New York.

Javandel, I., Doughty, D. and Tsang, C.-P. (1984) *Groundwater Transport: Handbook of Mathematical Models, American Geophysical Union, Water Research Monograph 10.* US Environmental Protection Agency, Washington.

Kitanidis, P. K. (1997) *Introduction to Geostatistics: With Applications in Hydrogeology,* 249 pp. Cambridge University Press, Cambridge.

Lowman, S.W. (1972) *Groundwater Hydraulics.* Geological Survey Professional Paper 708, US Government Printing Office, Washington.

Poehls, D.J. and Smith, G.J. *Encyclopedic Dictionary of Hydrogeology.* Academic Press is an imprint of Elsevier, Burlington, San Diego, London, Amsterdam, 2009.

Polubarinova-Kochina, P.Y. (1962) *Theory of Ground Water Movement.* Translated from Russian by J.M. Roger De Wiest. Princeton University Press, Princeton.

Price, M. (1985) *Introducing Groundwater.* Allen and Unwin, London.

Price, M. (1996) *Introducing Groundwater.* Chapman & Hall, London.

Remson, L., Hornberger, G.M. and Molz, E.J. (1971) *Numerical Methods in Subsurface Hydrology.* Wiley, New York.

Robins, N.S. (Ed.) (1998) *Groundwater Pollution, Aquifer Recharge and Vulnerability.* Special Publication No. 130. Geological Society, London.

Scheidegger, A.E. (1957) *The Physics of Flow through Porous Media.* The MacMillan Company, New York.

Schwartz, E.W. and Zhang, H. (2003) *Fundamentals of Groundwater.* John Wiley & Sons, New York.

Segol, G. (1994) *Classic Groundwater Simulations: Proving and Improving Numerical Models.* PTR Prentice Hall, Englewood Cliffs.

Spitz, K. and Moreno, J. (1996) *A Practical Guide to Groundwater and Solute Transport Modeling.* John Wiley & Sons, New York.

Todd, D.K. and Mays, L.W. (2004) *Groundwater Hydrology.* Third Edition. Wiley, New York.

USEPA (1986a) *RCRA Groundwater Monitoring Technical Enforcement Guidance Document. Office of Waste Programs Enforcement and Office of Solid Waste and Emergency Response, OSWER-9950.L.* USEPA, Washington.

USEPA (1986b) *Test Methods for Evaluating Solid Wastes. Office of Solid Waste and Emergency Response.* USEPA, Washington.

USEPA (1987) Alternate Concentration Limit Guidance, Interim Final EPA/530-SW-87-017. 114 p., July.

USEPA (1989a) *Risk Assessment Guidance for Superfund: Interim Final Guidance. Office of Emergency and Remedial Response (EPA/540/I-89/002).* USEPA, Washington.

USEPA (1989b) *Statistical Analysis of Groundwater Monitoring Data at RCRA Facilities, Interim Final Guidance. Office of Solid Waste.* USEPA, Washington.

USEPA (1989c) *Seminar Publication – Transport and Fate of Contaminants in the Subsurface. EPA/625/4-89/019.* USEPA, Washington.

USEPA (1989d) *Handbook of Suggested Practices for the Design and Installation of Groundwater Monitoring Wells. PB90-159-807.* USEPA, Washington.

USEPA (1990) *Handbook – Groundwater, Volume I: Groundwater and Contamination. EPA/625/6-90/016a.* USEPA, Washington.

USEPA (1991) *Handbook – Groundwater, Volume II: Methodology. EPA/625/6-90/016b.* USEPA, Washington.

USEPA (1992a) *RCRA Groundwater Monitoring: Draft Technical Guidance, Office of Solid Waste. EPA/530-4-93-001, PB93-139-350.* USEPA, Washington.

USEPA (1992b) *Statistical Analysis of Groundwater Monitoring Data at RCRA Facilities, Addendum Interim Final Guidance, Office of Solid Waste.* USEPA, Washington.

USEPA (1993a) *Office of Research and Development. Subsurface Characterization and Monitoring Techniques: A Desk Reference Guide, Volume I: Solids and Groundwater, Appendices A and B. U, EPA/625/R-93-003a.* Environmental Protection Agency, Washington.

USEPA (1993b) *Office of Research and Development. Subsurface Characterization and Monitoring Techniques: A Desk Reference Guide, Volume II: The Vadose Zone, Field Screening and Analytical Method – Appendices C and D, EPA/625/R-93/003i.* USEPA, Washington.

USEPA (1993c) *Technical Manual – Solid Waste Disposal Facility Criteria. EPA530-R-93-017, PB94-100-450.* USEPA, Washington.

USEPA (1994) *Handbook – Groundwater and Wellhead Protection. EPA/625/R-94/00-1.* USEPA, Washington.

USEPA (1995) *Groundwater Sampling – A Workshop Summary, November 30 to December 2, 1995, Dal Texas. EPA/600/R-94/205.* USEPA, Washington.

USEPA (2000) *Guidance for the Data Quality Objectives Process – EPA QA/G-4 EPA/600/R-96/05, 100 p., August.* USEPA, Washington.

USEPA (2002) *EPA Guidance for the Quality Assurance Project Plans – EPA QA/G-5 EPA/600/R-98/011, III p. December.* USEPA, Washington.

Verruijt, A. (1970) *The Theory of Groundwater Flow.* Macmillan, London.

Walton, W.C. (1970) *Groundwater Resource Evaluation.* McGraw Hill, New York.

Wang, E. and Anderson, M.P. (1995) *Introduction to Groundwater Modeling.* Academic Press, San Diego.

Index